Crystals and Crystal Structures

Richard J. D. Tilley
Emeritus Professor, University of Cardiff

JOHN WILEY & SONS, LTD

Copyright © 2006   John Wiley & Sons Ltd, The Atrium, Southern Gate, Chichester,
West Sussex PO19 8SQ, England

Telephone   (+44) 1243 779777

Email (for orders and customer service enquiries): cs-books@wiley.co.uk
Visit our Home Page on www.wileyeurope.com or www.wiley.com

Reprinted with corrections August 2007

Other Wiley Editorial Offices

John Wiley & Sons Inc., 111 River Street, Hoboken, NJ 07030, USA

Jossey-Bass, 989 Market Street, San Francisco, CA 94103-1741, USA

Wiley-VCH Verlag GmbH, Boschstr. 12, D-69469 Weinheim, Germany

John Wiley & Sons Australia Ltd, 42 McDougall Street, Milton, Queensland 4064, Australia

John Wiley & Sons (Asia) Pte Ltd, 2 Clementi Loop #02-01, Jin Xing Distripark, Singapore 129809

John Wiley & Sons Canada Ltd, 6045 Freemont Blvd, Mississauga, Ontario, Canada L5R 4J3

Wiley also publishes its books in a variety of electronic formats. Some content that appears in print may not
be available in electronic books.

Library of Congress Cataloging-in-Publication Data

Tilley, R. J. D.
    Crystals and crystal structure / Richard J. D. Tilley.
        p.   cm.
    Includes bibliographical references and index.
    ISBN 13: 978-0-470-01820-0 (cloth)
    ISBN 10: 0-470-01820-8 (cloth)
    ISBN 13: 978-0-470-01821-7 (pbk. : alk. paper)
    ISBN 10: 0-470-01821-6 (pbk. : alk. paper)
  1. Crystals.   2. Crystallography.   I. Title.

    QD905.2.T56 2006
    548–dc22   2006016562

British Library Cataloguing in Publication Data

A catalogue record for this book is available from the British Library

ISBN-13 978-0-470-01820-0 (HB)   ISBN-10 0-470-01820-8 (HB)
ISBN-13 978-0-470-01821-7 (PB)   ISBN-10 0-470-01821-6 (PB)

Typeset in 10.5/12.5 pt Times by Thomson Digital

Crystals and Crystal Structures

For Elizabeth, Gareth and Richard

Contents

7   The depiction of crystal structures   155

8   Defects, modulated structures and quasicrystals   187

Appendices   **215**

Preface

Crystallography – the study of crystals, their structures and properties – has antecedents predating modern science. Despite this long history, the subject remains dynamic and is changing rapidly. Protein crystallography is but one example. An understanding of many biological processes at a molecular level, already leading to a fundamental change in the treatment of many chronic illnesses, has been built upon protein crystal structure studies of awesome complexity. Other areas of rapid growth stand out. The development of computer software has revolutionised the way in which crystal structures can be depicted. The presentation of such structures as nets, for example, leads to images that compare in beauty with the best of abstract art.

Computers themselves, the microprocessors that do the work, and the displays that present the information, as well as the host of mobile devices that are essentially computers in one form or another, rely almost entirely upon a fundamental appreciation of crystal structures. Nanomaterials, fragments of material that are sometimes too small to organise into a crystal, have properties dominated by surfaces that are often best described as crystal planes. In both of these areas, a fundamental understanding of why crystalline and non-crystalline materials behave differently, something intimately connected with crystal structure, is vital.

It is thus apparent that crystallography plays an important part in a wide range of disciplines, including biology, chemistry, materials science and technology, mineralogy and physics, as well as engineering. Two further examples will suffice to demonstrate this. The scientific breakthroughs in the first half of the 20th century, leading to applications as diverse as nuclear power and semiconductor technology, were built, to a considerable extent, upon detailed understanding of metallic and inorganic crystal structures. In the latter half of the 20th century, molecular biology, now embedded in crystallography, has lead to a deep rooted change in medicine, as a result of crystal structure studies, starting with the structures of biologically important molecules such as insulin, via the epoch making determination of the crystal structure of DNA to recent studies of enormous protein structures.

With all of these aspects in mind, this book is designed as an introductory text for students and others that need to understand crystals and crystallography without necessarily becoming crystallographers. The aim is explain how crystal structures are described for anyone coming to this area of study for the first time. At the end of it, a student should be able to read scientific papers and articles describing a crystal structure, or use crystallographic databases, with complete confidence and understanding. It has grown out of lectures to undergraduate and postgraduate students in a wide range of disciplines, generally at the same introductory level. It is not designed to explain in detail how to determine crystal structures experimentally, although this aspect of the subject is reviewed in Chapter 6, and sources that provide this information are listed in the Bibliography. As far as practicable, mathematical descriptions have been employed with a light touch. However, the Bibliography cites references to areas where a sophisticated mathematical approach results in considerable generalisation.

The text is not focused upon one subset of crystal types, but covers both simple inorganic and organic structures and complex proteins. Furthermore, the book contains an introduction to areas of crystallography, such as modulated structures and quasicrystals, as well as protein crystallography itself, that are the subject of important and active research. With respect to the first two of these areas of research, the whole traditional framework of crystallography is being re-explored. In protein crystallography, not only are traditional crystallographic techniques being pushed to extremes, but the results of the crystallographic studies are being exploited for medical purposes with the utmost of rapidity. These and other areas of current crystallographic research cannot be ignored by students of science or engineering.

The book is organised into eight chapters. The first of these provides an introduction to the subject, and explains how crystal structures are described, using simple examples for this purpose. Chapter 2 introduces the important concept of lattices and provides formulae for the description and calculation of lattice and crystal geometry. Chapter 3 is concerned with symmetry in two dimensions, and introduces tiling, an idea of considerable crystallographic and general interest. Chapter 4 describes symmetry in three dimensions, and covers the important relationships between the symmetry of a crystal and its physical properties. Chapter 5 is focussed upon how to build or display a crystal structure given the information normally available in databases, viz. space group, unit cell and atom positions. A reader will be able to go from a database description of a structure to a model by following the steps outlined. Crystal structures are determined by diffraction of radiation. Chapter 6 outlines the relationship between

diffraction and structure, including information on X-ray, electron and neutron techniques. The all important 'phase problem' is discussed with respect to solving protein structures. Chapter 7 describes the principal ways in which structures are depicted in order to reveal structural relationships, or, in the case of proteins, biochemical reactivity. The final chapter, Chapter 8, is concerned with defects in crystals and recent developments in crystallography such as the recognition of incommensurately modulated crystals and quasicrystals. These materials require a replacement of classical crystallographic ideas by broader definitions to encompass these new structures.

Each chapter is prefaced by three 'Introductory Questions'. These are questions that have been asked by students in the past, and to some extent provide a focus for the student at the beginning of each chapter–something to look out for. The answers are given at the end of the chapter. Within each chapter, new crystallographic concepts, invariably linked to a definition of the idea, are given in **bold type** when encountered for the first time. To assist in understanding, all chapters are provided with a set of problems and exercises, designed to reinforce the concepts introduced. These are of two types. A multiple choice 'Quick Quiz' is meant to be tackled rapidly, and aims to reveal gaps in the reader's grasp of the preceding material. The 'Calculations and Questions' sections are more traditional, and contain numerical exercises that reinforce concepts that are often described in terms of mathematical relationships or equations. A bibliography provides students with sources for further exploration of the subject.

Crystallography is a three-dimensional subject. (Indeed, four or more dimensions are used in the mathematical description of incommensurate structures.) This implies that understanding will be assisted by any techniques that allow for three-dimensional representations. Computer graphics that can display crystal and molecular structures as virtual three-dimensional images are of particular importance. In some instances it can be said that these aids are indispensable if the spatial relationships between the components in a complex structure are to be understood. Indeed, in the complex chemistry of proteins and other large molecules, the structure, and hence the biological function cannot really be understood without the use of such aids. However, it is important not to overcomplicate matters. For example, overall protein structures are too complicated to visualise in terms of atom arrangements, and the global structure of a protein is usually represented in a stylised 'cartoon' form. Exactly the same approach should be used for other aspects of crystallography. Even the simplest of models are invaluable, and a small collection of geometric solids; spheres, cubes, octahedra, tetrahedra and so on, together with modelling clay, and matches or tooth picks, can be used to advantage. Generally, the simpler the depiction of a crystal structure, the easier it will be to remember it and to relate it to other structures and to biological, chemical and physical properties.

Having said that, access to computer software for the display of crystal structures will aid in understanding and is to be strongly encouraged. Information on sources of computer software is given in the Bibliography. Similarly, access to a crystallographic database is necessary for more advanced study or research. For this book, extensive use was made of the EPSRC's Chemical Database Service at Daresbury, UK. Further information on access to this resource is also given in the Bibliography.

This book could not have been compiled without considerable assistance. First and foremost, it is a pleasure to acknowledge the help of Dr. Andrew Slade and Mrs. Celia Carden, of John Wiley at Chichester, who have, as always, given sage advice and continual encouragement during the preparation of the manuscript, and Mr. Robert Hambrook, who transformed the manuscript into the book before you. In addition, a number of anonymous reviewers read early drafts of the manuscript and made invaluable comments that lead to clarification and the correction of errors. It is a pleasure to acknowledge the contributions made to my understanding of crystallography by three *ex* colleagues. Dr. R. Steadman, of the University of Bradford taught crystallography with flair, and produced student teaching materials that are as valuable now as when first drafted. Drs. G. Harburn and R. P. Williams of the University of Cardiff explained diffraction theory to me both expertly and patiently, and illuminated the areas of optical transforms and Fourier optics. In addition, I am particularly indebted to Professor A. H. White and Dr. B. W. Skelton of the Department of Chemistry, University of Western Australia, for the provision of Figure 6.16; Dr R. D. Tilley and Dr. J. H. Warner, Victoria University, Wellington, New Zealand, for the provision of Figure 6.24b; to Dr. D. Whitford, who kindly provided the clear and well designed protein structure diagrams shown in Figures 7.25 and 7.27 and Dr K. Saitoh for providing the elegant diffraction pattern in Figure 8.24. Mr Allan Coughlin and Mr Rolfe Jones in the University of Cardiff provided frequent help, encouragement and information during the writing of the book. The staff in the Trevithick Library, University of Cardiff, have been ever helpful, especially with respect to queries concerning obscure literature. The staff at the EPSRC's Chemical Database Service at Daresbury, UK, especially Dr. D. Parkin, were always on hand to give invaluable advice concerning use of both the databases of crystal structures and the software to visualise structures. Last, but by no means least, I thank my family for continual patience during the assembly of the book, and especially my wife Anne, for toleration of the routine that writing has demanded over the last couple of years.

Richard J. D. Tilley

1

Crystals and crystal structures

- *What is a crystal system?*

- *What are unit cells?*

- *What information is needed to specify a crystal structure?*

Crystals are solids that possess long-range order. The arrangement of the atoms at one point in a crystal is identical, (excepting localised mistakes or defects that can arise during crystal growth), to that in any other remote part of the crystal. Crystallography describes the ways in which the component atoms are arranged in crystals and how the long-range order is achieved. Many chemical (including biochemical) and physical properties depend on crystal structure and knowledge of crystallography is essential if the properties of materials are to be understood and exploited.

Crystallography first developed as an observational science; an adjunct to the study of minerals. Minerals were, (and still are), described by their habit, the shape that a mineral specimen may exhibit, which may vary from an amorphous mass to a well-formed crystal. Indeed, the

regular and beautiful shapes of naturally occurring crystals attracted attention from the earliest times, and the relationship between crystal shape and the disposition of crystal faces, the crystal morphology, was soon used in classification. At a later stage in the development of the subject, symmetry, treated mathematically, became important in the description of crystals. The actual determination of crystal structures, the positions of all of the atoms in the crystal, was a later level of refinement that was dependent upon the discovery and subsequent use of X-rays.

1.1 Crystal families and crystal systems

Careful measurement of mineral specimens allowed crystals to be classified in terms of six **crystal families**, called anorthic, monoclinic, orthorhombic, tetragonal, hexagonal and isometric. This classification has been expanded slightly by crystallographers into seven **crystal systems**. The crystal systems are sets of reference axes, which have a direction as well as a magnitude, and hence are vectors[1]. The crystal families and classes are given in Table 1.1.

---

[1]Vectors are set in **bold** typeface throughout this book.

Crystals and Crystal Structures.   Richard J. D. Tilley
© 2006 John Wiley & Sons, Ltd

Table 1.1   The seven crystal systems

Crystal family	Crystal system	Axial relationships
Isometric	Cubic	$a=b=c$, $\alpha=\beta=\gamma=90°$;
Tetragonal	Tetragonal	$a=b\neq c$, $\alpha=\beta=\gamma=90°$;
Orthorhombic	Orthorhombic	$a\neq b\neq c$, $\alpha=\beta=\gamma=90°$;
Monoclinic	Monoclinic	$a\neq b\neq c$, $\alpha=90°$, $\beta\neq90°$, $\gamma=90°$;
Anorthic	Triclinic	$a\neq b\neq c$, $\alpha\neq90°$, $\beta\neq90°$, $\gamma\neq90°$;
Hexagonal	Hexagonal	$a=b\neq c$, $\alpha=\beta=90°$, $\gamma=120°$;
	Trigonal or	$a=b=c$, $\alpha=\beta=\gamma$; or
	Rhombohedral	$a'=b'\neq c'$, $\alpha'=\beta'=90°$, $\gamma'=120°$;
		(hexagonal axes)

The three reference axes are labelled **a**, **b** and **c**, and the angles between the positive direction of the axes are α, β, and γ, where α lies between +**b** and +**c**, β lies between +**a** and +**c**, and γ lies between +**a** and +**b**, (Figure 1.1). The angles are chosen to be greater or equal to 90° except for the trigonal system, as described below. In figures, the **a**-axis is represented as projecting out of the plane of the page, towards the reader, the **b**-axis points to the right and the **c**-axis points towards the top of the page. This arrangement is a **right-handed coordinate system**.

Measurements on mineral specimens could give absolute values for the inter-axial angles, but only relative axial lengths could be derived. These lengths are written a, b and c.

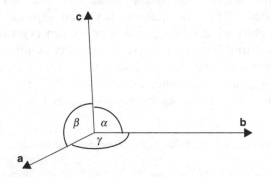

Figure 1.1   The reference axes used to characterise the seven crystal systems

The seven crystal systems are named according to the relationship between the axes and the inter-axial angles. The most symmetric of the crystal classes is the cubic or isometric system, in which the three axes are arranged at 90° to each other and the axial lengths are identical. These form the familiar Cartesian axes. The tetragonal system is similar, with mutually perpendicular axes. Two of these, usually designated **a** (=**b**), are of equal length, while the third, designated **c**, is longer or shorter than the other two. The orthorhombic system has three mutually perpendicular axes of different lengths. The monoclinic system is also defined by three unequal axes. Two of these, conventionally chosen as **a** and **c**, are at an oblique angle, β, while the third **c**, is normal to the plane containing **a** and **b**. The least symmetrical crystal system is the triclinic, which has three unequal axes at oblique angles. The hexagonal crystal system has two axes of equal length, designated **a** (=**b**), at an angle, γ, of 120°. The **c**-axis lies perpendicular to the plane containing **a** and **b**, and lies parallel to a six-fold axis of rotation symmetry, (see Chapter 4).

The trigonal system has three axes of equal length, each enclosing equal angles α (=$\beta=\gamma$), forming a rhombohedron. The axes are called rhombohedral axes, while the system name trigonal refers to the presence of a three-fold axis of rotation symmetry in the crystal (see Chapter 4). Crystals described in terms of rhombohedral axes are often more conveniently described in terms of a hexagonal set of axes. In this case, the hexagonal **c**-axis is parallel to the rhombohedral body diagonal, which is a three-fold axis of symmetry (Figure 1.2). The relationship between the two sets of axes is given by the *vector* equations:

$$\mathbf{a}_R = \tfrac{2}{3}\,\mathbf{a}_H + \tfrac{1}{3}\,\mathbf{b}_H + \tfrac{1}{3}\,\mathbf{c}_H$$
$$\mathbf{b}_R = -\tfrac{1}{3}\,\mathbf{a}_H + \tfrac{1}{3}\,\mathbf{b}_H + \tfrac{1}{3}\,\mathbf{c}_H$$
$$\mathbf{c}_R = -\tfrac{1}{3}\,\mathbf{a}_H - \tfrac{2}{3}\,\mathbf{b}_H + \tfrac{1}{3}\,\mathbf{c}_H$$
$$\mathbf{a}_H = \mathbf{a}_R - \mathbf{b}_R$$
$$\mathbf{b}_H = \mathbf{b}_R - \mathbf{c}_R$$
$$\mathbf{c}_H = \mathbf{a}_R + \mathbf{b}_R + \mathbf{c}_R$$

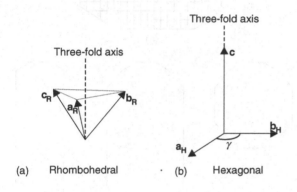

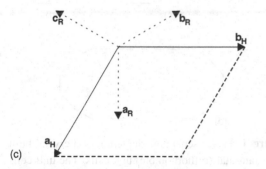

Figure 1.2 Rhombohedral and hexagonal axes: (a, b), axes in equivalent orientations with the trigonal 3-fold axis parallel to the hexagonal **c**-axis; (c) superposition of both sets of axes, projected down the hexagonal **c**-axis (= the rhombohedral 3-fold axis)

where the subscripts R and H stand for rhombohedral and hexagonal respectively. [Note that in these equations the vectors **a**, **b** and **c** are added vectorially, not arithmetically, (see Appendix 1)]. The arithmetical relationships between the axial lengths is given by:

$$a_H = 2\,a_R \sin\frac{\alpha}{2} \qquad\qquad c_H = a_R\sqrt{3 + 6\cos\alpha}$$

$$a_R = \frac{1}{3}\sqrt{3a_H^2 + c_H^2} \qquad \sin\frac{\alpha}{2} = \frac{3a_H}{2\sqrt{3a_H^2 + c_H^2}}$$

where the subscripts R and H stand for rhombohedral and hexagonal respectively.

1.2 Morphology and crystal classes

Observations of the shapes of crystals, the crystal **morphology**, suggested that the regular external form of a crystal was an expression of internal order. Among other observations, the cleavage of crystals, that is, the way in which they could be fractured along certain directions in such a manner that the two resultant fragments had perfect faces, suggested that all crystals might be built up by a stacking of infinitesimally small regular 'brick-like' elementary volumes, each unique to the crystal under consideration. These elementary volumes, the edges of which could be considered to be parallel to the axial vectors **a**, **b** and **c**, of the seven crystal systems, eventually came to be termed **morphological unit cells**. The relative axial lengths, *a*, *b* and *c* were taken as equal to the relative lengths of the unit cell sides. The values *a*, *b*, *c*, α, β and γ are termed the **morphological unit cell parameters**. [The absolute lengths of the axes, also written *a*, *b* and *c* or a_0, b_0 and c_0, determined by diffraction techniques, described below, yield the **structural unit cell** of the material. Unit cell parameters now refer only to these structural values.]

A central concept in crystallography is that the whole of a crystal can be built up by stacking identical copies of the unit cell in exactly the same orientation. That is to say, a crystal is characterised by both **translational** and **orientational** long-range order. The unit cells are displaced repeatedly in three dimensions, (**translational** long-range order), without any rotation or reflection, (**orientational** long-range order). This restriction leads to severe restrictions upon the shape (strictly speaking the symmetry) of a unit cell; a statement which will be placed on a firm footing in later chapters. The fact that some unit cell shapes are not allowed is easily demonstrated in two dimensions, as it is apparent that regular pentagons, for example, cannot pack together without leaving gaps (Figure 1.3). [A regular pentagon is a plane figure with five equal sides and five equal internal angles.]

Not only could unit cells be stacked by translation alone to yield the internal structure of the crystal, but, depending on the rate at which the unit cells were stacked in different directions, (i.e. the rate at which the crystal grew in three dimensions), different facets of the crystal became emphasised, while others were suppressed, so producing a variety of external shapes, or **habits**, (Figure 1.4),

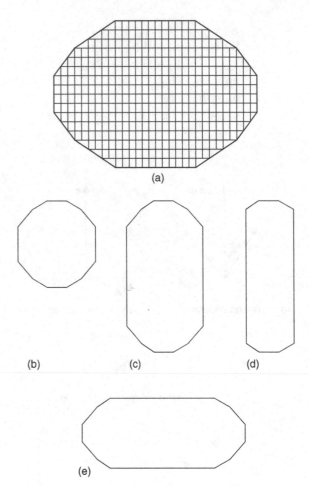

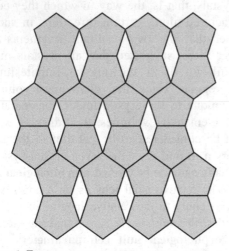

Figure 1.3 Irrespective of how they are arranged, regular pentagons cannot fill a plane completely; spaces always appear between some of the pentagons. Just one arrangement is drawn, others are possible

Figure 1.4 (a) schematic depiction of a crystal built of rectangular (orthorhombic) unit cells. The unit cells must be imagined to be much smaller than depicted, thus producing smooth facets. (b – e) different crystal habits derived by differing rates of crystal growth in various directions

thus explaining the observation that a single mineral could occur in differing crystal morphologies.

The faces of a crystal, irrespective of the overall shape of the crystal, could always be labelled with respect to the crystal axes. Each face was given a set of three integers, $(h \ k \ l)$, called **Miller indices**. These are such that the crystal face in question made intercepts on the three axes of a/h, b/k and c/l. A crystal face that intersected the axes in exactly the axial ratios was given importance as the **parametral plane**, with indices (111). [Miller indices are now used to label any plane, internal or external, in a crystal, as described in Chapter 2, and the nomenclature is not just confined to the external faces of a crystal.]

The application of Miller indices allowed crystal faces to be labelled in a consistent fashion. This, together with accurate measurements of the angles between crystal faces, allowed the morphology of crystals to be described in a reproducible way, which, in itself, lead to an appreciation of the symmetry of crystals. Symmetry was broken down into a combination of symmetry elements. These were described as mirror planes, axes of rotation, and so on, that, when taken in combination, accounted for the external shape of the crystal. The crystals of a particular mineral, regardless of its precise morphology, were always found to possess the same symmetry elements.

Symmetry elements are operators. That is, each one describes an operation, such as reflection. When these operations are applied to the crystal, the external form is reproduced. It was found that all crystals fell into one or another of 32 different groups of symmetry operations. These were called **crystal classes**. Each crystal class could be allocated to one of the six crystal families. These symmetry elements and the resulting crystal classes are described in detail in Chapters 3 and 4.

1.3 The determination of crystal structures

The descriptions above were made using optical techniques, especially optical microscopy. However, the absolute arrangement of the atoms in a crystal cannot be determined in this way. This limitation was overcome in the early years of the 20th century, when it was discovered that X-rays were scattered, or **diffracted**, by crystals in a way that could be interpreted to yield the absolute arrangement of the atoms in a crystal, the **crystal structure**. X-ray diffraction remains the most widespread technique used for structure determination, but diffraction of electrons and neutrons is also of great importance, as these reveal features that are complementary to those observed with X-rays.

The physics of diffraction by crystals has been worked out in detail. It is found that the incident radiation is scattered in a characteristic way, called a **diffraction pattern**. The positions and intensities of the diffracted beams are a function of the arrangements of the atoms in space and some other atomic properties, such as the atomic number of the atoms. Thus, if the positions and the intensities of the diffracted beams are recorded, it is possible to deduce the arrangement of the atoms in the crystal and their chemical nature. The determination of crystal structures by use of the diffraction of radiation is outlined in Chapter 6.

1.4 The description of crystal structures

The minimum amount of information needed to specify a crystal structure is the unit cell type, i.e., cubic, tetragonal, etc, the unit cell parameters and the positions of all of the atoms in the unit cell. The atomic contents of the unit cell are a simple multiple, Z, of the composition of the material.

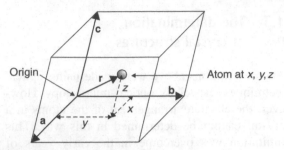

Figure 1.5   The position of an atom in a unit cell, x, y and z, is defined with respect to the directions of the unit cell edges. The numerical values of x, y and z are specified as fractions, (½, ¼, etc.) of the unit cell edges a, b and c

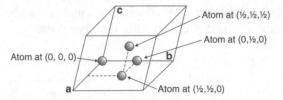

Figure 1.6   Atoms at positions 0, 0, 0; 0, ½, 0; ½, ½, 0; and ½, ½, ½ in a unit cell

The value of Z is equal to the number of **formula units** of the solid in the unit cell. Atom positions are expressed in terms of three coordinates, x, y, and z. These are taken as **fractions** of a, b and c, the unit cell sides, say ½, ½, ¼. The x, y and z coordinates are plotted with respect to the unit cell axes, not to a Cartesian set of axes, (Figure 1.5). The position of an atom can also be expressed as a vector, **r**:

$$\mathbf{r} = x\mathbf{a} + y\mathbf{b} + z\mathbf{c}$$

where **a**, **b** and **c** are the unit cell axes, (Figure 1.5).

An atom at a cell corner is given the coordinates (0, 0, 0). An atom at the centre of the face of a unit cell is given the coordinates (½, ½, 0) if it lies between the **a**- and **b**-axes, (½, 0, ½) if between the **a**- and **c**-axes, and (0, ½, ½) if between the **b**- and **c**-axes. An atom at the centre of a unit cell would have a position specified as (½, ½, ½), irrespective of the type of unit cell. Atoms at the centres of the cell edges are specified at positions (½, 0, 0), (0, ½, 0) or (0, 0, ½), for atoms on the **a**-, **b**- and **c**-axis, (Figure 1.6). Stacking of the unit cells to build a structure will ensure that an atom at the unit cell origin will appear at every corner, and atoms on unit cell edges or faces will appear on all of the cell edges and faces.

In figures, the conventional origin is placed at the left rear corner of the unit cell. The **a**- or x-axis is represented as projecting out of the plane of the page, towards the reader, the **b**- or y-axis points to the right and the **c**- or z-axis points towards the top of the page. In projections, the origin is set at the upper left corner of the unit cell projection. A frequently encountered projection is that perpendicular to the **c**-axis. In this case, the **a**- or x-axis is drawn pointing down, (from top to bottom of the page), and the **b**- or y-axis pointing to the right. In projections the x and y coordinates can be determined from the figure. The z position is usually given on the figure as a fraction.

A vast number of structures have been determined, and it is very convenient to group those with topologically identical structures together. On going from one member of the group to another, the atoms in the unit cell differ, reflecting a change in chemical compound, and the atomic coordinates and unit cell dimensions change slightly, reflecting the difference in atomic size, but relative atom positions are identical or very similar. Frequently, the group name is taken from the name of a mineral, as mineral crystals were the first solids used for structure determination. Thus, all crystals with a structure similar to that of sodium chloride, NaCl, (the mineral halite), are said to adopt the *halite* structure. These materials then all have a general

Table 1.2 Strukturbericht symbols and names for simple structure types

Symbol and name	Examples	Symbol and name	Examples
A1, cubic close-packed, copper	Cu, Ag, Au, Al	A2, body-centred cubic, iron	Fe, Mo, W, Na
A3, hexagonal close-packed, magnesium	Mg, Be, Zn, Cd	A4, diamond	C, Si, Ge, Sn
B1, halite, rock salt	NaCl, KCl, NiO, MgO	B2, caesium chloride	CsCl, CsBr, AgMg, BaCd
B3, zinc blende	ZnS, ZnSe, BeS, CdS	B4, wurtzite	ZnS, ZnO, BeO, CdS, GaN
C1, fluorite	CaF_2, BaF_2, UO_2, ThO_2	C4, rutile	TiO_2, SnO_2, MgF_2, FeF_2

formula MX, where M is a metal atom and X a non-metal atom, for example, MgO. Similarly, crystals with a structure similar to the rutile form of titanium dioxide, TiO_2, are grouped with the *rutile* structure. These all have a general formula MX_2, for example FeF_2. As a final example, compounds with a similar structure to the mineral fluorite, (sometimes called fluorspar), CaF_2, are said to adopt the *fluorite* structure. These also have a general formula MX_2, an example being UO_2. Examples of these three structures follow. Crystallographic details of a number of simple inorganic structures are given in Appendix 2. Some mineral names of common structures are found in Table 1.2 and Appendix 2.

A system of nomenclature that is useful for describing relatively simple structures is that originally set out in 1920, in Volume 1 of the German publication Strukturbericht. It assigned a letter code to each structure; A for materials with only one atom type, B for two different atoms, and so on. Each new structure was characterised further by the allocation of a numeral, so that the crystal structures of elements were given symbols A1, A2, A3 and so on. Simple binary compounds were given symbols B1, B2 and so on, and binary compounds with more complex structures C1, C2, D1, D2 and so on. As the number of crystal structures and the complexity displayed increased, the system became unworkable. Nevertheless, the terminology is still used, and is useful for the description of simple structures. Some Strukturbericht symbols are given in Table 1.2.

1.5 The cubic close-packed (A1) structure of copper

A number of elemental metals crystallise with the cubic A1 structure, also called the copper structure.

Unit cell: cubic

Lattice parameter for copper[2], $a = 0.3610$ nm.

Z = 4 Cu

Atom positions: 0, 0, 0; ½, ½, 0;
  0, ½, ½; ½, 0, ½;

There are four copper atoms in the unit cell, (Figure 1.7). Besides some metals, the noble gases, Ne(s), Ar(s), Kr(s), Xe(s), also adopt this structure in the solid state. This structure is often called the face-centred cubic (*fcc*) structure or the cubic close-packed (*ccp*) structure, but the Strukturbericht symbol, A1 is the most compact notation. Each atom has 12 nearest neighbours, and if the atoms are supposed to be hard touching spheres, the fraction of the volume occupied is

[2]Lattice parameters and interatomic distances in crystal structures are usually reported in Ångström units, Å, in crystallographic literature. 1 Å is equal to 10^{-10} m, that is, 10 Å = 1 nm. In this book, the SI unit of length, nm, will be used most often, but Å will be used on occasion, to conform with crystallographic practice.

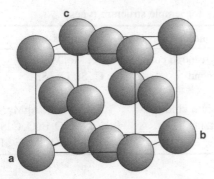

Figure 1.7 The cubic unit cell of the A1, copper, structure

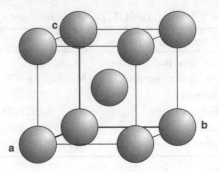

Figure 1.8 The cubic unit cell of the A2, tungsten, structure

0.7405. More information on this structure is given in Chapter 7.

1.6 The body-centred cubic (A2) structure of tungsten

A second common structure adopted by metallic elements is that of the cubic structure of tungsten, W.

Unit cell: cubic

Lattice parameter for tungsten, $a = 0.316$ nm.

Z = 2 W

Atom positions: 0, 0, 0; ½, ½, ½;

There are two tungsten atoms in the body-centred unit cell, one at (0, 0, 0) and one at the cell centre, (½, ½, ½), (Figure 1.8). This structure is often called the body-centred cubic (*bcc*) structure, but the Strukturbericht symbol, A2, is a more compact designation. In this structure, each atom has 8 nearest neighbours and 6 next nearest neighbours at only 15% greater distance. If the atoms are supposed to be hard touching spheres, the fraction of the volume

occupied is 0.6802. This is less than either the A1 structure above or the A3 structure that follows, both of which have a volume fraction of occupied space of 0.7405. The A2 structure is often the high temperature structure of a metal that has an A1 close-packed structure at lower temperatures.

1.7 The hexagonal (A3) structure of magnesium

The third common structure adopted by elemental metals is the hexagonal magnesium, Mg, structure.

Unit cell: hexagonal

Lattice parameters for magnesium,
$a = 0.321$ nm, $c = 0.521$ nm

Z = 2 Mg

Atom positions: 0, 0, 0; ⅓, ⅔, ½;

There are two atoms in the unit cell, one at (0, 0, 0) and one at (⅓, ⅔, ½). [The atoms can also be placed at ⅔, ⅓, ¼; ⅓, ⅔, ¾, by changing the unit cell origin. This is preferred for some

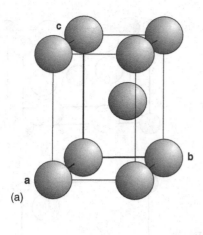

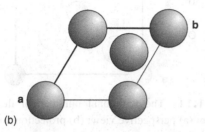

Figure 1.9 The hexagonal unit cell of the A3, magnesium, structure: (a) perspective view; (b) projection down the **c**-axis

purposes]. The structure is shown in perspective, (Figure 1.9a) and projected down the **c**-axis, (Figure 1.9b). This structure is often referred to as the hexagonal close-packed (*hcp*) structure. If the atoms are supposed to be hard touching spheres, the fraction of the volume occupied is 0.7405, equal to that in the A1 structure of copper, and the ratio of the lattice parameters, c/a, is equal to $\sqrt{8}/\sqrt{3}$, $= 1.633$. The ideal volume, V, of the unit cell, equal to the area of the base of the unit cell multiplied by the height of the unit cell, is:

$$V - \frac{\sqrt{3}}{2}a^2c = 0.8660\,a^2c$$

More information on this structure, and the relationship between the A1 and A3 structures, is given in Chapter 7.

1.8 The *halite* structure

The general formula of crystals with the *halite* structure is *MX*. The mineral halite, which names the group, is sodium chloride, NaCl, also called rock salt.

Unit cell: cubic.

Lattice parameter for halite, $a = 0.5640$ nm.

Z = 4 {NaCl}

Atom positions: Na: ½, 0, 0;   0, 0, ½;
0, ½, 0;   ½, ½, ½
Cl: 0, 0, 0;   ½, ½, 0;
½, 0, ½;   0, ½, ½

There are four sodium and four chlorine atoms in the unit cell. For all materials with the *halite* structure, Z = 4. In this structure, each atom is surrounded by six atoms of the opposite type at the corners of a regular octahedron (see Chapter 7). A perspective view of the halite structure is shown in Figure 1.10a, and a projection, down the **c**-axis, in Figure 1.10b.

This structure is adopted by many oxides, sulphides, halides and nitrides with a formula *MX*.

1.9 The *rutile* structure

The general formula of crystals with the *rutile* structure is MX_2. The mineral rutile, which names the group, is one of the structures adopted by titanium dioxide, TiO_2. [The other common form

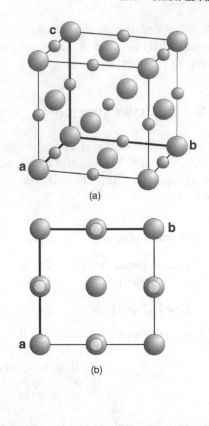

(a)

(b)

Na

Cl

Figure 1.10   The cubic unit cell of the B1, *halite*, structure: (a) perspective view; (b) projection down the **c**-axis

of TiO_2 encountered is called anatase. Other structures for TiO_2 are also known.]

Unit cell: tetragonal.

Lattice parameters for rutile, $a = 0.4594$ nm, $c = 0.2959$ nm.

$Z = 2 \{TiO_2\}$

Atom positions: Ti: 0, 0, 0;   ½, ½, ½
O: $^3/_{10}$, $^3/_{10}$, 0; $^4/_5$, $^1/_5$, ½;
$^7/_{10}$, $^7/_{10}$, 0; $^1/_5$, $^4/_5$, ½

There are two molecules of TiO_2 in the unit cell, that is, for all materials that adopt the *rutile*

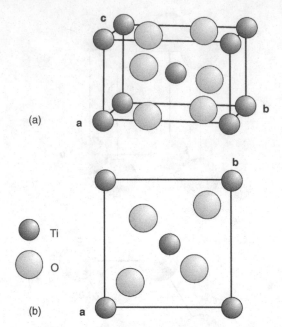

(a)

Ti

O

(b)

Figure 1.11   The tetragonal unit cell of the *rutile* structure: (a) perspective view; (b) projection down the **c**-axis

structure, $Z = 2$. In this structure, each titanium atom is surrounded by six oxygen atoms at the corners of an octahedron. A perspective view of the rutile structure is shown in Figure 1.11a, and a projection, down the **c**-axis, in Figure 1.11b.

This structure is relatively common and adopted by a number of oxides and fluorides with a formula MX_2.

1.10   The *fluorite* structure

The general formula of crystals with the *fluorite* structure is MX_2. The mineral fluorite, calcium fluoride, CaF_2, which names the group, is sometimes also called fluorspar.

Unit cell: cubic.

Lattice parameter for fluorite,
$a = 0.5463$ nm.

$Z = 4 \{CaF_2\}$

Atom positions:

Ca: 0, 0, 0; ½, ½, 0; 0, ½, ½; ½, 0, ½
F: ¼, ¾, ¼; ¼, ¾, ¾; ¼, ¼, ¼; ¼, ¼, ¾;
¾, ¼, ¼; ¾, ¼, ¾; ¾, ¾, ¼; ¾, ¾, ¾

There are four calcium and eight fluorine atoms in the unit cell. The number of molecules of CaF_2 in the unit cell is four, so that, for all *fluorite* structure compounds, $Z = 4$. In this structure, each calcium atom is surrounded by eight fluorine atoms at the corners of a cube. Each fluorine atom is surrounded by four calcium atoms at the vertices of a tetrahedron (see also Chapter 7). A perspective view of the structure is shown in Figure 1.12a, and a projection of the structure down the **c**-axis in Figure 1.12b.

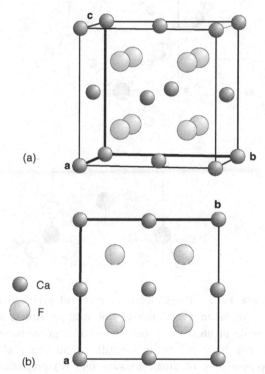

Ca
F

Figure 1.12 The cubic unit cell of the *fluorite* structure: (a) perspective view; (b) projection down the **c**-axis

This structure is adopted by a number of oxides and halides of large divalent cations of formula MX_2.

1.11   The structure of urea

The structures of molecular crystals tend to have a different significance to those of inorganic and mineral structures. Frequently, the information of most importance is the molecular geometry, and how the molecules are arranged in the crystallographic unit cell is often of less interest. To introduce the changed emphasis when dealing with molecular materials, the crystal structure of the organic compound urea is described. Urea is a very simple molecule, with a formula CH_4N_2O. The unit cell is small and of high symmetry. It was one of the earliest organic structures to be investigated using the methods of X-ray crystallography, and in these initial investigations the data was not precise enough to locate the hydrogen atoms. [The location of hydrogen atoms in a structure remains a problem to present times, see also Chapters 6 and 7.] The crystallographic data for urea is[3]

Unit cell: tetragonal.

Lattice parameters for urea,
$a = 0.5589$ nm, $c = 0.46947$ nm.

$Z = 2 \{CH_4N_2O\}$

Atom positions:
C1:	0, 0.5000, 0.3283
C2:	0.5000, 0, 0.6717
O1:	0.5000, 0, 0.4037
N1:	0.1447, 0.6447, 0.1784
N2:	0.8553, 0.3553, 0.1784
N3:	0.6447, 0.8553, 0.8216
N4:	0.3553, 0.1447, 0.8216

[3]Data adapted from: V. Zavodnik, A. Stash, V. Tsirelson, R. de Vries and D. Feil, Acta Crystallogr., B**55**, 45 (1999).

H1:   0.2552, 0.7552, 0.2845
H2:   0.1428, 0.6428, 0.9661
H3:   0.8448, 0.2448, 0.2845
H4:   0.8572, 0.3572, 0.9661
H5:   0.7552, 0.7448, 0.7155
H6:   0.2448, 0.2552, 0.7155
H7:   0.6429, 0.8571, 0.0339
H8:   0.3571, 0.1428, 0.0339

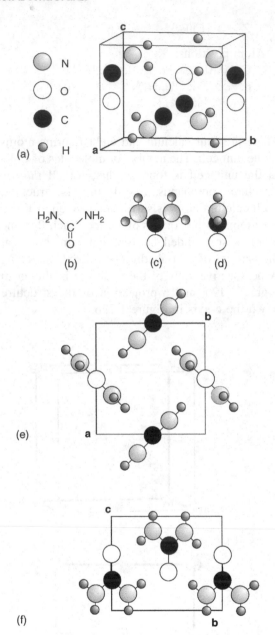

Notice that atoms of the same chemical type are numbered sequentially. The number of molecules of urea in the unit cell is two, so that $Z = 2$. The atoms in a unit cell, (including hydrogen), are shown in Figure 1.13a. This turns out to be not very helpful, and an organic chemist would have difficulty in recognising it as urea. This is because the molecules lie along the unit cell sides, so that a whole molecule is not displayed in the unit cell, only molecular fragments. [The unit cell is chosen in this way because of symmetry constraints, described in the following chapters.] The chemical structural formula for urea is drawn in Figure 1.13b, and this is compared to a molecule of urea viewed front on (Figure 1.13c), and edge on (Figure 1.13d) extracted from the crystallographic data. The crystal structure is redrawn in Figures 1.13e, f with the atoms linked to form molecules. This latter depiction now agrees with chemical intuition, and shows how the molecules are arranged in space.

Note that the list of atoms in the unit cell is becoming lengthy, albeit that this is an extremely simple structure. The ways used by crystallographers to reduce these lists to manageable proportions, by using the symmetry of the crystal, is explained in later chapters.

1.12   The density of a crystal

The theoretical density of a crystal can be found by calculating the mass of all the atoms in the unit

Figure 1.13   The structure of urea: (a) perspective view of the tetragonal unit cell of urea; (b) structural formula of urea; (c) a 'ball and stick' representation of urea 'face on', as in (b); (d) a 'ball and stick' representation of urea 'sideways on'; (e) projection of the structure along the **c**-axis; (f) projection of the structure down the **a**-axis

cell. The mass of an atom, m_A, is its molar mass (grams mol^{-1}) divided by the Avogadro constant, $N_A, (6.02214 \times 10^{23}\, mol^{-1})$

$$m_A = \text{molar mass}/N_A \quad \text{(grams)}$$
$$= \text{molar mass}/(1000 \times N_A) \text{ (kilograms)}$$

The total mass of all of the atoms in the unit cell is then

$$n_1\, m_1 + n_2\, m_2 + n_3\, m_3 \ldots /(1000 \times N_A)$$

where n_1 is the number of atoms of type 1, with a molar mass of m_1, and so on. This is written in a more compact form as

$$\sum_{i=1}^{q} n_i m_i /(1000 \times N_A)$$

where there are q different atom types in the unit cell. The density, ρ, is simply the total mass is divided by the unit cell volume, V:

$$\rho = \left\{ \sum_{i=1}^{q} n_i m_i /(1000 \times N_A) \right\}/V$$

For example, the theoretical density of halite is calculated in the following way. First count the number of different atom types in the unit cell. To count the number of atoms in a unit cell, use the information:

an atom within the cell counts as 1

an atom in a face counts as ½

an atom on an edge counts as ¼

an atom on a corner counts as ⅛

A quick method to count the number of atoms in a unit cell is to displace the unit cell outline to remove all atoms from corners, edges and faces. The atoms remaining, which represent the unit cell contents, are all within the boundary of the unit cell and count as 1.

The unit cell of the halite structure contains 4 sodium (Na) and 4 chlorine (Cl) atoms. The mass of the unit cell, m, is then given by:

$$m = [(4 \times 22.99) + (4 \times 35.453)]/1000 \times N_A$$
$$= 3.882 \times 10^{-25}\, kg$$

Where 22.99 g mol^{-1} is the molar mass of sodium, 35.453 g mol^{-1} is the molar mass of chlorine, and N_A is the Avogadro constant, $6.02214 \times 10^{23}\, mol^{-1}$.

The volume, V, of the cubic unit cell is given by a^3, thus:

$$V = (0.5640 \times 10^{-9})^3\, m^3$$
$$= 1.79406 \times 10^{-28}\, m^3.$$

The density, ρ, is given by the mass m divided by the volume, V:

$$\rho = 3.882 \times 10^{-25}\, kg/1.79406 \times 10^{-28}\, m^3$$
$$= 2164\, kg\, m^{-3}$$

The measured density is 2165 kg m^{-3}. The theoretical density is almost always slightly greater than the measured density because real crystals contain defects that act so as to reduce the total mass per unit volume.

Answers to introductory questions

What is a crystal system?

A crystal system is a set of reference axes, used to define the geometry of crystals and crystal structures. There are seven crystal systems, cubic, tetragonal, orthorhombic, monoclinic, triclinic, hexagonal and trigonal. As the crystal systems are sets of reference axes, they have a direction as well as a magnitude, and hence are vectors. They must be specified by length and interaxial angles.

The three reference axes are labelled **a**, **b** and **c**, and the angles between the positive direction of the axes as α, β, and γ, where α lies between +**b** and +**c**, β lies between +**a** and +**c**, and γ lies between +**a** and +**b**. The angles are chosen to be greater or equal to 90° except for the trigonal system. In figures, the **a**-axis is represented as projecting out of the plane of the page, towards the reader, the **b**-axis points to the right and the **c**-axis points towards the top of the page.

What are unit cells?

All crystals can be built by the regular stacking of a small volume of material called the unit cell. The edges of the unit cell are generally taken to be parallel to the axial vectors **a**, **b** and **c**, of the seven crystal systems. The lengths of the unit cell sides are written a, b and c, and the angles between the unit cell edges are written, α, β and γ. The collected values a, b, c, α, β and γ for a crystal structure are termed the unit cell or lattice parameters.

What information is needed to specify a crystal structure?

The minimum amount of information needed to specify a crystal structure is the unit cell type, i.e., cubic, tetragonal, etc, the unit cell parameters and the positions of all of the atoms in the unit cell. The atomic contents of the unit cell are a simple multiple, Z, of the composition of the material. The value of Z is equal to the number of formula units of the solid in the unit cell. Atom positions are expressed in terms of three coordinates, x, y, and z. These are taken as fractions of a, b and c, the unit cell sides, for example, ½, ½, ¼.

Problems and exercises

Quick quiz

1 The number of crystal systems is:
   (a) 5
   (b) 6
   (c) 7

2 The angle between the **a**- and **c**-axes in a unit cell is labelled:
   (a) α
   (b) β
   (c) γ

3 A tetragonal unit cell is defined by:
   (a) a = b = c, $\alpha = \beta = \gamma = 90°$
   (b) a = b $\neq$ c, $\alpha = \beta = \gamma = 90°$
   (c) a $\neq$ b $\neq$ c, $\alpha = \beta = \gamma = 90°$

4 A crystal is built by the stacking of unit cells with:
   (a) Orientational and translational long-range order
   (b) Orientational long-range order
   (c) Translational long-range order

5 Miller indices are used to label
   (a) Crystal shapes

(b) Crystal faces

(c) Crystal sizes

6 Crystal structures are often determined by the scattering of:
(a) Light
(b) Microwaves
(c) X-rays

7 In crystallography the letter Z specifies:
(a) The number of atoms in a unit cell
(b) The number of formula units in a unit cell
(c) The number of molecules in a unit cell

8 The position of an atom at the corner of a monoclinic unit cell is specified as:
(a) 1, 0, 0
(b) 1, 1, 1
(c) 0, 0, 0

9 The number of atoms in the unit cell of the halite structure is:
(a) 2
(b) 4
(c) 8

10 When determining the number of atoms in a unit cell, an atom in a face counts as:
(a) ½
(b) ¼
(c) ⅛

Calculations and Questions

1.1 The rhombohedral unit cell of arsenic, As, has unit cell parameters $a_R = 0.412$ nm, $\alpha = 54.17°$. Use graphical vector addition (Appendix 1) to determine the equivalent hexagonal lattice parameter a_H. Check your answer arithmetically, and calculate a value for the hexagonal lattice parameter c_H.

1.2 Cassiterite, tin dioxide, SnO_2, adopts the *rutile* structure, with a tetragonal unit cell, lattice parameters, $a = 0.4738$ nm, $c = 0.3187$ nm, with Sn atoms at: 0, 0, 0; ½, ½, ½; and O atoms at: $^3/_{10}$, $^3/_{10}$, 0; $^4/_5$, $^1/_5$, ½; $^7/_{10}$, $^7/_{10}$, 0; $^1/_5$, $^4/_5$, ½. Taking one corner of the unit cell as origin, determine the atom positions in nm and calculate the unit cell volume in nm^3. Draw a projection of the structure down the **c**-axis, with a scale of 1 cm = 0.1 nm.

1.3 The structure of $SrTiO_3$, is cubic, with a lattice parameter $a = 0.3905$ nm. The atoms are at the positions: Sr: ½, ½, ½; Ti: 0, 0, 0; O: ½, 0, 0; 0, ½, 0; 0, 0, ½. Sketch the unit cell. What is the number of formula units in the unit cell? [This structure type is an important one, and belongs to the *perovskite* family.]

1.4 (a) Ferrous fluoride, FeF_2, adopts the tetragonal *rutile* structure, with lattice parameters $a = 0.4697$ nm, $c = 0.3309$ nm. The molar masses are Fe, 55.847 g mol^{-1}, F, 18.998 g mol^{-1}. Calculate the density of this compound. (b) Barium fluoride, BaF_2, adopts the cubic *fluorite* structure, with lattice parameter $a = 0.6200$ nm. The molar masses are Ba, 137.327 g mol^{-1}, F, 18.998 g mol^{-1}. Calculate the density of this compound.

1.5 Strontium chloride, $SrCl_2$, adopts the *fluorite* structure, and has a density of 3052 kg m^{-3}. The molar masses of the atoms are Sr, 87.62 g mol^{-1}, Cl, 35.45 g mol^{-1}. Estimate the lattice parameter, a, of this compound.

1.6 Molybdenum, Mo, adopts the A2 (tungsten) structure. The density of the metal is

$10222 \text{ kg mol}^{-1}$ and the cubic lattice parameter is $a = 0.3147 \text{ nm}$. Estimate the molar mass of molybdenum.

1.7 A metal difluoride, MF_2, adopts the tetragonal *rutile* structure, with lattice parameters $a = 0.4621 \text{ nm}$, $c = 0.3052 \text{ nm}$ and density 3148 kg m^{-3}. The molar mass of fluorine, F, is $18.998 \text{ g mol}^{-1}$. Estimate the

molar mass of the metal and hence attempt to identify it.

1.8 The density of anthracene, $C_{14}H_{10}$, is 1250 kg m^{-3} and the unit cell volume is $475.35 \times 10^{-30} \text{ m}^3$. Determine the number of anthracene molecules, Z, which occur in a unit cell. The molar masses are: C, 12.011 g mol^{-1}, H, 1.008 g mol^{-1}.

2

Lattices, planes and directions

- *How does a crystal lattice differ from a crystal structure?*

- *What is a primitive unit cell?*

- *What are Miller-Bravais indices used for?*

The development of the idea of a lattice was among the earliest mathematical explorations in crystallography. Crystal structures and crystal lattices are different, although these terms are frequently (and incorrectly) used as synonyms. A crystal **structure** is built of **atoms**. A crystal **lattice** is an infinite pattern of **points**, each of which must have the same surroundings in the same orientation. A lattice is a mathematical concept.

All crystal structures can be built up from a lattice by placing an atom or a group of atoms at each lattice point. The crystal structure of a simple metal and that of a complex protein may both be described in terms of the same lattice, but whereas the number of atoms allocated to each lattice point is often just one for a simple metallic crystal, it may easily be thousands for a protein crystal.

2.1 Two-dimensional lattices

In two dimensions, if any lattice point is chosen as the origin, the position of any other lattice point is defined by the vector $\mathbf{P}(uv)$:

$$\mathbf{P}(uv) = u\mathbf{a} + v\mathbf{b} \qquad (2.1)$$

where the vectors $\mathbf{a}$ and $\mathbf{b}$ define a parallelogram and u and v are integers. The parallelogram is the **unit cell** of the lattice, with sides of length a and b. The coordinates of the lattice points are indexed as u, v, (Figure 2.1). Standard crystallographic terminology writes negative values as $\bar{u}$ and $\bar{v}$, (pronounced u bar and v bar). To agree with the convention for crystal systems given in Table 1.1, it is usual to label the angle between the lattice vectors, γ. The **lattice parameters** are the lengths of the axial vectors and the angle between them, a, b and γ. The choice of the vectors $\mathbf{a}$ and $\mathbf{b}$, which are called the **basis vectors** of the lattice, is completely arbitrary, and any number of unit cells can be constructed. However, for crystallographic purposes it is most convenient to choose as small a unit cell as possible and one that reveals the symmetry of the lattice.

Despite the multiplicity of possible unit cells, only five unique two-dimensional or **plane lattices**

Crystals and Crystal Structures.   Richard J. D. Tilley
© 2006 John Wiley & Sons, Ltd

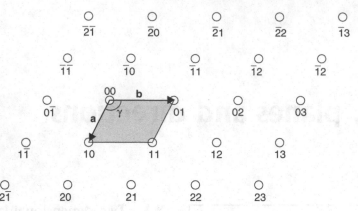

Figure 2.1   Part of an infinite lattice: the numbers are the indices, u, v, of each lattice point. The unit cell is shaded. Note that the points are exaggerated in size and do not represent atoms

are possible, (Figure 2.2). Those unit cells that contain only one lattice point are called **primitive** cells, and are labelled p. They are normally drawn with a lattice point at each cell corner, but it is easy

to see that the unit cell contains just one lattice point by mentally displacing the unit cell outline slightly. There are four primitive plane lattices, oblique, (*mp*), (Figure 2.2a), rectangular, (*op*),

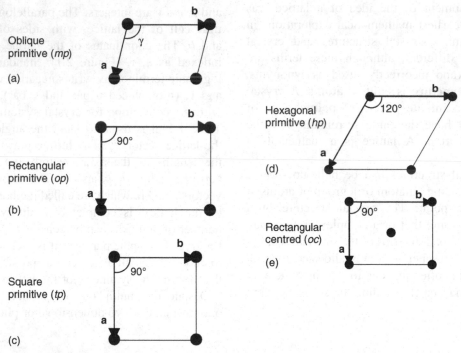

Figure 2.2   The unit cells of the five plane lattices: (a) oblique, (*mp*); (b) rectangular, (*op*); (c) square, (*tp*); (d) hexagonal, (*hp*); (e) rectangular centred, (*oc*)

(Figure 2.2b), square, (*tp*), (Figure 2.2c), and hexagonal, (*hp*), (Figure 2.2d).

The fifth plane lattice contains one lattice point at each corner and one in the unit cell centre. Such unit cells are called **centred** cells, and are labelled *c*. In this particular case, as the lattice is rectangular it is designated *oc*, (Figure 2.2e). The unit cell contains two lattice points, as can be verified by displacing the unit cell outline slightly. It is easily seen that this lattice could also be drawn as a primitive lattice, (Figure 2.3). This latter lattice, known as a rhombic (*rp*) lattice, has the two basis vectors of equal length, and an interaxial angle, γ, different from 90°. In terms of the basis vectors of the centred *oc* cell, **a** and **b**, the basis vectors of the rhombic *rp* cell, **a**′ and **b**′ are:

$$\mathbf{a}' = \tfrac{1}{2}(\mathbf{a} + \mathbf{b}); \quad \mathbf{b}' = \tfrac{1}{2}(-\mathbf{a} + \mathbf{b})$$

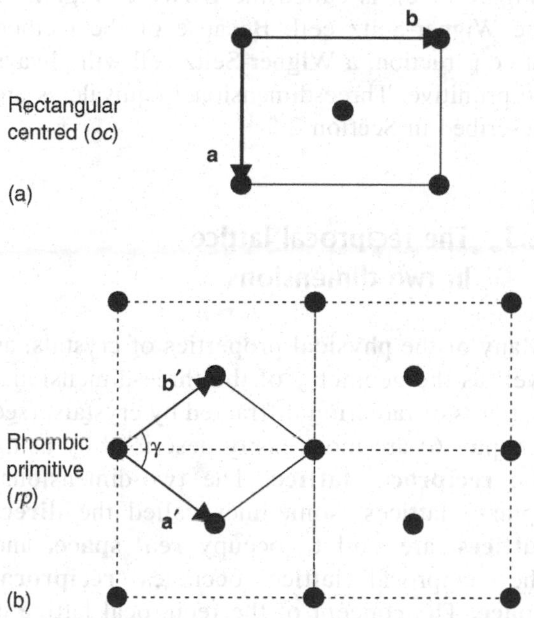

Rectangular centred (*oc*)

(a)

Rhombic primitive (*rp*)

(b)

Figure 2.3 The relationship between the rectangular (*oc*) lattice (a) and (b) the equivalent rhombic primitive (*rp*) designation of the same lattice

Table 2.1 The five plane lattices

Crystal system (Lattice type)	Lattice symbol	Lattice parameters
Oblique	*mp*	$a \neq b,\ \gamma \neq 90°$
Rectangular primitive	*op*	$a \neq b,\ \gamma = 90°$
Rectangular centred	*oc*	$a \neq b,\ \gamma = 90°$
	rp	$a' = b',\ \gamma \neq 90°$
Square	*tp*	$a = b,\ \gamma = 90°$
Hexagonal	*hp*	$a = b,\ \gamma = 120°$

Note that this is a vector equation and the terms are to be added vectorially (see Appendix 1). The five plane lattices are summarised in Table 2.1.

Although the lattice points in any primitive cell can be indexed in accordance with equation (2.1), using integer values of *u* and *v*, this is not possible with the *oc*-lattice. For example, taking the basis vectors as the unit cell sides, the coordinates of the two lattice points in the unit cell are 0, 0 and ½, ½. For crystallographic purposes it is better to choose a basis that reflects the symmetry of the lattice rather than stick to the rigid definition given by equation (2.1). Because of this, lattices in crystallography are defined in terms of **conventional crystallographic bases** (and hence **conventional crystallographic unit cells**). In this formalism, the definition of equation (2.1) is relaxed, so that the coefficients of each vector, *u*, *v*, must be either **integral** or **rational**. (A rational number is a number that can be written as *a/b* where both *a* and *b* are integers.) The *oc* lattice makes use of this definition, as the rectangular nature of the lattice is of prime importance. Remember though, that the *oc* and *rp* definitions are simply alternative descriptions of the same array of points. The lattice remains unique. It is simply necessary to use the description that makes things easiest.

It is an axiomatic principle of crystallography that a lattice cannot take on the symmetry of a regular pentagon. It is easy to see this. Suppose

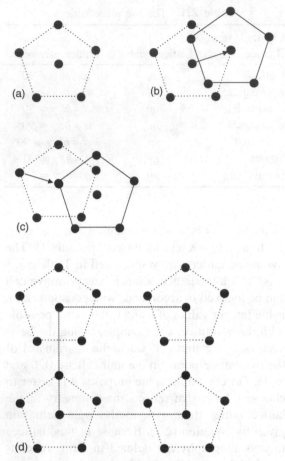

Figure 2.4 Lattices with pentagonal symmetry cannot form: (a) a fragment of a 'lattice' with pentagonal symmetry; (b, c) displacement of the fragment by a lattice vector does not extend the lattice; (d) fragments with pentagonal symmetry can be part of a pattern by placing each on a lattice point

that a lattice fragment is drawn in which a lattice point is surrounded by five others arranged at the vertices of a regular pentagon, (Figure 2.4a). Now each point in a lattice must have the exactly the same surroundings as any other lattice point. Displacement of the fragment by a lattice vector (Figure 2.4b,c) will show that some points are seen to be closer than others, which means that the construction

is not a lattice. However, the fragment can form part of a pattern by placing the fragment itself upon each point of a lattice, in this case an *op* lattice (Figure 2.4d).

2.2   Unit cells

The unit cells described above are conventional crystallographic unit cells. However, the method of unit cell construction described is not unique. Other shapes can be found that will fill the space and reproduce the lattice. Although these are not often used in crystallography, they are encountered in other areas of science. The commonest of these is the **Wigner-Seitz** cell.

This cell is constructed by drawing a line from each lattice point to its nearest neighbours, (Figure 2.5a). A second set of lines is then drawn normal to the first, through the midpoints, (Figure 2.5b). The polygon so defined, (Figure 2.5c), is called the **Dirichlet region** or the Wigner-Seitz cell. Because of the method of construction, a Wigner-Seitz cell will always be primitive. Three-dimensional equivalents are described in Section 2.5.

2.3   The reciprocal lattice in two dimensions

Many of the physical properties of crystals, as well as the geometry of the three-dimensional patterns of radiation diffracted by crystals, (see Chapter 6) are most easily described by using the **reciprocal lattice**. The two-dimensional (plane) lattices, sometimes called the **direct lattices**, are said to occupy **real space**, and the reciprocal lattice occupies **reciprocal space**. The concept of the reciprocal lattice is straightforward. (Remember, the reciprocal lattice is simply another lattice.) It is defined in terms of two basis vectors labelled **a*** and **b***.

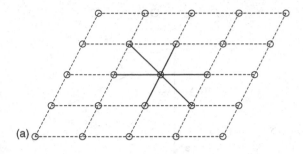

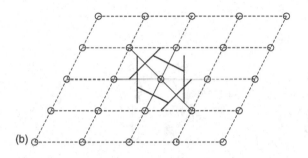

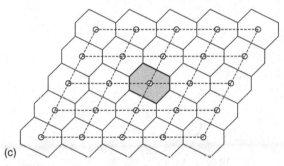

Figure 2.5 The construction of a Wigner–Seitz cell or Dirichlet region: (a) draw a line from each lattice point to its nearest neighbours; (b) draw a set of lines normal to the first, through their mid-points; (c) the polygon formed, (shaded) is the cell required

The direction of these vectors is perpendicular to the end faces of the direct lattice unit cell. The lengths of the basis vectors of the reciprocal lattice are the inverse of the perpendicular distance from the lattice origin to the end faces of the direct lattice unit cell. For the square and rectan-gular plane lattices, this is simply the inverse of the lattice spacing:

$$a^* = 1/a, \quad b^* = 1/b$$

For the oblique and hexagonal plane lattices, these are given by:

$$a^* = 1/d_{10}, \quad b^* = 1/d_{01}$$

where the perpendicular distances between the rows of lattice points is labelled d_{10} and d_{01}.

The construction of a reciprocal plane lattice is simple and is illustrated for the oblique plane (*mp*) lattice. Draw the plane lattice and mark the unit cell (Figure 2.6a). Draw lines perpendicular to the two sides of the unit cell. These lines give the axial directions of the reciprocal lattice basis vectors, (Figure 2.6b). Determine the perpendicular distances from the origin of the direct lattice to the end faces of the unit cell, d_{10} and d_{01} (Figure 2.6c). The inverse of these distances, $1/d_{10}$ and $1/d_{01}$, are the reciprocal lattice axial lengths, a^* and b^*.

$$1/d_{10} = a^*$$
$$1/d_{01} = b^*$$

Mark the lattice points at the appropriate reciprocal distances, and complete the lattice, (Figure 2.6d). Note that in this case, as in all real and reciprocal lattice pairs, the vector joining the origin of the reciprocal lattice to a lattice point *hk* is perpendicular to the (*hk*) planes in the real lattice and of length $1/d_{hk}$, (Figure 2.6e).

It will be seen that the angle between the reciprocal axes, **a*** and **b***, is $(180 - \gamma) = \gamma*$, when the angle between the direct axes, **a** and **b**, is γ. It is thus simple to construct the reciprocal lattice by drawing **a*** and **c*** at an angle of $(180 - \gamma)$ and marking out the lattice with the appropriate spacing a^* and b^*.

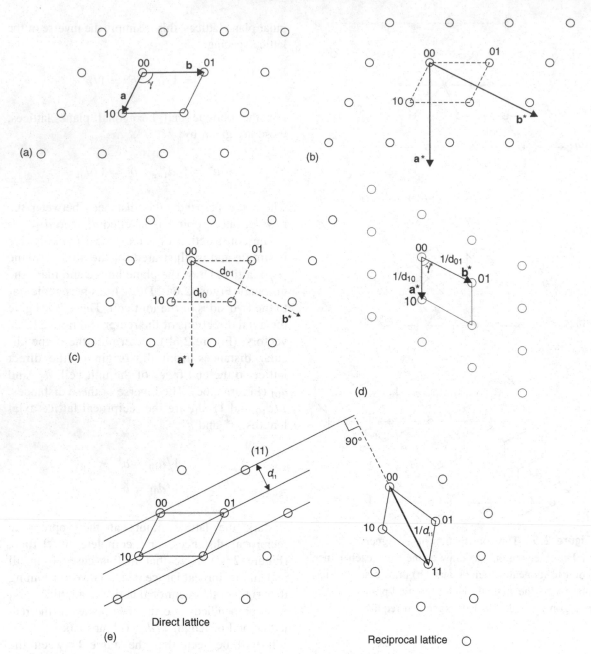

Figure 2.6 The construction of a reciprocal lattice: (a) draw the plane lattice and mark the unit cell; (b) draw lines perpendicular to the two sides of the unit cell to give the axial directions of the reciprocal lattice basis vectors; (c) determine the perpendicular distances from the origin of the direct lattice to the end faces of the unit cell, d_{10} and d_{01}, and take the inverse of these distances, $1/d_{10}$ and $1/d_{01}$, as the reciprocal lattice axial lengths, a^* and b^*; (d) mark the lattice points at the appropriate reciprocal distances, and complete the lattice; (e) the vector joining the origin of the reciprocal lattice to a lattice point hk is perpendicular to the (hk) planes in the real lattice and of length $1/d_{hk}$

Sometimes it can be advantageous to construct the reciprocal lattice of the centred rectangular *oc*-lattice using the primitive unit cell. In this way it will be found that the primitive reciprocal lattice so formed can also be described as a centred rectangular lattice. This is a general feature of reciprocal lattices. Each direct lattice generates a reciprocal lattice of the same type, i.e. *mp* → *mp*, *oc* → *oc*, etc. In addition, the reciprocal lattice of a reciprocal lattice is the direct lattice.

A construction in reciprocal space identical to that used to delineate the Wigner-Seitz cell in direct space gives a cell known as the first **Brillouin zone**, (Figure 2.7). The first Brillouin zone of a lattice is thus a primitive cell.

2.4 Three-dimensional lattices

Three-dimensional lattices use the same nomenclature as the two-dimensional lattice described above. If any lattice point is chosen as the origin, the position of any other lattice point is defined by the vector **P**(*uvw*):

$$P(uvw) = u\mathbf{a} + v\mathbf{b} + w\mathbf{c}$$

where **a**, **b**, and **c** are the basis vectors, and *u*, *v* and *w* are positive or negative integers or rational numbers. As before, there are any number of ways of choosing **a**, **b** and **c**, and crystallographic convention is to choose vectors that are small and reveal the underlying symmetry of the lattice. The parallelepiped formed by the three basis vectors **a**, **b** and **c**, defines the unit cell of the lattice, with edges of length *a*, *b*, and *c*. The numerical values of the unit cell edges and the angles between them are collectively called the lattice parameters of the unit cell. It follows from the above description that the unit cell is not unique and is chosen for convenience and to reveal the underlying symmetry of the crystal.

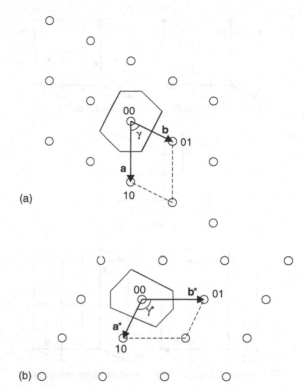

Figure 2.7 The first Brillouin zone of a reciprocal lattice: (a) the real lattice and Wigner-Seitz cell; (b) the reciprocal lattice and first Brillouin zone. The zone is constructed by drawing the perpendicular bisectors of the lines connecting the origin, 00, to the nearest neighbouring lattice points, in an identical fashion to that used to obtain the Wigner-Seitz cell in real space

There are only 14 possible three-dimensional lattices, called **Bravais** lattices (Figure 2.8). Bravais lattices are sometimes called direct lattices. Bravais lattices are defined in terms of conventional crystallographic bases and cells, (see Section 2.1). The rules for selecting the preferred lattice are determined by the symmetry of the lattice, (see Chapters 3, 4 for information on symmetry). In brief, the *main conditions* are:

(i) The three basis vectors define a right-handed coordinate system, that is, **a** (or *x*)

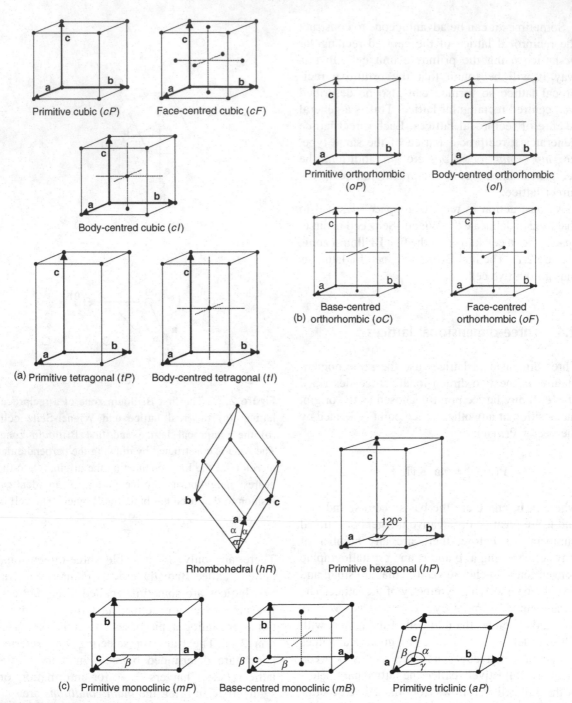

Figure 2.8   The 14 Bravais lattices. Note that the lattice points are exaggerated in size and are not atoms. The monoclinic lattices have been drawn with the **b**-axis vertical, to emphasise that it is normal to the plane containing the **a**- and **c**-axes

points out of the page, **b** (or y) points to the right and **c** (or z) is vertical.

(ii) The **a**, **b** and **c** basis vectors for a cubic lattice are parallel to the three four-fold symmetry axes.

(iii) The basis vector **c** for the hexagonal lattice lies parallel to the unique six-fold symmetry axis; **a** and **b** are along two-fold symmetry axes perpendicular to **c** and at 120° to each other.

(iv) The basis vector **c** for the tetragonal lattice is taken along the unique four-fold symmetry axis; **a** and **c** lie along two-fold symmetry axes perpendicular to each other and **c**.

(v) The basis vectors **a**, **b** and **c** for an orthorhombic crystal lie along three mutually perpendicular two-fold symmetry axes.

(vi) The unique symmetry direction in monoclinic lattices is conventionally labelled **b**; **a** and **c** lie in the lattice net perpendicular to **b** and include an oblique angle.

(vii) A rhombohedral lattice is described in two ways. If described in terms of a hexagonal lattice, **c** lies along the three-fold symmetry axis, with **a** and **c** chosen as for the hexagonal system. In terms of rhombohedral axes, **a**, **b** and **c** are the shortest non-coplanar lattice vectors symmetrically equivalent with respect to the three-fold axis.

(viii) A triclinic cell is chosen as primitive.

The smallest unit cell possible for any of the lattices, the one that contains just one lattice point, is the primitive unit cell. A primitive unit cell, usually drawn with a lattice point at each corner, is labelled P. All other lattice unit cells contain more than one lattice point. A unit cell with a lattice point at each corner and one at the centre of the unit cell, (thus containing two lattice points in total), is called a **body-centred** unit cell, and labelled I. A unit cell with a lattice point in the middle of each face, thus containing four lattice points, is called a **face-centred** unit cell, and labelled F. A unit cell that has just one of the faces of the unit cell centred, thus containing two lattice points, is labelled **A-face-centred**, A, if the centred faces cut the **a**-axis, **B-face-centred**, B, if the centred faces cut the **b**-axis and **C-face-centred**, C, if the centred faces cut the **c**-axis. The 14 Bravais lattices are summarised in Table 2.2. Note that all of the non-primitive Bravais lattices can be described in terms of alternative primitive unit cells.

The rhombohedral primitive lattice is often described in terms of a hexagonal lattice, but the lattice points can also be indexed in terms of a cubic face-centred lattice in the unique case in which the rhombohedral angle α_r is exactly equal to 60°, (also see Section 4.5).

As in two-dimensions, a lattice with a three-dimensional unit cell derived from regular pentagons, such as an icosahedron, cannot be constructed.

2.5 Alternative unit cells

As outlined above, a number of alternative unit cells can be described for any lattice. The most widely used is the Wigner-Seitz cell, constructed in three dimensions in an analogous way to that described in Section 2.2. The Wigner-Seitz cell of a body-centred cubic Bravais I lattice is drawn in Figure 2.9a,b. It has the form of a truncated octahedron, centred upon any lattice point, with the square faces lying on the cube faces of the Bravais lattice unit cell. The Wigner-Seitz cell of a face-centred cubic Bravais F lattice is drawn in Figure 2.9c,d. The polyhedron is a regular

Table 2.2 Bravais lattices

Crystal system	Lattice symbol	Lattice parameters
Triclinic	aP	$a \neq b \neq c$, $\alpha \neq 90°$, $\beta \neq 90°$, $\gamma \neq 90°$;
Monoclinic primitive	mP	$a \neq b \neq c$, $\alpha = 90°$, $\beta \neq 90°$, $\gamma = 90°$;
Monoclinic centred	mC	
Orthorhombic primitive	oP	$a \neq b \neq c$, $\alpha = \beta = \gamma = 90°$
Orthorhombic C-face-centred	oC	
Orthorhombic body-centred	oI	
Orthorhombic face-centred	oF	
Tetragonal primitive	tP	$a = b \neq c$, $\alpha = \beta = \gamma = 90°$
Tetragonal body-centred	tI	
Trigonal (Rhombohedral)	hR	$a = b = c$, $\alpha = \beta = \gamma$ (primitive cell); $a' = b' \neq c'$, $\alpha' = \beta' = 90°$, $\gamma' = 120°$ (hexagonal cell)
Hexagonal primitive	hP	$a = b \neq c$, $\alpha = \beta = 90°$, $\gamma = 120°$
Cubic primitive	cP	$a = b = c$, $\alpha = \beta = \gamma = 90°$
Cubic body-centred	cI	
Cubic face-centred	cF	

rhombic dodecahedron. Note that it is displaced by ½ **a** with respect to the crystallographic cell, and is centred upon the lattice point marked * in Figure 2.9c. The lattice points labelled A, B, C help to make the relationship between the two cells clearer. Recall that a Wigner-Seitz cell is always primitive. Other unit cells are sometimes of use when crystal structures are discussed, (see Chapter 7).

2.6 The reciprocal lattice in three dimensions

As with the two-dimensional lattices, the three-dimensional (Bravais) lattices, the direct lattices, are said to occupy real space, and the reciprocal

lattices occupy reciprocal space. Similarly, a reciprocal lattice is simply a lattice. It is defined in terms of three basis vectors **a***, **b*** and **c***. The direction of these vectors is perpendicular to the end faces of the direct lattice unit cell. This means that a direct axis will be perpendicular to a reciprocal axis if they have different labels, that is, **a** is perpendicular to **b*** and **c***. The reciprocal lattice axes are parallel to the direct lattice axes for cubic, tetragonal and orthorhombic direct lattices, that is, **a** is parallel to **a***, **b** is parallel to **b*** and **c** is parallel to **c***.

A direct lattice of a particular type, (triclinic, monoclinic, orthorhombic, etc.), will give a reciprocal lattice cell of the same type, (triclinic, monoclinic, orthorhombic, etc.). The reciprocal lattice of the cubic F direct lattice is a cubic I lattice and

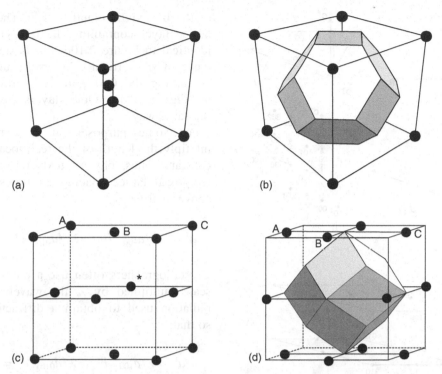

Figure 2.9 Wigner-Seitz cells: (a) the body-centred cubic lattice; (b) the Wigner-Seitz cell of (a); (c) the face-centred cubic lattice; (d) the Wigner-Seitz cell of (c). The face-centred cubic lattice point marked * forms the central lattice point in the Wigner-Seitz cell

the reciprocal lattice of the cubic I lattice is a cubic F lattice.

The reciprocal lattice of a reciprocal lattice is the direct lattice.

The lengths of the basis vectors of the reciprocal lattice, a^*, b^* and c^*, are the inverse of the perpendicular distance from the lattice origin to the end faces of the direct lattice unit cell, d_{100}, d_{010} and d_{001}, that is:

$$a^* = 1/d_{100}, \ b^* = 1/d_{010}, \ c^* = 1/d_{001}$$

For cubic, tetragonal and orthorhombic lattices, these are equivalent to:

$$a^* = 1/a, \ b^* = 1/b, \ c^* = 1/c$$

The construction of a three-dimensional reciprocal lattice is similar to that for a plane lattice, although more complex in that the third dimension is less easy to portray. The construction of the reciprocal lattice of a P monoclinic lattice is described in Figure 2.10. The direct lattice has a lozenge-shaped unit cell, with the **b**-axis normal to the **a**- and **c**-axes, (Figure 2.10a). To construct the sheet containing the **a*** and **c*** axes, draw the direct lattice unit cell projected down **b**, and draw normals to the end faces of the unit cell, (Figure 2.10b). These give the directions of the reciprocal lattice **a*** and **c*** axes. The reciprocals of the perpendicular distances from the origin to the faces of the unit cell give the axial lengths, (Figure 2.10c). These allow the reciprocal lattice plane to be drawn, (Figure 2.10d).

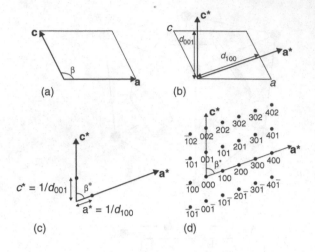

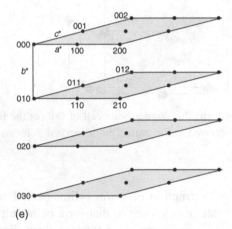

Figure 2.10 The construction of a reciprocal lattice: (a) the **a**–**c** section of the unit cell in a monoclinic (*mP*) direct lattice; (b) reciprocal lattice axes lie perpendicular to the end faces of the direct cell; (c) reciprocal lattice points are spaced $a^* = 1/d_{010}$ and $c^* = 1/d_{001}$; (d) the lattice plane is completed by extending the lattice; (e) the reciprocal lattice is completed by adding layers above and below the first plane

The b-axis is normal to **a** and **c**, and so the **b*** axis is parallel to **b** and normal to the section containing **a*** and **c***, drawn. The length

of the **b*** axis is equal to $1/b$. The reciprocal lattice layer containing the 010 point is then identical to Figure 2.10d, but is stacked a distance of **b*** vertically below it, and the layer containing the $0\bar{1}0$ point is vertically above it, (Figure 2.10e). Other layers then follow in the same way.

For some purposes, it is convenient to multiply the length of the reciprocal axes by a constant. Thus, physics texts frequently use a reciprocal lattice spacing 2π times that given above, that is:

$$a^* = 2\pi/d_{100}, \ b^* = 2\pi/d_{010}, \ c^* = 2\pi/d_{001}.$$

Crystallographers often use a reciprocal lattice scale multiplied by λ, the wavelength of the radiation used to obtain a diffraction pattern, so that:

$$a^* = \lambda/d_{100}, \ b^* = \lambda/d_{010}, \ c^* = \lambda/d_{001}.$$

As with two-dimensional lattices, the procedure required to construct the (primitive) Wigner-Seitz cell in the direct lattice yields a cell called the first Brillouin zone when applied to the reciprocal lattice. The lattice that is reciprocal to a real space face-centred cubic *F* lattice is a body-centred cubic *I* lattice. The (primitive) Wigner-Seitz cell of the cubic body-centred *I* lattice, (Figure 2.9a), a truncated octahedron, (Figure 2.9b), is therefore identical in shape to the (primitive) first Brillouin zone of a face-centred cubic *F* lattice. In the same way, the lattice that is reciprocal to a real space body-centred cubic *I* lattice is a face-centred cubic *F* lattice. The Wigner-Seitz cell of the real space face-centred cubic *F* lattice, (Figure 2.9c), a regular rhombic dodecahedron, (Figure 2.9d), is identical in shape to the first Brillouin zone of a body-centred centred cubic *I* lattice, (Table 2.3).

Table 2.3 Reciprocal and real space cells

Lattice	Direct lattice	Reciprocal lattice
plane	oblique *mp*: Wigner-Seitz cell Figure 2.7a	oblique *mp*: 1st Brillouin zone Figure 2.7b
cubic 3-d	*F* lattice: Wigner-Seitz cell is a rhombic dodecahedron, Figure 2.9d.	*I* lattice: 1st Brillouin zone is a truncated octahedron, Figure 2.9b
	I lattice: Wigner-Seitz cell is a truncated octahedron, Figure 2.9b	*F* lattice: 1st Brillouin zone is a rhombic dodecahedron, Figure 2.9d

2.7   Lattice planes and Miller indices

As described in Chapter 1, the facets of a well-formed crystal or internal planes through a crystal structure are specified in terms of **Miller Indices**, *h*, *k* and *l*, written in round brackets, (*hkl*). The same terminology is used to specify planes in a lattice.

Miller indices, (*hkl*), represent not just one plane, but the set of all identical parallel lattice planes. The values of *h*, *k* and *l* are the reciprocals of the fractions of a unit cell edge, *a*, *b* and *c* respectively, intersected by an appropriate plane. This means that a set of planes that lie parallel to a unit cell edge is given the index 0 (zero) regardless of the lattice geometry. Thus a set of planes that pass across the ends of the unit cells, cutting the **a**-axis at a position 1 *a*, and parallel to the **b**- and **c**-axes of the unit cell has Miller indices (100), (Figure 2.11a,b). The same principles apply to the other planes shown. The set of planes that lies parallel to the **a**- and **c**-axes, and intersecting the end of each unit cell at a position 1 *b* have Miller indices (010), (Figure 2.11c,d). The set of planes that lies parallel to the **a**- and **b**-axes, and intersecting the end of each unit cell at a position 1 *c* have Miller indices (001), (Figure 2.11e,f). Planes cutting both the **a**-axis and **b**-axis at 1 *a* and 1 *b* will be (110) planes, (Figure 2.11 g,h), and planes cutting the **a**-, **b**- and **c**-axes at 1 *a*, 1 *b* and 1 *c* will be (111).

Remember that the Miller indices refer to a family of planes, not just one. For example, Figure 2.12 shows part of the set of (122) planes, which cut the unit cell edges at 1 *a*, ½ *b* and ½ *c*.

The Miller indices for lattice planes can be determined using a simple method. For generality, take a triclinic lattice, and imagine a set of planes parallel to the **c**-axis, intersecting parallel sheets of lattice points as drawn in Figure 2.13. The Miller indices of this set of planes are determined by travelling along the axes in turn and counting the number of spaces between planes encountered on passing from one lattice point to the next. Thus, in Figure 2.13, three spaces are crossed on going from one lattice point to another along the **a**-direction, so that *h* is 3. In travelling along the **b**-direction, two spaces are encountered in going from one lattice point to another, so that the value of *k* is 2. As the planes are supposed to be parallel to the **c**-axis, the value of *l* is 0. The planes have Miller indices (320). For non-zero values of *l*, repeat the sketch so as to include **a**- and **c**- or **b**- and **c**-, and repeat the procedure.

The planes described in Figures 2.11, 2.12 and 2.13 are crossed in travelling along positive directions of the axes. Some planes may intersect one of the axes in a positive direction and the other in a negative direction. Negative intersections are written with a negative sign over the index, $\bar{h}$, (pronounced h bar), $\bar{k}$, (pronounced k

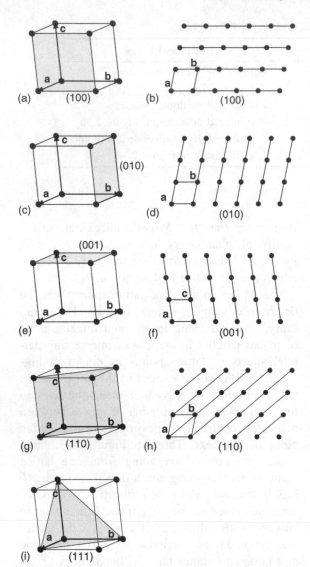

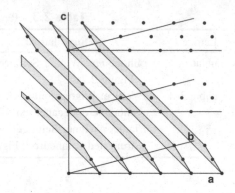

Figure 2.12   Part of the set of (122) lattice planes

Miller indices (110). The plane shown as a dotted line in Figure 2.14b cuts the **a**-axis at $-1a$ and the **b**-axis at $+1b$, and the planes have Miller indices $(\bar{1}10)$, (pronounced one bar, one, zero). The plane shown as a dotted line in Figure 2.14c cuts the **a**-axis at $+1a$ and the **b**-axis at $-1b$, and the planes have Miller indices $(1\bar{1}0)$, (pronounced one, one bar, zero). Finally the plane shown as a dotted line in Figure 2.14d cuts the **a**-axis in $-1a$ and the **b**-axis in $-1b$, so the planes have Miller indices $(\bar{1}\bar{1}0)$, (pronounced one bar, one bar, zero). Note that a $(\bar{1}\,\bar{1}\,0)$ plane is identical to a (110) plane, as the position of the axes is arbitrary; they can be placed anywhere on the diagram.

Figure 2.11   Miller indices of lattice planes: (a, b) (100); (c, d) (010); (e, f) (001), (g, h) (110); (i) (111)

bar) and $\bar{l}$, (pronounced 1 bar). For example, there are four planes related to (110), three of which involve travelling in a negative axial direction in order to count the spaces between the planes encountered, (Figure 2.14). The plane shown as a dotted line in Figure 2.14a cuts the **a**-axis at $+1a$ and the **b**-axis at $+1b$, and the planes have

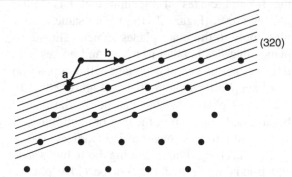

Figure 2.13   Determination of Miller indices: count the spaces crossed on passing along each cell edge, to give (320)

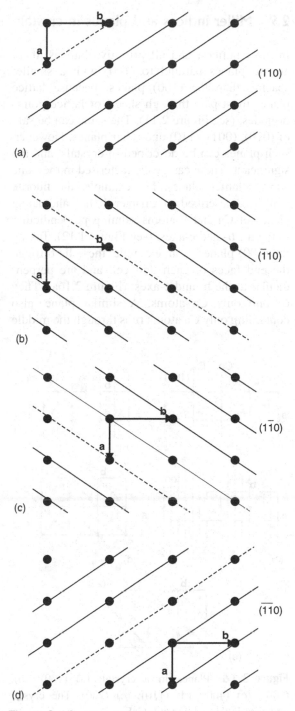

Figure 2.14 Negative Miller indices: (a) (110); (b) ($\bar{1}$10); (c) (1$\bar{1}$0); (d) ($\bar{1}\bar{1}$0)

Similarly, planes with Miller indices (1$\bar{1}$0) are identical to ($\bar{1}$10) planes.

In the three-dimensional direct and reciprocal lattice pairs, the vector joining the origin of the reciprocal lattice to a lattice point hkl is perpendicular to the (hkl) planes in the real lattice and of length $1/d_{hkl}$. An alternative method of constructing the reciprocal lattice is thus to draw the normals to the relevant (hkl) planes on the direct lattice and plot the reciprocal lattice points along lines parallel to these normals at separations of $1/d_{hkl}$. This method is often advantageous when constructing the reciprocal lattices for all of the face- or body-centred direct lattices. The same is true for the planar direct and reciprocal lattice pairs. The vector joining the origin of the reciprocal lattice to a lattice point hk is perpendicular to the (hk) planes in the real lattice and of length $1/d_{hk}$.

In lattices of high symmetry, there are always some sets of (hkl) planes that are identical from the point of view of symmetry. For example, in a cubic lattice, the (100), (010) and (001) planes are identical in every way. Similarly, in a tetragonal lattice, (110) and ($\bar{1}$10) planes are identical. Curly brackets, {hkl}, designate these related planes. Thus, in the cubic system, the symbol {100} represents the three sets of planes (100), (010), and (001), the symbol {110} represents the six sets of planes (110), (101), (011), ($\bar{1}$10), ($\bar{1}$01), and (0$\bar{1}$1), and the symbol {111} represents the four sets (111), (11$\bar{1}$), (1$\bar{1}$1) and ($\bar{1}$11).

2.8 Hexagonal lattices and Miller-Bravais indices

The Miller indices of planes in hexagonal lattices can be ambiguous. For example, three sets of planes lying parallel to the **c**-axis, which is imagined to be perpendicular to the plane of the diagram, are shown in (Figure 2.15). These planes have Miller indices A, (110), B, (1$\bar{2}$0) and C, ($\bar{2}$10). Although these Miller indices seem to refer

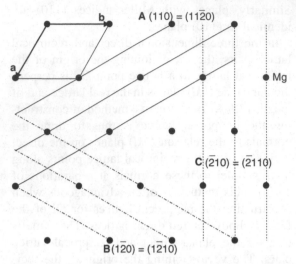

Figure 2.15 Miller-Bravais indices in hexagonal lattices. The three sets of identical planes marked have different Miller indices but similar Miller-Bravais indices

to different types of plane, clearly the three planes are identical, and all are equivalent to the planes through the 'long' diagonal of the unit cell. In order to eliminate this confusion, four indices, $(hkil)$, are often used to specify planes in a hexagonal crystal. These are called **Miller-Bravais indices** and are only used in the hexagonal system. The index i is given by:

$$h + k + i = 0, \text{ or } i = -(h+k)$$

In reality this third index is not needed, as it is simply derived from the known values of h and k. However, it does help to bring out the relationship between the planes. Using four indices, the planes are A, $(11\bar{2}0)$, B, $(1\bar{2}10)$ and C, $(\bar{2}110)$. Because it is a redundant index, the value of i is sometimes replaced by a dot, to give indices $(hk.l)$. This nomenclature emphasises that the hexagonal system is under discussion without actually including a value for i.

2.9 Miller indices and planes in crystals

In most lattices, and all primitive lattices, there are no planes parallel to (100) with a smaller spacing than the (100) planes, because lattice planes must pass through sheets of lattice points or nodes, (see Figure 2.11). The same can be said of (010), (001), (110) and other planes. However, such planes can be described in crystals, and are significant. These can be characterised in the same way as lattice planes. For example, the fluorite structure, described in Chapter 1, has alternating planes of Ca and F atoms running perpendicular to the **a**-, **b**- and **c**-axes, (see Figure 1.12). Taking the (100) planes as an example, these lie through the end faces of each unit cell and are perpendicular to the **b**- and **c**-axes, (Figure 2.16a). They contain only Ca atoms. A similar plane, also containing only Ca atoms runs through the middle

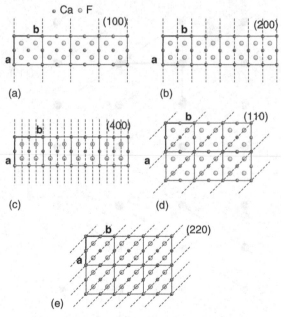

Figure 2.16 Planes in a crystal: (a) (100); (b), (200); (c) (400); (d) (110); (e) (220). The crystal structure is that of fluorite, CaF_2

of the cell. Using the construction described earlier, (Figure 2.13), these can be indexed as (200), as two interplanar spaces are crossed in moving one lattice parameter distance, (Figure 2.16b). Thus all the set of (200) planes contain only Ca atoms. The set of parallel planes with half the spacing of the (200) set will be indexed as (400), as four spaces are crossed in moving one lattice parameter distance. However, these planes are not all identical, as some contain only F atoms and others only Ca atoms, (Figure 2.16c). The Miller indices of (110), (Figure 2.16d) and (220), (Figure 2.16e) planes can be determined in the same way, by counting the spaces crossed in moving along one unit cell length in each direction. The atomic composition of both sets of planes is the same.

Any general set of planes parallel to the **b**- and **c**-axes, and so only cutting the **a** cell edge is written (h00). Any general set of planes plane parallel to the **a**- and **c**-axes, and so only cutting the **b** cell-edge has indices (0k0) and any general plane parallel to the **a**- and **b**-axes, and so cutting the **c** cell-edge has indices (00l). Planes that cut two edges and parallel to a third are described by indices (hk0), (0kl) or (h0l). Planes that are at an angle to all three axes have indices (hkl). Negative intersections and symmetrically equivalent planes are defined using the same terminology as described in Section 2.7, above.

2.10 Directions

The response of a crystal to an external stimulus such as a tensile stress, electric field, and so on, is frequently dependent upon the direction of the applied stimulus. It is therefore important to be able to specify directions in crystals in an unambiguous fashion. Directions are written generally as [uvw] and are enclosed in square brackets. Note that the symbol [uvw] includes all parallel directions, just as (hkl) specifies a set of parallel planes.

The three indices u, v, and w define the coordinates of a point within the lattice. The index u gives the coordinates in terms of the lattice repeat a along the **a**-axis, the index v gives the coordinates in terms of the lattice repeat b along the **b**-axis and the index w gives the coordinates in terms of the lattice repeat c along the **c**-axis. The direction [uvw] is simply the vector pointing from the origin to the lattice point with coordinates u, v, w. The direction [230], with $u = 2$ and $v = 3$, is drawn in Figure 2.17a. Remember, though, that, any parallel direction shares the symbol [uvw], because the origin of the coordinate system is not fixed and can always be moved to the starting point of the vector, (Figure 2.17b). (A North wind is always a North wind, regardless of where you stand.)

The direction [100] is parallel to the **a**-axis, the direction [010] is parallel to the **b**-axis and [001] is parallel to the **c**-axis. The direction [111] lies along the body diagonal of the unit cell. Negative values of u, v and w are written $\bar{u}$, (pronounced

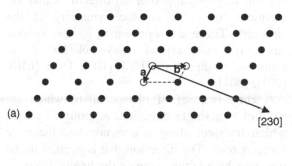

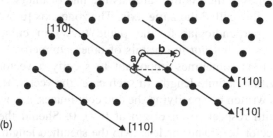

Figure 2.17 Directions in a lattice: (a) [230]; (b) [110]

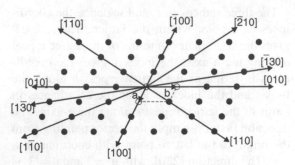

Figure 2.18   Directions in a lattice

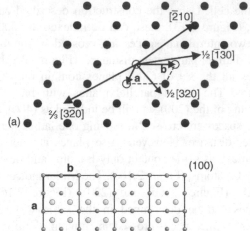

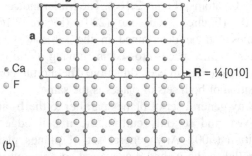

Figure 2.19   Vectors: (a) vectors in a lattice; (b) the displacement of part of a crystal of fluorite, CaF_2, by a vector, $\mathbf{R} = \frac{1}{4}$ [010] with respect to the other half

u bar), $\bar{v}$, (pronounced v bar) and $\bar{w}$, (pronounced w bar). Further examples of directions in a lattice are illustrated in Figure 2.18. Because directions are vectors, [uvw] is not identical to [$\bar{u}\,\bar{v}\,\bar{w}$], in the same way that the direction 'North' is not the same as the direction 'South'.

Directions in a crystal are specified in the same way. In these instances the integers u, v, w are applied to the unit cell vectors, **a**-, **b**- and **c**-. As with Miller indices, it is sometimes convenient to group together all directions that are identical by virtue of the symmetry of the structure. These are represented by the notation ⟨uvw⟩. In a cubic crystal the symbol ⟨100⟩ represents the six directions [100], [$\bar{1}$00], [010], [0$\bar{1}$0], [001], [00$\bar{1}$].

A **zone** is a set of planes, all of which are parallel to a single direction. A group of planes which intersect along a common line therefore forms a zone. The direction that is parallel to the planes, which is the same as the line of intersection, is called the **zone axis**. The zone axis [uvw] is perpendicular to the plane (uvw) in cubic crystals but *not* in crystals of other symmetry.

It is sometimes important to specify a vector with a definite length. In such cases the vector, **R**, is written by specifying the end coordinates, u, v w, with respect to an origin at 0, 0, 0. Should the vector be greater or less than the specified length, it is prefixed by the appropriate scalar multiplier,

(see Appendix 1), in accordance with normal vector arithmetic, (Figure 2.19a).

Crystals often contain planar boundaries which separate two parts of a crystal that are not in perfect register. Vectors are used to define the displacement of one part with respect to the other. For example, a fault involving the displacement of one part of a crystal of fluorite, CaF_2, by ¼ of the unit cell edge, i.e. $\mathbf{R} = \frac{1}{4}$ [010] with respect to the other part[1], is drawn in (Figure 2.19b).

As with Miller indices, directions in hexagonal crystals are sometimes specified in terms of a

---

[1]In cubic crystals, a vector such as that describing the boundary drawn in Figure 2.19b is frequently denoted as ¼ **a** [010]. Similarly, a vector ⅔ [$\bar{3}\bar{1}$0] may be written as ⅔ **a** [$\bar{3}\bar{1}$0]. This notation is confusing, and is best avoided.

four-index system, $[u'v'tw']$ called **Weber indices**. The conversion of a three-index set to a four-index set is given by the following rules.

$$[uvw] \rightarrow [u'v'tw']$$

$$u' = n(2u - v)/3$$

$$v' = n(2v - u)/3$$

$$t = -(u' + v')$$

$$w' = nw$$

In these equations, n is a factor *sometimes* needed to make the new indices into smallest integers. Thus the direction [001] always transforms to [0001]. The three equivalent directions in the basal (0001) plane of a hexagonal crystal structure such as magnesium, (Figure 2.20), are obtained by using the above transformations. The correspondence is:

$$[100] = [2\bar{1}\bar{1}0]$$
$$[010] = [\bar{1}2\bar{1}0]$$
$$[\bar{1}10] = [\bar{1}\bar{1}20]$$

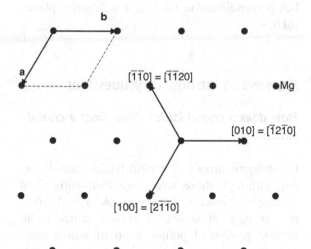

Figure 2.20 Directions in a hexagonal lattice

The relationship between directions and planes depends upon the symmetry of the crystal. In cubic crystals, (and *only* cubic crystals), the direction [hkl] is normal to the plane (hkl).

2.11 Lattice geometry

The most important metrical properties of lattices and crystals for everyday crystallography are given below. These are expressed most compactly using vector notation, but are given here 'in longhand', without derivation, as a set of useful tools.

The volume of the reciprocal unit cell, V* is given by:

$$V^* = 1/V$$

where V is the volume of the direct unit cell. The direction $[uvw]$ lies in the plane (hkl) when:

$$hu + kv + lw = 0$$

The intersection of two planes, $(h_1k_1l_1)$ and $(h_2k_2l_2)$ is the direction $[uvw]$, where:

$$u = k_1l_2 - k_2l_1$$
$$v = h_2l_1 - h_1l_2$$
$$w = h_1k_2 - h_2k_1$$

Three planes, $(h_1k_1l_1)$, $(h_2k_2l_2)$ and $(h_3k_3l_3)$ form a zone when:

$$h_1(k_2l_3 - l_2k_3) - k_1(h_2l_3 - l_2h_3)$$
$$+ l_1(h_2k_3 - k_2h_3) = 0$$

The plane $(h_3k_3l_3)$ belongs to the same zone as $(h_1k_1l_1)$ and $(h_2k_2l_2)$ when:

$$h_3 = mh_1 \pm nh_2; k_3 = mk_1 \pm nk_2; l_3 = ml_1 \pm ml_2$$

where m and n are integers.

Table 2.4 Interplanar spacing d_{hkl}

System	$1/(d_{hkl})^2$
cubic	$[h^2 + k^2 + l^2]/a^2$
tetragonal	$[(h^2 + k^2)/a^2] + [l^2/c^2]$
orthorhombic	$[h^2/a^2] + [k^2/b^2] + [l^2/c^2]$
monoclinic	$[h^2/a^2 \sin^2 \beta] + [k^2/b^2] + [l^2/c^2 \sin^2 \beta] - [(2hl \cos \beta)/(ac \sin^2 \beta)]$
triclinic*	$[1/V^2]\{[S_{11}h^2] + [S_{22}k^2] + [S_{33}l^2] + [2S_{12}hk] + [2S_{23}kl] + [2S_{13}hl]\}$
hexagonal	$[\tfrac{4}{3}][(h^2 + hk + k^2)/a^2] + [l^2/c^2]$
rhombohedral	$\{[(h^2 + k^2 + l^2)\sin^2 \alpha + 2(hk + kl + hl)(\cos^2 \alpha - \cos \alpha)]/[a^2(1 - 3\cos^2 \alpha + 2\cos^3 \alpha)]\}$

* V = unit cell volume.
$S_{11} = b^2c^2 \sin^2 \alpha$ $S_{22} = a^2c^2 \sin^2 \beta$ $S_{33} = a^2b^2 \sin^2 \gamma$
$S_{12} = abc^2(\cos \alpha \cos \beta - \cos \gamma)$ $S_{23} = a^2bc(\cos \beta \cos \gamma - \cos \alpha)$
$S_{13} = ab^2c(\cos \gamma \cos \alpha - \cos \beta)$

The three directions $[u_1v_1w_1]$, $[u_2v_2w_2]$ and $[u_3v_3w_3]$ lie in one plane when

$$u_1(v_2w_3 - w_2v_3) - v_1(u_2w_3 - w_2u_3)$$
$$+ w_1(u_2v_3 - v_2u_3) = 0$$

Two directions $[u_1v_1w_1]$ and $[u_2v_2w_2]$ lie in a single plane (hkl) when:

$$h = v_1w_2 - v_2w_1$$
$$k = u_2w_1 - u_1w_2$$
$$l = u_1v_2 - u_2v_1$$

The reciprocal lattice vector

$$\mathbf{r} = u\mathbf{a}^* + v\mathbf{b}^* + w\mathbf{c}^*$$

lies perpendicular to the direct lattice planes (uvw), and the direct lattice vector

$$\mathbf{R} = h\mathbf{a} + k\mathbf{b} + l\mathbf{c}$$

lies perpendicular to the reciprocal lattice planes (hkl).

Table 2.5 Unit cell volume, V

System	V
Cubic	a^3
tetragonal	a^2c
orthorhombic	abc
monoclinic	$abc \sin \beta$
triclinic	$abc\sqrt{(1 - \cos^2 \alpha - \cos^2 \beta - \cos^2 \gamma + 2\cos\alpha \cos\beta \cos\gamma)}$
hexagonal	$[\sqrt{(3)}/2][a^2c] \approx 0.866a^2c$
rhombohedral	$a^3\sqrt{(1 - 3\cos^2 \alpha + 2\cos^3 \alpha)}$

Answers to introductory questions

How does a crystal lattice differ from a crystal structure?

Crystal structures and crystal lattices are different, although these terms are frequently (and incorrectly) used as synonyms. A crystal structure is built of atoms. A crystal lattice is an infinite pattern of points, each of which must have the same surroundings in the same orienta-

Table 2.6 Interplanar angle φ

System	$\cos \varphi$
cubic	$[h_1 h_2 + k_1 k_2 + l_1 l_2]/\{[h_1^2 + k_1^2 + l_1^2][h_2^2 + k_2^2 + l_2^2]\}^{1/2}$
tetragonal	$\dfrac{\dfrac{h_1 h_2 + k_1 k_2}{a^2} + \dfrac{l_1 l_2}{c^2}}{\left[\left(\dfrac{h_1^2 + k_1^2}{a^2} + \dfrac{l_1^2}{c^2}\right)\left(\dfrac{h_2^2 + k_2^2}{a^2} + \dfrac{l_2^2}{c^2}\right)\right]^{1/2}}$
orthorhombic	$\dfrac{\dfrac{h_1 h_2}{a^2} + \dfrac{k_1 k_2}{b^2} + \dfrac{l_1 l_2}{c^2}}{\left[\left(\dfrac{h_1^2}{a^2} + \dfrac{k_1^2}{b^2} + \dfrac{l_1^2}{c^2}\right)\left(\dfrac{h_2^2}{a^2} + \dfrac{k_2^2}{b^2} + \dfrac{l_2^2}{c^2}\right)\right]^{1/2}}$
monoclinic*	$d_1 d_2 \left(\dfrac{h_1 h_2}{a^2 \sin^2 \beta} + \dfrac{k_1 k_2}{b^2} + \dfrac{l_1 l_2}{c^2 \sin^2 \beta} - \dfrac{(l_1 h_2 + l_2 h_1)\cos\beta}{ac \sin^2 \beta}\right)$
triclinic*	$\dfrac{d_1 d_2}{V^2}[S_{11} h_1 h_2 + S_{22} k_1 k_2 + S_{33} l_1 l_2 + S_{23}(k_1 l_2 + k_2 l_1) + S_{13}(l_1 h_2 + l_2 h_1) + S_{12}(h_1 k_2 + h_2 k_1)]$
hexagonal	$\dfrac{h_1 h_2 + k_1 k_2 + \frac{1}{2}(h_1 k_2 + h_2 k_1) + \dfrac{3a^2 l_1 l_2}{4c^2}}{\left[\left(h_1^2 + k_1^2 + h_1 k_1 + \dfrac{3a^2 l_1^2}{4c^2}\right)\left(h_2^2 + k_2^2 + h_2 k_2 + \dfrac{3a^2 l_2^2}{4c^2}\right)\right]^{1/2}}$
rhombohedral*	$\dfrac{d_1 d_2\{(h_1 h_2 + k_1 k_2 + l_1 l_2)\sin^2\alpha + [h_1(k_2 + l_2) + k_1(h_2 + l_2) + l_1(h_2 + k_2)\cos\alpha(\cos\alpha - 1)]\}}{a^2(1 - 3\cos^2\alpha + 2\cos^3\alpha)}$

* $V =$ unit cell volume, d_1 is the interplanar spacing of $(h_1 k_1 l_1)$ and d_2 is the interplanar spacing of $(h_2 k_2 l_2)$ and
$S_{11} = b^2 c^2 \sin^2\alpha$ $S_{22} = a^2 c^2 \sin^2\beta$ $S_{33} = a^2 b^2 \sin^2\gamma$
$S_{12} = abc^2(\cos\alpha\cos\beta - \cos\gamma)$ $S_{23} = a^2 bc(\cos\beta\cos\gamma - \cos\alpha)$
$S_{13} = ab^2 c(\cos\gamma\cos\alpha - \cos\beta)$

tion. A lattice is a mathematical concept. There are only 5 different two-dimensional (planar) lattices and 14 different three-dimensional (Bravais) lattices.

All crystal structures can be built up from the Bravais lattices by placing an atom or a group of atoms at each lattice point. The crystal structure of a simple metal and that of a complex protein may both be described in terms of the same lattice, but whereas the number of atoms allocated to each lattice point is often just one for a simple metallic crystal, it may easily be thousands for a protein crystal.

What is a primitive unit cell?

A primitive unit cell is a lattice unit cell that contains only one lattice point. The four primitive plane lattice unit cells are labelled p: oblique, (mp), rectangular, (op), square, (tp) and hexagonal, (hp). They are normally drawn with a lattice point at each cell corner, but it is easy to see that the unit cell contains just one lattice point by mentally displacing the unit cell outline slightly. There are five primitive Bravais lattices, labelled *P*: triclinic, (*aP*), monoclinic primitive, (*mP*), tetragonal primitive, (*tP*), hexagonal primitive, (*hP*) and cubic primitive, (*cP*). In addition the trigonal lattice, when referred to rhombohedral axes, has a primitive unit cell, although the lattice is labelled *hR*.

What are Miller-Bravais indices used for?

The facets of a well-formed crystal or internal planes through a crystal structure are specified in terms of Miller Indices, *h*, *k* and *l*, written in round brackets, (*hkl*). Miller indices, (*hkl*), represent not just one plane, but the set of all identical parallel planes.

The Miller indices of planes in crystals with a hexagonal unit cell can be ambiguous. In order to eliminate this confusion, four indices, (*hkil*), are often used to specify planes in a hexagonal crystal. These are called Miller-Bravais indices and are only used in the hexagonal system. The index *i* is given by:

$$h + k + i = 0, \text{ or } i = -(h + k)$$

In reality this third index is not needed, as it is simply derived from the known values of *h* and *k*. However, it does help to bring out relationships between planes that are not obvious when using just three indices. Because it is a redundant index, the value of *i* is sometimes replaced by a dot, to give indices (*hk.l*). This nomenclature emphasises that the hexagonal system is under discussion without actually including a value for *i*.

Problems and exercises

Quick quiz

1 A lattice is:
  (a) A crystal structure
  (b) An ordered array of points
  (c) A unit cell

2 The basis vectors in a lattice define:
  (a) The crystal structure
  (b) The atom positions
  (c) The unit cell

3 The number of different two-dimensional plane lattices is:

  (a) 5
  (b) 6
  (c) 7

4 A rectangular primitive plane lattice has lattice parameters:
  (a) $a \neq b$, $\gamma = 90°$
  (b) $a = b$, $\gamma = 90°$
  (c) $a \neq b$, $\gamma \neq 90°$

5 The number of Bravais lattices is:
  (a) 12
  (b) 13
  (c) 14

6 An orthorhombic body centred Bravais lattice has lattice parameters:
   (a) $a, b, c, \alpha = \beta = \gamma = 90°$
   (b) $a(=b), c, \alpha = \beta = \gamma = 90°$
   (c) $a(=b), c, \alpha = \beta = 90°, \gamma = 120°$

7 A face centred (F) lattice unit cell contains:
   (a) One lattice point
   (b) Two lattice points
   (c) Four lattice points

8 A unit cell with a lattice point at each corner and one at the centre of the cell is labelled:
   (a) B
   (b) C
   (c) I

9 The notation [uvw] means:
   (a) A single direction in a crystal
   (b) A set of parallel directions in a crystal
   (c) A direction perpendicular to a plane (uvw)

10 The notation {hkl} represents:
   (a) A set of directions that are identical by virtue of the symmetry of the crystal
   (b) A set of planes that are identical by virtue of the symmetry of the crystal
   (c) Both a set of planes or directions that are identical by virtue of the symmetry of the crystal

Calculations and Questions

2.1 Several patterns of points are shown in the figure below. Assuming these to be infinite in extent, which of them are plane lattices? For those that are lattices, name the lattice type.

2.2 Draw the plane direct and reciprocal lattice for:

(a) an oblique lattice with parameters $a = 8$ nm, $b = 12$ nm, $\gamma = 110°$

(b) a rectangular centred lattice with parameters $a = 10$ nm, $b = 14$ nm

(c) the rectangular lattice in (b) drawn as a primitive lattice

Confirm that the reciprocal lattices in (b) and (c) are identical and rectangular centred.

2.3 Sketch the direct and reciprocal lattice for

(a) a primitive monoclinic Bravais lattice with $a = 15$ nm, $b = 6$ nm, $c = 9$ nm, $\beta = 105°$

(b) a primitive tetragonal Bravais lattice with $a = 7$ nm, $c = 4$ nm

2.4 Index the lattice planes drawn in the figure below. The **c**-axis in all lattices is normal to the plane of the page and hence the index l is 0 in all cases.

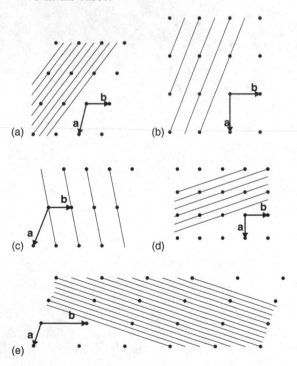

2.5 Index the lattice planes drawn in the figure below using Miller – Bravais ($hkil$) and Miller indices (hkl). The lattice is hexagonal with the **c**-axis is normal to the plane of the page and hence the index l is 0 in all cases.

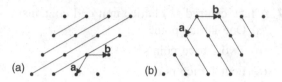

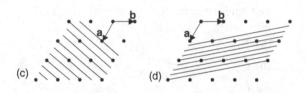

2.6 Give the indices of the directions marked on the figure below. In all cases the axis not shown is perpendicular to the plane of the paper and hence the index w is 0 in all cases.

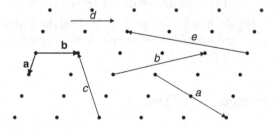

2.7 Along what direction $[uvw]$ do the following pairs of planes $(h_1k_1l_1)$ and $(h_2k_2l_2)$ intersect? (a) (110), ($1\bar{1}0$); (b) ($\bar{2}10$), (011); (c) (111), (100); (d) (212), ($1\bar{2}1$).

2.8 Calculate the interplanar spacing, d_{hkl}, for: (a) (111), cubic, $a = 0.365$ nm; (b) (210), tetragonal, $a = 0.475$ nm, $c = 0.235$ nm; (c) (321) orthorhombic, $a = 1.204$ nm, $b = 0.821$ nm, $c = 0.652$ nm; (d) (222), monoclinic, $a = 0.981$ nm, $b = 0.365$ nm, $c = 0.869$ nm, $\beta = 127.5°$; (e) (121), hexagonal, $a = 0.693$ nm, $c = 1.347$ nm.

3

Two-dimensional patterns and tiling

- *What is a point group?*
- *What is a plane group?*
- *What is an aperiodic tiling?*

At the end of chapter 1, an inherent difficulty became apparent. How is it possible to conveniently specify a crystal structure in which the unit cell may contain hundreds or even thousands of atoms? In fact, crystallographers make use of the symmetry of crystals to reduce the list of atom positions to reasonable proportions. However, the application of symmetry to crystals has far more utility than this accountancy task. The purpose of this chapter is to introduce the notions of symmetry, starting with two-dimensional patterns.

This aspect of crystallography, the mathematical description of the arrangement of arbitrary objects in space, was developed in the latter years of the 19th century. It went hand in hand with the observational crystallography sketched at the beginning of Chapter 1. They are separated here solely for ease of explanation.

3.1 The symmetry of an isolated shape: point symmetry

Everyone has an intuitive idea of symmetry, and it seems reasonable to consider that the letters **A** and **C** are equally symmetrical, and both more so than the letter **G**. Similarly, a square seems more symmetrical than a rectangle. Symmetry is described in terms of transformations that leave an object apparently unchanged. These transformations are mediated by **symmetry elements** and the action of transformation by a symmetry element is called a **symmetry operation**. Typical symmetry elements include mirrors and axes and the operations associated with them are reflection and rotation.

A consideration of the symmetry of an isolated object like the letter **A**, suggests that it can be divided into two identical parts by a **mirror**, denoted *m* in text, and drawn as a continuous line in a diagram (Figure 3.1a). The same can be said of the letter **C**, although this time the mirror is horizontal, (Figure 3.1b). An equilateral triangle contains three mirror lines (Figure 3.2a). However, it is also easy to see that the triangle can be thought to contain an **axis of rotation**, lying through the centre of the triangle and

Crystals and Crystal Structures.   Richard J. D. Tilley
© 2006 John Wiley & Sons, Ltd

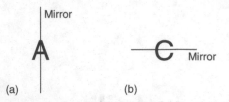

Figure 3.1   Symmetry of letters: (a) A; (b) C. The symmetry element in each case is a mirror, represented by a line

normal to the plane of the paper. The operation associated with this axis is rotation of the object in a **counter-clockwise** manner[1] through an angle of $(360/3)°$, which regenerates the initial shape each time. The axis is called a **triad** axis or a **three-fold** axis of rotation, and it is represented on a figure by the symbol ▲, (Figure 3.2b).

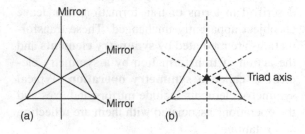

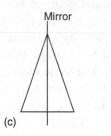

Figure 3.2   Symmetry of triangles: (a) mirrors in an equilateral triangle; (b) triad axis in an equilateral triangle; (c) mirror in an isosceles triangle

[1]Clearly whether the rotation takes place in a clockwise or anticlockwise direction is irrelevant in many cases. Conventionally an anticlockwise direction is always chosen.

Note that a general (non-equilateral) triangle does not possess this combination of symmetry elements. An isosceles triangle, for example, has only one mirror line present (Figure 3.2c) and a scalene triangle has none. It is thus reasonable to say that the equilateral triangle is more symmetric than an isosceles triangle, which is itself more symmetrical than a scalene triangle. We are thus lead to the idea that more symmetrical objects contain more symmetry elements than less symmetrical ones.

The most important symmetry operators for a planar shape consist of the mirror operator and an *infinite number* of rotation axes. Note that a mirror is a symmetry operator that can change the **handedness** or **chirality** of an object, that is, a left hand is transformed into a right hand by reflection. Two mirror image objects cannot be superimposed simply by rotation in the plane, just as a right hand glove will not fit a left hand. The only way in which the two figures can be superimposed is by lifting one from the page, (i.e. using a third dimension), and turning it over. The chirality introduced by mirror symmetry has considerable implications for the physical properties of both isolated molecules and whole crystals. It is considered in more detail in later Chapters.

The rotation operator 1, a **monad** rotation axis, implies that no symmetry is present, because the shape has to be rotated counter-clockwise by $(360/1)°$ to bring it back into register with the original, (Figure 3.3a). There is no graphical symbol for a monad axis. A **diad** or two-fold rotation axis is represented in text as 2, and in drawings by the symbol ◖. Rotation in a counter-clockwise manner by $360°/2$, $(180°)$, around a diad axis returns the shape to the original configuration (Figure 3.3b). A **triad** or three-fold rotation axis is represented in text by 3 and on figures by ▲. Rotation counter-clockwise by $360°/3$, $(120°)$, around a triad axis returns the shape to the original configuration (Figure 3.3c). A **tetrad** or four-fold rotation axis is represented in text by

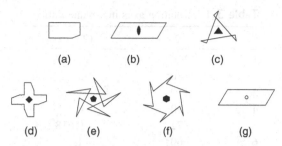

Figure 3.3 Rotation axes: (a) monad, (no symbol); (b) diad, two-fold; (c) triad, three-fold; (d) tetrad, four-fold; (e) pentad, five-fold; (f) hexad, six-fold; (g) centre of symmetry, equivalent to (b)

by 4 and on figures by ◆. Rotation counter-clock-wise by 360°/4, (90°), about a tetrad axis returns the shape to the original configuration (Figure 3.3d). The **pentad** or five-fold rotation axis, represented in text by 5 and on figures by ⬟, is found in regular pentagons. Rotation counter-clockwise by 360°/5, (72°), about a pentad axis returns the shape to the original configuration (Figure 3.3e). A **hexad** or six-fold rotation axis is represented in text by 6 and on figures by ⬣. Rotation counter-clockwise by 360°/6, (60°), around a hexad axis returns the shape to the original configuration (Figure 3.3f). In addition, there are an infinite number of other rotation symmetry operators, such as seven-fold, eight-fold and higher axes.

An important symmetry operator is the **centre of symmetry** or **inversion centre**, represented in text by $\bar{1}$, (pronounced 'one bar'), and in drawings by °. This operation is an inversion through a point in the shape, so that any object at a position (x, y) with respect to the centre of symmetry is paired with an identical object at $(-x, -y)$, written $(\bar{x}, \bar{y})$, (pronounced 'x bar', 'y bar'). In two dimensions, the presence of a centre of symmetry is equivalent to a diad axis, (Figure 3.3g).

When the various symmetry elements present in a shape are applied, it is found that one point is left unchanged by the transformations and when they are drawn on a figure, they all

pass through this single point. The collection of symmetry elements of an isolated shape is called the **point group** and the combination of elements is called the **general point symmetry** of the shape. (Here the word *group* is mathematically precise, and the results of symmetry operations can be related to each other using the formalism of group theory. This aspect will not be covered here.) The symmetry elements that characterise the point group are collected together into a **point group symbol**. Crystallographers generally use **International** or **Hermann-Mauguin** symbols for this purpose, and this practice is followed here. The point symmetry is described by writing the rotation axes present, followed by the mirror planes. Thus, the equilateral triangle has the point symmetry, 3*m*. Alternatively, it is said to belong to the **point group**, 3*m*. Note that the order in which the symmetry operators are written is specific and is described in detail below. An alternative nomenclature for point group symmetry is that proposed by **Schoenflies**. This is generally used by chemists, and is described in Appendix 3.

3.2 Rotation symmetry of a plane lattice

As described in Chapter 1, a crystal is defined by the fact that the whole structure can be built by the regular stacking of a unit cell that is translated but neither rotated nor reflected. The same is true for two-dimensional 'crystals' or patterns. This imposes a limitation upon the combinations of symmetry elements that are compatible with the use of unit cells to build up a two-dimensional pattern or a three-dimensional crystal. To understand this, consider the rotational symmetry of the five unique plane lattices, described in the previous chapter.

Suppose that a rotational axis of value n is normal to a plane lattice. It is convenient, (but not

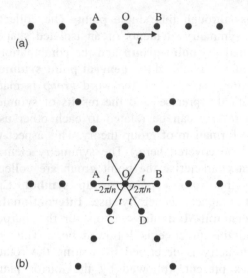

(a)

(b)

Figure 3.4 Lattice rotation: (a) a lattice row, A – O – B, each node of which is separated by a vector **t**; (b) the same row rotated by $\pm 2\pi/n$

Table 3.1   Rotation axes in a plane lattice

n	$2\pi/n$	$2\cos(2\pi/n) = m$
1	360°	2
2	180°	−2
3	120°	−1
4	90°	0
5	72°	0.618[*]
6	60°	1
7	51.43°	1.244[*]

[*]Not allowed

mandatory) to locate it at a lattice point, O, part of the row of nodes A-O-B, each of which is separated by the lattice vector **t**, of length t, where t is simply the (scalar) length of vector **t** (Figure 3.4a). The rotation operation will generate two new rows of points, lying at angles $\pm 2\pi/n$ to the original row, containing points C and D, (Figure 3.4b). As all points in a lattice are identical, C and D must lie on a lattice row parallel to that containing A and B. The separation of C and D must be an integral number, m, of the distance t. By simple geometry, the distance CD is given by $2t \cos(2\pi/n)$. Hence, for a lattice to exist:

$$2t\cos(2\pi/n) = mt$$
$$2\cos(2\pi/n) = m\text{(integer)}$$

The solutions to this equation, for the lowest values of n, are given in Table 3.1. This shows that the only rotation axes that can occur in a lattice are 1, 2, 3, 4, and 6. The fivefold rotation axis and all axes with n higher than 6, *cannot occur* in a plane lattice.

This derivation provides a formal demonstration of the fact mentioned in Chapters 1 and 2, that a unit cell in a lattice or a pattern cannot possess five-fold symmetry. The same will be found to be true for three-dimensional lattices and crystals, described in Chapter 4. (Here it is pertinent to note that five-fold symmetry can occur in two- and three-dimensional patterns, if the severe constraints imposed by the mathematical definition of a lattice are relaxed slightly, as described in below in Section 3.7 and Chapter 8, Section 8.9.

Note that the fact that a unit cell cannot show overall five-fold rotation symmetry does not mean that a unit cell cannot contain a pentagonal arrangement of atoms. Indeed, pentagonal coordination of metal atoms is commonly found in inorganic crystals and pentagonal ring structures occur in organic molecules. However, the pentagonal groups must be arranged within the unit cell to generate an overall symmetry appropriate to the lattice type, described in the following section, (see, e.g. Figure 2.4).

3.3   The symmetry of the plane lattices

The symmetry properties inherent in patterns are important in crystallography. The simplest of these patterns are the plane lattices described in Chapter 2. The symmetry elements that can be found in the plane lattices are the allowed rotation axes just described, together with mirror planes. The way in which these symmetry elements are arranged will provide an introduction to the way in which symmetry elements are disposed in more complex patterns.

Consider the situation with an oblique primitive (*mp*) lattice, (Figure 3.5a). The symmetry of the unit cell is consistent with the presence of a diad axis, which can be placed conveniently

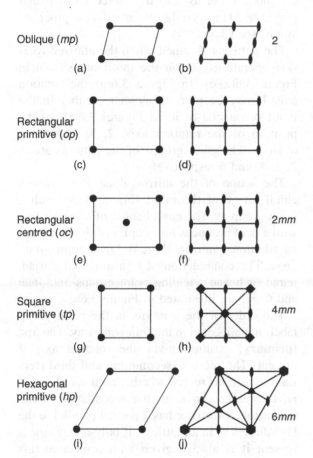

Oblique (*mp*)

(a)   (b)   2

Rectangular primitive (*op*)

(c)   (d)

Rectangular centred (*oc*)

(e)   (f)   2*mm*

Square primitive (*tp*)

(g)   (h)   4*mm*

Hexagonal primitive (*hp*)

(i)   (j)   6*mm*

Figure 3.5 Symmetry of the plane lattices: (a, b) oblique primitive, *mp*, 2; (c, d) rectangular primitive, *op*, 2*mm*; (e, f) rectangular centred, *oc*, 2*mm*; (g, h) square, *tp*, 4*mm*; (i, j) hexagonal primitive, *hp*, 6*mm*

through a lattice point, at the origin of the unit cell, without loosing generality. This means that there must be a diad through every lattice point. However, the presence of this axis also forces diad axes to be placed at the centre of each of the unit cell sides, and also one at the centre of the cell, (Figure 3.5b). The symmetry operations must leave every point in the lattice identical, and so the lattice symmetry is also described as the **lattice point symmetry**. The lattice point symmetry of the *mp* lattice is given the symbol 2.

The other lattice symmetries are described in a similar way. The rectangular primitive (*op*) lattice, (Figure 3.5c), also has lattice symmetry consistent with a diad, located for convenience at the origin of the unit cell, which, of necessity, then repeats at each lattice point. This, as before, generates diads at the centre of each of the unit cell sides and one at the unit cell centre, (Figure 3.5d). However, the lattice also shows mirror symmetry, not apparent in the oblique lattice. These mirrors lie along the unit cell edges and this necessitates the introduction of parallel mirror lines half-way along each unit cell side. The mirrors are represented as lines on Figure 3.5d and the point symmetry is given the symbol 2*mm*. (The order of the symbols for this and the other point symmetry groups listed is explained below).

The rectangular centred (*oc*) lattice, (Figure 3.5e), also has the same diads and mirrors as the *op* lattice, located in the same positions, (Figure 3.5f). The presence of the lattice centring, however, forces the presence of additional diads between the original set. Nevertheless, the point symmetry does not change, compared to the *op* lattice, and remains 2*mm*.

The square (*tp*) lattice, (Figure 3.5g), has, as principle symmetry element, a tetrad rotation axis through the lattice point at the unit cell origin, which necessitates a tetrad through each lattice point. This generates additional diads at the centre of each unit cell side, and another tetrad at the cell centre. These in turn necessitate the presence of mirror lines as shown, (Figure 3.5h). The lattice point symmetry symbol is 4*mm*.

Finally, the hexagonal primitive (*hp*) lattice, (Figure 3.5i), has a hexad rotation axis at each lattice point. This generates diads and triads as shown. In addition, there are six mirror lines through each lattice point. In other parts of the unit cell, two mirror lines intersect at diads and three mirror lines intersect at triads, (Figure 3.5j). The lattice point symmetry is described by the symbol 6*mm*.

Note that this emphasis on the symmetry of the lattices has a profound effect on thinking about structures. Up to now, the lattice type has been characterised by the lattice parameters, the lengths of the lattice vectors and the inter-vector angles. Now, however, it is possible to define a lattice in terms of the symmetry rather than the dimensions. In fact this is the norm in crystallography. As we will see, the crystal system of a phase is allocated in terms of symmetry and not lattice parameters. To pre-empt future chapters somewhat, consider the definition of a monoclinic unit cell in terms of the lattice parameters $a \neq b \neq c$; $\alpha = 90°$, $\beta \neq 90°$, $\gamma = 90°$ (see Chapter 1). The unit cell may still be regarded as monoclinic (not orthorhombic) even if the angle β is 90°, provide that the symmetry of the structure complies with that expected of a monoclinic unit cell.

3.4 The ten plane crystallographic point symmetry groups

The general point groups described in Section 3.1 are not subject to any limitations. The point groups obtained by excluding all rotation operations incompatible with a lattice are called the **crystallographic plane point groups**. These are formed, therefore, by combining the rotation axes 1, 2, 3, 4, and 6, with mirror symmetry. When the above symmetry elements are combined, it is found that there are just *ten* (two-dimensional) crystallographic plane point groups, 1, 2, *m*, 2*mm*, 4, 4*mm*, 3, 3*m*, 6 and 6*mm*.

The application of these symmetry elements is identical to that depicted in Figure 3.3. However, as we are ultimately aiming to explain patterns in two and three dimensions rather than solid shapes, it is convenient to illustrate the symmetry elements with respect to a unit of pattern called a **motif**, placed at a general position with respect to the rotation axis, (Figure 3.6). In this example the motif is an asymmetric 'three atom planar molecule'. (In crystals, the motif is a group of atoms, see Chapter 5).

The patterns obtained when the allowed rotation operations act on the motif are shown in Figure 3.6(a–e). In Figure 3.6(a) the rotation axis 1 may be placed anywhere, either in the motif or outside of it. In Figures 3.6 (b–e) the position of the rotation axis, 2, 3, 4 or 6, is obvious. The point groups of the patterns are 1, 2, 3, 4, and 6 respectively.

The action of the mirror alone is to form a chiral image of the motif (Figure 3.6f), with a point group *m*. The combination of a mirror line with a diad produces four copies of the motif and an additional mirror, (Figure 3.6g), point group 2*mm*. The combination of a mirror with a triad, tetrad or hexad, yielding point groups 3*m*, 4*mm* and 6*mm*, are illustrated in Figure 3.6h, i, j.

The order of the symbols in the point group labels are allocated in the following way. The first (**primary**) position gives the rotation axis if present. The second (**secondary**) and third (**tertiary**) positions record whether a mirror element, *m*, is present. An *m* in the secondary position means that the mirror has a normal parallel to the [10] direction, in all lattices. If only one mirror is present it is always given with respect to this direction. An *m* in the secondary position has a normal parallel to [01] in a rectangular unit cell, and to [1$\bar{1}$] in both a square and a hexagonal unit cell, (Table 3.2).

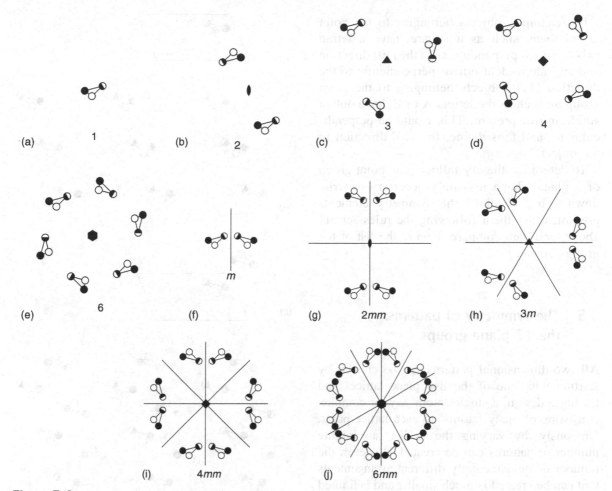

Figure 3.6 The ten plane crystallographic point groups: (a) 1; (b) 2; (c) 3; (d) 4; (e) 6; (f) *m*; (g) 2*mm*; (h) 3*m*; (i) 4*mm*; (j) 6*mm*

Table 3.2 The order of two-dimensional point group and plane group symbols

Lattice	Primary	Secondary	Tertiary
		Position in Hermann-Mauguin symbol	
Oblique, *mp*	Rotation point	–	–
Rectangular, *op*, *cp*	Rotation point	[10]	[01]
Square, *tp*	Rotation point	[10] = [01]	[1$\bar{1}$] = [11]
Hexagonal, *hp*	Rotation point	[10] = [01] = [$\bar{1}\bar{1}$]	[1$\bar{1}$] = [12] = [$\bar{2}\bar{1}$]

For example, objects belonging to the point group 4*mm*, such as a square, have a tetrad axis, a mirror perpendicular to the [10] direction and an independent mirror perpendicular to the direction [1$\bar{1}$]. Objects belonging to the point group *m*, such as the letters **A** or **C**, have only a single mirror present. This would be perpendicular to, and thus define, the [10] direction in the object.

To determine the crystallographic point group of a planar shape it is only necessary to write down a list of all of the symmetry elements present, order them following the rules set out above, and then compare them to the list of ten groups given.

3.5 The symmetry of patterns: the 17 plane groups

All two-dimensional patterns can be created by starting with one of the five plane lattices and adding a design, a single 'atom' say, or complex consisting of many 'atoms' to each lattice point. Obviously, by varying the design, an infinite number of patterns can be created. However, the number of fundamentally different arrangements that can be created is much smaller and is limited by the symmetries that have been described above.

Perhaps the simplest way to start deriving these patterns is to combine the five lattices with the patterns generated by the ten point groups. For example, take as motif the triangular 'molecule' represented by the point group 1, (Figure 3.6a), and combine this with the oblique *mp* lattice, (Figure 3.5a). The 'molecule' can be placed anywhere in the unit cell and the rotation axis (1) can conveniently pass through a lattice point. The resultant pattern, (Figure 3.7a, b), is labelled *p*1, and is called a **plane group**. The initial letter describes the lattice type, (primitive), and the number part

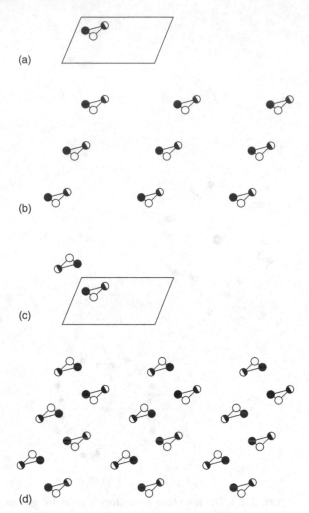

Figure 3.7 The plane groups *p*1 and *p*2: (a) the motif of point group 1 added to the plane lattice *mp*; (b) the pattern formed by repetition of (a), representative of plane group *p*1; (c) the motif in point group 2 added to the plane lattice *mp*; (d) the pattern formed by repetition of (c), representative of plane group *p*2

gives information on the symmetry operators present, in this case just the monad axis, 1. In agreement with the designation primitive, there is just one lattice point, and so one motif, per unit cell.

Remember that the motif can have any shape whatever. For example a motif with a five fold axis, say a regular pentagon, or a starfish, is allowed. However, the only pattern that can result, when this motif is combined with an *mp* lattice, is that given by the plane group *p*1, (see e.g. Figure 2.4). That is, the plane group is defined by the overall symmetry of the pattern, not by the symmetry present in the motif.

It is also possible to combine the same *mp* lattice with the arrangement representing point group 2. In this case, it is convenient to place the diad axis through the lattice point at the origin, (Figure 3.7c). Note that the presence of the diad makes the presence of two triangular 'molecules' mandatory, and it is convenient to consider the overall pattern to be generated by the addition of the whole group associated with point group 2, that is, two triangular 'molecules' and the rotation diad, to each lattice point. The pattern generated, (Figure 3.7d), is labelled *p*2, where the initial letter relates to the primitive lattice and the label 2 indicates the presence of the diad axis. In agreement with the designation of primitive, there is only one lattice point in the unit cell, although this is associated with two triangular 'molecules' and a diad axis.

The arrangements equivalent to the next two point groups, *m* and 2*mm* cannot be combined with the *mp* lattice because the mirrors are incompatible with the symmetry of the lattice, and to form new patterns it is necessary to combine these with the *op* and *oc* lattices. When building up these patterns a new type of symmetry element is revealed. This is easy to see if the point group *m* is combined with firstly the *op* lattice and then with the *oc* lattice.

Consider the *op* lattice, (Figure 3.8a). The symmetry operator *m* replicates the triangular motif associated with each lattice point to yield a design consisting of two mirror image 'molecules'. Note that the mirror line running through the origin automatically generates another parallel mirror through the centre of the unit cell,

but the motif is not replicated further, (Figure 3.8b). The same procedure applied to the *oc* lattice (Figure 3.8c) gives a different result, as the triangular 'molecule' labelled A associated with a lattice point at the origin must also be associated with the lattice point at the centre of the unit cell, to give A', (which is identical to A), (Figure 3.8d). It is now apparent that the triangular 'molecules' A and B (Figure 3.8d) are related by the operation of a combination of a reflection and a translation, called **glide**. The glide operator appears as a result of combining the translation properties of the lattice with the mirror operator. The new symmetry element, drawn as a dashed line on figures, is the **glide line**.

The translation component of the glide must be subject to the limitations of the underlying lattice, just as the rotation axes are. Following initial reflection of the motif (Figure 3.9a) the glide line will lie parallel to a lattice row of the underlying plane lattice (Figure 3.9b). The translation of the motif, the glide vector **t**, will be parallel to the glide line and also to a lattice row of the underlying plane lattice. Suppose that the lattice repeat along the glide direction is the vector **T**. In order to fulfil the repeat constraints of the lattice, twice the vector **t** must bring the motif back to a new lattice position a distance of $p\mathbf{T}$ from the original lattice point (Figure 3.9b). It is possible to write:

$$2\mathbf{t} = p\mathbf{T}$$

where p is an integer. Thus:

$$\mathbf{t} = (p/2)\mathbf{T} = 0\mathbf{T}, \mathbf{T}/2, 3\mathbf{T}/2, \ldots$$

Now of these options, only a displacement of $\mathbf{T}/2$ is distinctive. The glide displacement is therefore restricted to one half of the lattice repeat. Thus displacement parallel to the unit cell **a**-axis is

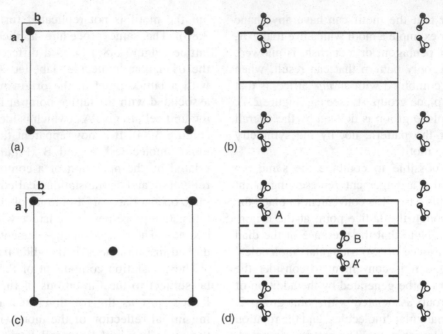

Figure 3.8   The plane groups *pm* and *cm*: (a) the plane lattice *op*; (b) the pattern formed by adding the motif of point group *m* to the lattice in (a), representative of plane group *pm*; (b) the plane lattice *oc*; (d) the pattern formed by adding the motif of plane group *m* to the lattice in (c), representative of plane group *cm*. Mirror lines are heavy, and glide lines in (d) are heavy dashed

limited to **a**/2, and parallel to the **b**-axis is limited to **b**/2. Similar relationships hold for displacements in diagonal directions. Note that just as the operation of a mirror through the cell origin automatically generates a parallel mirror through the cell centre, a glide line through the cell origin automatically generates another parallel glide line through the unit cell centre.

In building up these patterns, it will become apparent that point groups with a tetrad axis can only be combined with a square (*tp*) lattice, and those with triad or hexad axes can only be combined with a hexagonal (*hp*) lattice.

When all the translations inherent in the five plane lattices are combined with the sym-

metry elements found in the ten point groups plus the glide line, just 17 different arrangements, called the 17 (two-dimensional) **plane groups**, are found, (Table 3.3). All two-dimensional crystallographic patterns must belong to one or other of these groups and be described by one of 17 unit cells.

Each two-dimensional plane group is given a symbol that summarises the symmetry properties of the pattern. The symbols have a similar meaning to those of the point groups. The first letter gives the lattice type, primitive (*p*) or centred (*c*). A rotation axis, if present, is represented by a number, 1, (monad), 2, (diad), 3, (triad), 4, (tetrad) and 6, (hexad), and this is given second place in the symbol. Mirrors (*m*) or glide lines (*g*)

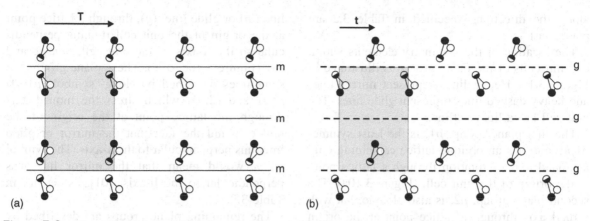

Figure 3.9 The glide operation: (a) reflection across a mirror line; (b) translation parallel to the mirror plane by a vector **t**, which is constrained to be equal to **T**/2, where **T** is the lattice repeat vector parallel to the glide line. The lattice unit cell is shaded

Table 3.3 Plane (two-dimensional) symmetry groups

System	Lattice	Point group	Plane group	Number[*]	Figure
Oblique	*mp*	1	*p*1	1	3.10a
		2	*p*2	2	3.10b
Rectangular	*op, oc*	*m, 2mm*	*pm* (*p*1*m*1)	3	3.10c
			pg (*p*1*g*1)	4	3.10d
			cm (*c*1*m*1)	5	3.10e
			*p*2*mm*	6	3.10f
			*p*2*mg*	7	3.10g
			*p*2*gg*	8	3.10h
			*c*2*mm*	9	3.10i
Square	*tp*	4, 4*mm*	*p*4	10	3.10j
			*p*4*mm*	11	3.10k
			*p*4*gm*	12	3.10l
Hexagonal	*hp*	3, 3*m*, 6, 6*mm*	*p*3	13	3.10m
			*p*3*m*1	14	3.10n
			*p*31*m*	15	3.10o
			*p*6	16	3.10p
			*p*6*mm*	17	3.10q

[*]Each plane group has a number allocated to it, found in the International Tables for Crystallography (see Bibliography).

along the directions specified in Table 3.2 are placed last.

The location of the symmetry elements within the unit cells of the plane groups is illustrated in Figure 3.10. Heavy lines represent mirror lines and heavy dashed lines represent glide lines. The unit cell has a light outline.

The first plane group, $p1$, is the least symmetrical, and has an oblique lattice containing only a monad axis of rotation through a lattice point at the origin of the unit cell, (Figure 3.10a). The second plane group, $p2$, is also oblique, but with a diad axis through a lattice point at the origin of the unit cell, (Figure 3.10b). The other plane groups with a single rotation axis are $p4$, $p3$ and $p6$. In all of these, the rotation axis passes through a lattice point at the origin of the unit cell, (Figure 3.10 j, m, p).

A number of plane groups do not include rotation axes. These are the groups with a mirror line (m) or glide line, (g), through a lattice point at the origin of the unit cell running perpendicular to the **a**-axis of the unit cell, pm, pg and cm, (Figure 3.10c, d, e). These plane groups are sometimes described by 'long' symbols $p1m1$, $p1g1$ and $c1m1$, which stress the monad axis through the lattice point at the origin of the unit cell and the fact that the mirror or glide line runs perpendicular to the **a**-axis. The symbol $p11m$ would mean that the mirror line was perpendicular to the **b**-axis, [01], as set out in Table 3.2.

The remaining plane groups are described by a string of four symbols, typified by $p2mm$ and $p2mg$. The first two symbols give the lattice and the rotation axis present. The significance of the order of the last two symbols is given in Table 3.2. Thus, the meaning of the symbol $p2mm$, which is derived from the *op* lattice, is that the lattice is primitive, there is a diad axis

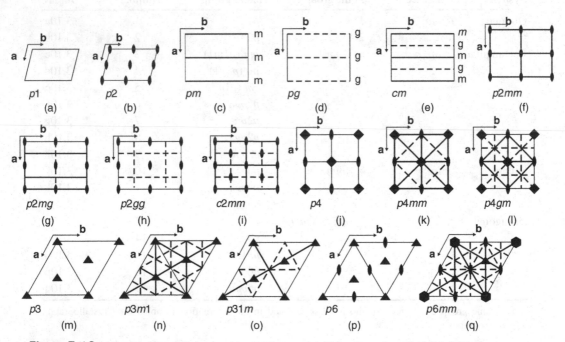

Figure 3.10   The location of the symmetry elements within the unit cells of the 17 plane groups

through the lattice point at the origin of the unit cell, and a mirror line perpendicular to the **a**-axis, [10], and a mirror line perpendicular to the **b**-axis, [01], (Figure 3.10f). The meaning of the symbol *p2mg* is that the lattice is primitive, there is a diad axis through the lattice point at the origin of the unit cell, and a mirror line perpendicular to the **a**-axis, [10], and a glide line perpendicular to the **b**-axis, [01], (Figure 3.10g). The positions of the mirror and glide lines are often through the origin of the unit cell, but not always. They occur in positions that are compatible with the rotation axis at the origin. The addition of a monad axis, 1, is often used as a place holder, to ensure that the mirror or glide line is correctly placed, as described above for the groups *p1m1* and *p11m*. Thus the plane groups *p31m* and *p3m1* are both derived from *p3* by the addition of mirrors. The group *p3m1* has a mirror perpendicular to the **a**-axis, [10], while the group *p31m* has a mirror perpendicular to [1$\bar{1}$].

To determine the plane group of a pattern, write down a list of all of the symmetry elements that it posses, (not always easy), order them and then compare the list to the symmetry elements associated with the point groups given in Table 3.3.

3.6 Two-dimensional 'crystal structures'

In describing a crystal structure, it is necessary to list the positions of all the atoms in the unit cell, and even in two-dimensional cases this can be a considerable chore. However, all structures must be equivalent to one of the 17 patterns described by the plane groups, and the task of listing atom positions can be reduced greatly by making use of the symmetry elements apparent in the appropriate plane group. To aid in this, each plane group is described in terms of a pair of standard dia-

grams. The first of these shows all of the symmetry elements in the unit cell, marked as in Figure 3.10. The accompanying diagram plots the various locations of a motif, initially placed in a **general position** in the unit cell, forced by the operation of the symmetry elements present, (also see Section 3.7). Conventionally a motif is represented by a **circle**. Sometimes this diagram also shows the position of the motif in parts of the neighbouring unit cells if that clarifies the situation.

Take the simplest plane group, *p1*, (number 1), as an example. Recall that there is only one lattice point in a primitive (*p*) unit cell. The motif can be placed anywhere in the unit cell. The unit cell has no symmetry elements present, (Figure 3.11a), and hence the motif is not replicated, (Figure 3.11b). Note that the cell is divided up into quarters, and motifs in the surrounding unit cells are also shown to demonstrate the pattern of repetition. Although the unit cell contains just one lattice point, the motif

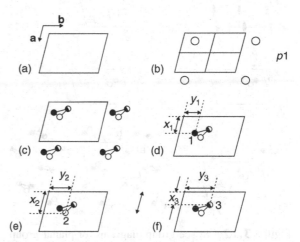

Figure 3.11 Space group diagrams for planar group *p1*: (a) symmetry elements; (b) equivalent general positions; (c) position of a triangular molecule motif; (d–f) specification of the coordinates (*x*, *y*) of an atom

associated with the lattice point can contain any number of atoms. It is convenient to take the triangular 'molecule' used previously as the motif, (Figure 3.11c).

The **atom positions** in *any* unit cell, (not just in plane group $p1$), x and y, are always defined in directions parallel to the unit cell edges. The three atoms in the 'molecule' motif will have atom positions (x_1, y_1), (x_2, y_2) and (x_3, y_3), defined in this way, (Figure 3.11d-f). Because of the absence of symmetry elements, there is no alternative but to list the (x, y) coordinates of each atom.

The standard crystallographic diagrams for plane group $p2$ are drawn in Figure 3.12 a, b. The plane group $p2$ contains a diad axis at the origin of the unit cell, (Figure 3.12a). This means that an atom at (x_1, y_1) will be repeated at a position, $-x_1$, $-y_1$, written $(\bar{x}, \bar{y})$, (pronounced x bar, y bar), which is equivalent to the position $(1-x)$, $(1-y)$, (Figure 3.12b–e). If the motif is a group of atoms, the positions of each atom in the new group will be related to those of the starting group by the same rotation symmetry. Note that although there are more diad axes present, they do not produce more atoms in the unit cell. The pattern produced by extending the number of unit cells visible is drawn in Figure 3.12f.

The results of the operation of the symmetry elements present in the plane groups pm and pg are summarised on Figure 3.13a–f. In plane group pm, the mirror lines lie perpendicular to the **a**-axis (Figure 3.13a). The consequence of the mirror is to transform the motif into its mirror image. This form cannot be superimposed upon the original simply be moving it around on the plane of the page. In standard crystallographic diagrams, a chiral relationship with a normal motif is indicated by placing a **comma within the circular motif** symbol, (Figure 3.13 b). The relationship between the two forms is clearer when the motif used is the triangular 'molecule' (Figure 3.13 c). Plane group pg has glide lines lying perpendicular to the **a**-axis (Figure 3.13 d). The glide operation involves reflection, and so the two forms of the motif are chiral pairs, (Figure 3.13 e, f).

The defining symmetry element in the plane group $p4$ is a tetrad axis, which is conveniently located at the corner of a unit cell. The operation of the symmetry elements comprising the plane group $p4$, (Figure 3.14a), will cause a motif at a general position to be reproduced four times in the unit cell, (Figure 3.14b). The positions of the atoms generated by the tetrad axis are derived as follows. An atom is placed at a general position (x, y), (Figure 3.14c). Successive counter- clockwise rotations of 90°, (Figures 3.13d–f), generate the remaining three. The equivalent atoms, transposed into just one unit cell are shown on Figure 3.14g.

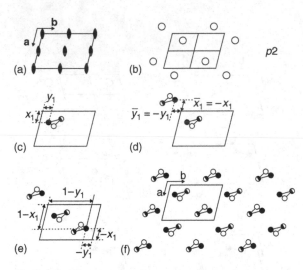

Figure 3.12 Space group diagrams for planar group $p2$: (a) symmetry elements; (b) equivalent general positions; (c–e) specification of the coordinates (x, y) of an atom in a general position; (f) the overall pattern

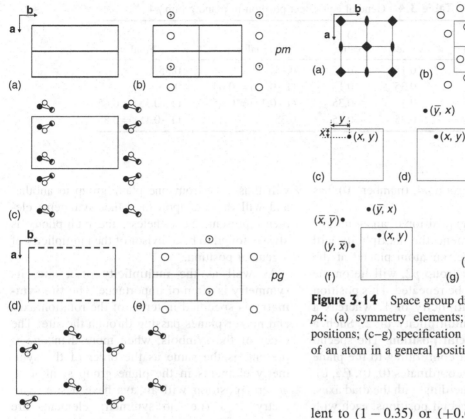

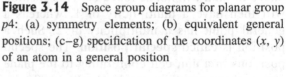

Figure 3.13 Space group diagrams for planar groups *pm* and *pg*: (a) symmetry elements present in group *pm*; (b) equivalent general positions for group *pm*; (c) an overall pattern consistent with group *pm*; (d) symmetry elements present in group *pg*; (b) equivalent general positions for group *pg*; (c) an overall pattern consistent with group *pg*

Figure 3.14 Space group diagrams for planar group *p4*: (a) symmetry elements; (b) equivalent general positions; (c–g) specification of the coordinates (x, y) of an atom in a general position

The positions are described in crystallographic texts as (x, y), $(\bar{y}, x)$, $(\bar{x}, \bar{y})$, $(y, \bar{x})$. This terminology can be misleading. It means this. If the first position chosen, (x, y) is, say, given by $x = 0.15$, $y = 0.35$, then the point $(\bar{y}, x)$, has an x coordinate of (-0.35), [which is equiva-

lent to $(1 - 0.35)$ or $(+0.65)$], and a y coordinate of 0.15. The four coordinates equivalent to the positions (x, y), $(\bar{y}, x)$, $(\bar{x}, \bar{y})$, $(y, \bar{x})$, are set out in Table 3.4 and can be visualised from Figure 3.14.

3.7 General and special positions

In the previous section, an atom or a motif was allocated to a position, (x, y), which was anywhere in the unit cell. Such a position is called a **general position**, and the other positions generated by the symmetry operators present are called **general equivalent positions**. The number of general equivalent positions is given by the **multiplicity**. Thus, the general position (x, y) in the plane group *p2*, (number 2) has a multiplicity

Table 3.4   General equivalent position in plane group $p4$

Position	x value	y value	Final x	Final y
(x, y)	0.15	0.35	0.15	0.35
$(\bar{y}, x)$	−0.35	0.15	$(1-0.35) = 0.65$	0.15
$(\bar{x}, \bar{y})$	−0.15	−0.35	$(1-0.15) = 0.85$	$(1-0.35) = 0.65$
$(y, \bar{x})$	0.35	−0.15	0.35	$(1-0.15) = 0.85$

of 2 and in the plane group $p4$, (number 10), has a multiplicity of 4.

If, however, the (x, y) position of an atom falls upon a symmetry element, the multiplicity will decrease. For example, an atom placed at the unit cell origin in the group $p2$, will lie on the diad axis, and will not be repeated. This position will then have a multiplicity of 1, whereas a general point has a multiplicity of 2. Such a position is called **special position**. The special positions in a unit cell conforming to the plane group $p2$ are found at coordinates (0, 0), (½, 0), (0, ½) and (½, ½), coinciding with the diad axes. There are four such special positions, each with a multiplicity of 1.

In a unit cell conforming to the plane group $p4$, there are special positions associated with the tetrad axes, at the cell origin, (0, 0), and at the cell centre, (½, ½), and associated with diad axes at positions (½, 0) and (0, ½), (see Figure 3.14a). The multiplicity of an atom located on the tetrad axis at the cell origin will be one, as it will not be repeated anywhere else in the cell by any symmetry operation. The same will be true of an atom placed on the tetrad axis at the cell centre. If an atom is placed on a diad axis, say at (½, 0), another must occur at (0, ½). The multiplicity of an atom on a diad will thus be 2.

Note that the multiplicity of a diad in plane group $p2$ is different from that of a diad in plane group $p4$. The multiplicity of a special position

will thus vary from one plane group to another and will depend upon the other symmetry elements present. Nevertheless, the multiplicity is always found to be a divisor of the multiplicity of a general position.

As well as the multiplicity of a site, its symmetry is often of importance. The **site symmetry** is specified in terms of the rotation axes and mirror planes passing through the site. The order of the symbols, when more than one is present, is the same as the order of the symmetry elements in the plane group symbol. A general position will always have a site symmetry of 1 (i.e. no symmetry elements are present). The site symmetry of an atom at a diad in plane group $p2$ or $p4$ will be 2, corresponding to the diad axis itself. The site symmetry of an atom on a tetrad axis in plane group $p4$ will be 4. When a position is not involved in site symmetry, the place is represented by a dot, as a place holder. Thus in plane group $p2mm$ (Figure 3.10f), the site symmetry of an atom on either of the two mirrors perpendicular to the **a**-axis is given as $.m.$, and an atom on a mirror perpendicular to the **b**-axis is given as $..m$. An atom on any of the diads is also on a mirror perpendicular to the **a**-axis and to a mirror perpendicular to the **b**-axis, hence the site symmetry is written as $2mm$.

Both the general and special positions in a unit cell are generally listed in a table that gives the multiplicity of each position, the

Table 3.5 Positions in the plane group *p*2

Multiplicity	Wyckoff letter	Site symmetry	Coordinates of equivalent positions
2	*e*	1	(1) x, y; (2) $\bar{x}, \bar{y}$
1	*d*	2	½, ½
1	*c*	2	½, 0
1	*b*	2	0, ½
1	*a*	2	0, 0

coordinates of the equivalent positions, and the site symmetry. In these tables, the general position is listed first, and the special positions are listed in order of decreasing multiplicity and increasing point symmetry. These are also accompanied by a **Wyckoff letter** or **symbol** listed upwards from the site with the lowest multiplicity, in alphabetical order. The two following examples, Table 3.5 for the plane group *p*2, and Table 3.6 for plane group *p*4, give more detail.

From the forgoing, it is seen that in order to specify a two-dimensional crystal structure, all that needs to be specified are (i), the lattice parameters, (ii), the plane group symbol, (iii), a list of atom types together with the Wyckoff letter and the atomic coordinates. For complex crystals, it simplifies the amount of information to be recorded enormously.

Table 3.6 Positions in the plane group *p*4

Multiplicity	Wyckoff letter	Point symmetry	Coordinates of equivalent positions
4	*d*	1	(1) x, y; (2) $\bar{x}, \bar{y}$; (3) ($\bar{y}$, x); (4) (y, $\bar{x}$)
2	*c*	2..	½, 0; 0, ½
1	*b*	4..	½, ½
1	a	4..	0, 0

3.8 Tesselations

A tesselation is a pattern of tiles that covers an area with no gaps or overlapping pieces. Obviously the unit cells associated with the five plane lattices will tesselate a plane periodically. In recent years, there has been considerable interest in the mathematics of such pattern forming, much of which is of relevance to crystallography. There are an infinite number of ways in which a surface may be tiled, and although tiling theory is generally restricted to using identically shaped pieces (tiles), the individual pieces can be rotated or reflected, as well as translated, whereas in crystallography, only translation is allowed.

Regular tesselations or **tilings** are those in which every tile has the same shape, and every intersection of the tiles has the same arrangement of tiles around it. There are only three regular tesselations, derived from the three regular polygons: equilateral triangles, squares or hexagons, (Figure 3.15). No other regular polygon will completely tile a plane without gaps appearing.

A **periodic tesselation** or **tiling** is one in which a region (i.e. a unit cell) can be outlined that tiles the plane by translation, that is by moving it laterally, but without rotating it or reflecting it. The rules for the production of a periodic tiling are less exacting than those to produce a regular tiling. This can be illustrated by a periodic tesselation using *equilateral pentagons*, (Figure 3.16a). Note that the *equilateral* pentagon is distinct from the *regular* pentagon, because, although it has five sides of equal length, the internal angles differ from 72°. The tesselation can only be completed if some of the pentagonal tiles are rotated or reflected. The tesselation is periodic, however, because a unit of structure, (a unit cell), consisting of four pentagonal tiles, (Figure 3.16b), will tesselate the plane by translation alone.

A **non-periodic (aperiodic) tesselation or tiling** is one that cannot be created by the repeated

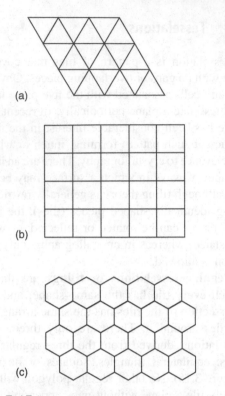

(a)

(b)

(c)

Figure 3.15 The regular tesselations: (a) equilateral triangles; (b) squares; (c) regular hexagons

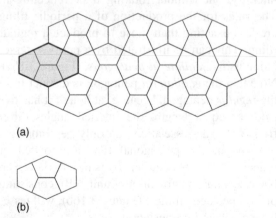

(a)

(b)

Figure 3.16 A periodic tiling using equilateral (not regular) pentagons (a) and the unit of structure (b) which will tile the area by translation alone. This tiling has been known and used since ancient times

translation of a small region (a unit cell) of the pattern. There are many of these including, for example, spiral tilings. One family of non-periodic tesselation patterns of particular relevance to crystallography are called **Penrose tilings**. These tilings can be constructed from two tile shapes called kites and darts, (Figure 3.17). These are derived from a rhombus with each edge equal to the 'Golden Ratio' of $(1 + \sqrt{5})/2 = 1.61803398\ldots$ in length, (Figure 3.17a). The rhombus is divided into two along the long diagonal in the golden ratio of $(1 + \sqrt{5})/2 = 1.61803398\ldots$ and the point so found is joined to the obtuse corners of the rhombus. This creates the kite, (Figure 3.17b) and dart, (Figure 3.17c). Each line segment of

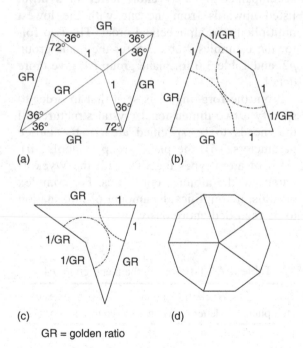

(a)

(b)

(c)

(d)

GR = golden ratio

Figure 3.17 Kites and darts: (a) a rhombus with each edge equal to the golden ratio, GR; (b) a kite, with arcs that force an aperiodic Penrose tiling when matched; (c) a dart, with arcs that force an aperiodic Penrose tiling when matched; (d) five kites make up a decagon

the kites and darts is then found to be 1 or the golden ratio. A kite is one-fifth of a decagon, (Figure 3.17d). The rhombus itself tessellates periodically, because it is just an oblique unit cell. However, if ways of joining the kites and darts are restricted, the pattern formed is non-periodic or aperiodic.

The procedure by which the non-periodicity is forced upon the tesselation is to join the tiles in a particular fashion. An attractive way of labelling the tiles is to draw circular arcs of two colours or patterns, cutting the edges in the golden ratio, (Figure 3.17b, c). The tesselation is then carried out by making sure that when tiles are joined, the circular arcs of the same type are continuous. A Penrose tiling constructed in this way, the 'infinite sun' pattern is drawn in Figure 3.18. Any region of the pattern appears to show five-fold rotational symmetry, and this remains true for any area of the tiling, no matter how large the pattern becomes. However, extending the drawing reveals that the pattern itself never quite manages to repeat.

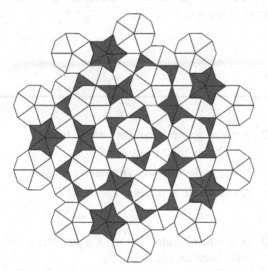

Figure 3.18 The 'infinite sun' pattern of Penrose tiles

Although, at first glance, it seems that a unit cell can be found, it turns out that this is never possible, Details of how to construct Figure 18 and other Penrose tilings is given in the Bibliography.

This has startling consequences for crystallography. A Penrose tiling pattern appears to have a five-fold or perhaps 10-fold rotational symmetry, a feature not allowed in the 'classical' derivation of patterns given above. More surprisingly, solids that are analogous to three-dimensional Penrose tilings have been discovered. These break the fundamental rules of crystallography by having unit cells with (apparently) five-fold and higher rotational symmetry. The discovery of these compounds threw the world of crystallography into fevered activity as scientists tried to explain these enigmatic phases. The materials, now called **quasicrystals**, are described in Chapter 8, Section 8.9.

Answers to introductory questions

What is a point group?

A point group describes the symmetry of an isolated shape.

When the various symmetry elements that are present in a two-dimensional (planar) shape are applied in turn, it is seen that one point is left unchanged by the transformations. When these elements are drawn on a figure, they all pass through this single point. For this reason, the combination of operators is called the general point symmetry of the shape. There are no limitations imposed upon the symmetry operators that are allowed, and in particular, five-fold rotation axes are certainly allowed, and are found in many natural objects, such as star-fish or flowers. There are many general point groups.

The scaffolding that underlies any two-dimensional pattern is a plane lattice. The point groups obtained by excluding all rotation operators incompatible with a plane lattice are called the crystallographic plane point groups. These are formed by combining the rotation operators 1, 2, 3, 4, and 6, with mirror symmetry. There are just 10 crystallographic point groups associated with two-dimensional shapes or 'crystals'.

What is a plane group?

A plane group describes the symmetry of a two-dimensional repeating pattern.

A way of making such a repeating pattern is to combine the ten (two-dimensional) crystallographic plane point groups with the 5 (two-dimensional) plane lattices. The result is 17 different arrangements, called the 17 (two-dimensional) plane groups. All two-dimensional crystallographic patterns must belong to one or other of these groups. These 17 are the only crystallographic patterns that can be formed in a plane. Obviously, by varying the motif, an infinite number of designs can be created, but they will all be found to possess one of the 17 unit cells described.

What is an aperiodic tiling?

A tesselation or tiling is a pattern of tiles that covers an area with no gaps or overlapping pieces. There are an infinite number of ways in which a surface may be tiled, and although tiling theory is generally restricted to using identically shaped pieces (tiles), the individual pieces can be rotated or reflected, as well as translated, whereas in crystallography, only translation is allowed.

A periodic tesselation or tiling is one in which a region, (a unit cell), can be outlined that tiles the plane by translation, that is by moving it laterally, but without rotating it or reflecting it. Contrariwise, an aperiodic (non-periodic) tesselation or tiling is one that cannot be created by the translation of a small region of the pattern without rotation or reflection. In effect this means that the pattern does not possess a unit cell. Although any small area may suggest a unit cell, when the pattern is examined over large distances it is found that repetition does not occur. The 'unit cell' is infinitely long. Perhaps the best known family of aperiodic tesselation patterns are called Penrose tilings.

Problems and exercises

Quick quiz

1 One of the following operations is NOT a symmetry element:
   (a) Rotation
   (b) Reflection
   (c) Translation

2 A point group is another name for:
   (a) The symmetry of a pattern
   (b) A collection of symmetry elements
   (c) A single symmetry element

3 A tetrad operator involves a counter clockwise rotation of:
   (a) 180°

(b) 90°

(c) 60°

4 One of the following rotation axes is not compatible with a crystal:
(a) 4-fold
(b) 6-fold
(c) 8-fold

5 The number of two-dimensional crystallographic point symmetry groups is:
(a) 17
(b) 10
(c) 5

6 The number of plane groups is:
(a) 17
(b) 10
(c) 5

7 A glide operator consists of:
(a) Reflection plus rotation
(b) Rotation plus translation
(c) Reflection plus translation

8 If the position $(x, y) = (0.25, 0.80)$, the position $(\bar{y}, x)$ is:
(a) $(-0.80, 0.25)$
(b) $(-0.25, 0.80)$
(c) $(0.80, -0.25)$

9 If the (x, y) position of an atom falls upon a symmetry element, the multiplicity will:
(a) Increase
(b) Decrease
(c) Remain unchanged

10 A Penrose tiling can be described as:
(a) Regular
(b) Periodic
(c) Aperiodic

Calculations and Questions

3.1 Write out the symmetry elements present in the 'molecules' in the figure below, and the corresponding point group of each one. Assume that the molecules are planar, exactly as drawn, and not three-dimensional. The shapes are: (a) pentagonal C_5H_5, in ferrocene; (b) linear, CS_2; (c) triangular, SO_2; (d) square, XeF_4; (e) planar, C_4H_4, ethane.

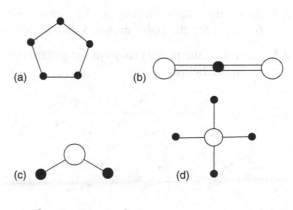

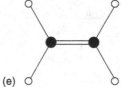

3.2 Write out the symmetry elements present in the two 'crossword' blanks drawn in the figure below, and the corresponding point group of each pattern.

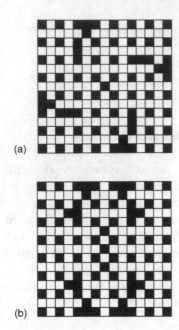

(a)

(b)

3.3 Using the motif in Figure 3.6, draw the diagrams for the point groups 5 and 5*m*.

3.4 Determine the plane groups of the patterns in the figure below.

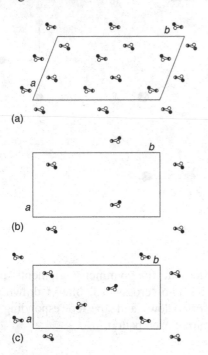

(a)

(b)

(c)

3.5 Determine the plane groups of the patterns below.

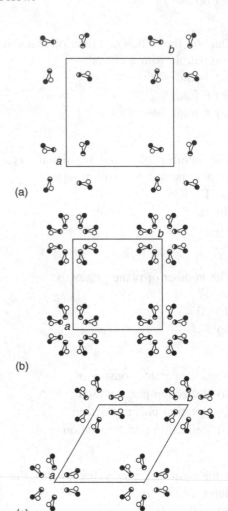

(a)

(b)

(c)

3.6 The general equivalent positions in the plane group *c2mm* are given by:

(1) x, y (2) $\bar{x}, \bar{y}$

(3) $\bar{x}, y$ (4) $x, \bar{y}$

(5) $\frac{1}{2} + x, \frac{1}{2} + y$ (6) $\frac{1}{2} + \bar{x}, \frac{1}{2} + \bar{y}$

(7) $\frac{1}{2}+\bar{x}, \frac{1}{2}+y$ (8) $\frac{1}{2}+x, \frac{1}{2}+\bar{y}$

The multiplicity of the site is 8. What are the coordinates of all of the equivalent atoms within the unit cell, if one atom is found at (0.125, 0.475)?

3.7 The general equivalent positions in the plane group *p4gm* are given by:

(1) x, y (2) $\bar{x}, \bar{y}$

(3) $\bar{y}, x$ (4) $y, \bar{x}$

(5) $\bar{x}+\frac{1}{2}, y+\frac{1}{2}$ (6) $x+\frac{1}{2}, \bar{y}+\frac{1}{2}$

(7) $y+\frac{1}{2}, x+\frac{1}{2}$ (8) $\bar{y}+\frac{1}{2}, \bar{x}+\frac{1}{2}$

The multiplicity of the site is 8. What are the coordinates of all of the equivalent atoms within the unit cell, if one atom is found at (0.210, 0.395)?

4

Symmetry in three dimensions

- *What is a rotoinversion axis?*

- *What relates a crystal class to a crystal-lographic point group?*

- *What are enantiomorphic pairs?*

This chapter follows, in outline, the format of Chapter 3. In this chapter, though, three-dimensional crystals form the subject matter, and the two-dimensional concepts already presented will be taken into this higher dimension.

4.1 The symmetry of an object: point symmetry

Any solid object can be classified in terms of the collection of symmetry elements that can be attributed to the shape. The combinations of allowed symmetry elements form the **general three-dimensional point groups** or **non-crystallographic three-dimensional point groups**. The symmetry operators are described here by the International or Hermann-Mauguin symbols.

The alternative Schoenflies symbols, generally used by chemists, are given in Appendix 3.

Many of the symmetry elements described with respect to planar shapes also apply to three-dimensional solids. For example, the rotation axes are the same and the mirror line becomes a mirror plane. As mentioned in Chapter 3, a mirror is a symmetry operator that changes the handedness or chirality of the object. A **chiral species** is one that cannot be superimposed upon its mirror image. That is, a left hand is transformed into a right hand by reflection. Objects, particularly molecules or crystals that display handedness are termed **enantiomorphs**, or an **enantiomorphic pair**. Such objects cannot be superimposed one upon the other, in the same way that a right-hand and left-hand glove cannot be superimposed one on the other. The amino acid alanine, $CH_3CH(NH_2)COOH$, is an example of a chiral molecule, which can exist in a 'right handed' or 'left handed' form, (Figure 4.1). [The crystal structure of alanine is described in more detail in Chapter 5.] Enantiomorphic objects are structurally identical in every way except for the change from right-handed to left-handed. The chemical and physical properties of these compounds are also identical except that a chiral or enantiomorphic object displays the additional

Crystals and Crystal Structures. Richard J. D. Tilley
© 2006 John Wiley & Sons, Ltd

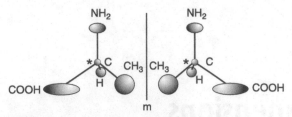

Figure 4.1 The enantiomorphic (mirror image) forms of the molecule alanine, $CH_3CH(NH_2)COOH$: (a) the naturally occurring form, (S)-$(+)$-alanine; (b) the synthetic form (R)-$(-)$-alanine. An enantiomorphic molecule cannot be superimposed upon its mirror image

Figure 4.2 Rotation axes in nature: (a) approximately five-fold, but really only a monad is present; (b) five-fold; (c) five fold plus mirrors; (d) eight-fold

property of **optical activity**, described below, (Section 4.11.) as well as different biological or pharmocological activity.

The most important symmetry operators for a solid object are identical to those described in Chapter 3. The mirror operator, denoted m in text, is drawn as a continuous line in a diagram. The rotation operator 1, a monad rotation axis, implies that no symmetry is present, because the shape has to be rotated counter clockwise by $(360/1)°$ to bring it back into register with the original. A diad or two-fold rotation axis is represented in text as 2, and in drawings by the symbol. Rotation in a counter-clockwise manner by $360°/2$, that is, $180°$, around a diad axis returns the shape to the original configuration. A triad or three-fold rotation axis, is represented in text by 3 and on figures by ▲. Rotation counter clockwise by $360°/3$, that is, $120°$, around a triad axis returns the shape to the original configuration. A tetrad or four-fold rotation axis is represented in text by 4 and on figures by ◆. Rotation counter clockwise by $360°/4$, that is, $90°$, about a tetrad axis returns the shape to the original configuration. The pentad or five-fold rotation axis, represented in text by 5 and on figures by ⬟, is found in starfish and many other natural objects. Rotation counter clockwise by $360°/5$, that is $72°$, about a pentad axis returns the shape to the original configuration. A hexad or

six-fold rotation axis is represented in text by 6 and on figures by ⬣. Rotation counter clockwise by $360°/6$, that is, $60°$, around a hexad axis returns the shape to the original configuration. In addition, there are an infinite number of other rotation symmetry operators, such as seven-fold, eight-fold and higher axes. Many of these are found in nature, (Figure 4.2).

A symmetry operator that takes on increased importance in solids compared to plane figures is the centre of symmetry or inversion centre, represented in text by $\bar{1}$, (pronounced 'one bar'), and in drawings by °. An object possess a centre of symmetry at position $(0, 0, 0)$ if any point at position (x, y, z) is accompanied by an equivalent point located at $(-x, -y, -z)$ written $(\bar{x}, \bar{y}, \bar{z})$, (pronounced 'x bar, y bar, z bar'), (Figure 4.3a). Note that a centre of symmetry inverts an object so that a lower face becomes an upper face, and *vice versa*, (Figure 4.3b). The centre of symmetry of a regular octahedral molecule lies at the central metal atom (Figure 4.3c). Any line that passes through the centre of symmetry of

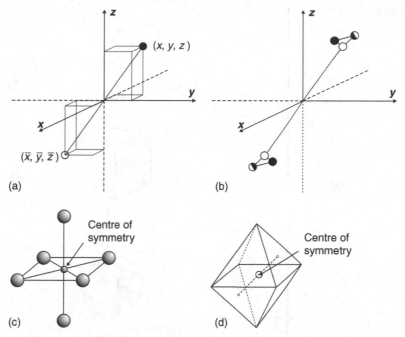

Figure 4.3 The centre of symmetry operator: (a) any point at position (x, y, z) is accompanied by an equivalent point located at $(-x, -y, -z)$ or $(\bar{x}, \bar{y}, \bar{z})$; (b) a centre of symmetry inverts an object so that a lower face becomes an upper face; (c) the centre of symmetry of a regular octahedral molecule lies at the central metal atom; (d) any line that passes through the centre of symmetry of a solid connects equivalent faces

a solid connects equivalent faces or objects (Figure 4.3d).

4.2 Axes of inversion: rotoinversion

Apart from the symmetry elements described in Chapter 3 and above, an additional type of rotation axis occurs in a solid that is not found in planar shapes, the **inversion axis**, $\bar{n}$, (pronounced 'n bar'). The operation of an inversion axis consists of a rotation combined with a centre of symmetry. These axes are also called **improper rotation axes**, to distinguish them from the ordinary **proper rotation axes** described above. The symmetry operation of an improper rotation axis is that of **rotoinversion**. Two solid objects related by the operation of an inversion axis are found to be enantiomorphous.

The operation of a two-fold improper rotation axis, $\bar{2}$, (pronounced 'two bar'), is drawn in Figure 4.4. The initial atom position (Figure 4.4a), is rotated 180° counter clockwise (Figure 4.4b) then inverted through the centre of symmetry, (Figure 4.4c). It is seen that the operation is identical to that of a mirror plane, (Figure 4.4d), and this latter designation is used in preference to that of the improper axis.

The operation of some of the other improper rotation axes can be illustrated with respect to the five **Platonic solids**, the regular tetrahedron, regular octahedron, regular icosahedron, regular cube and regular dodecahedron. These polyhedra have regular faces and vertices, and each has

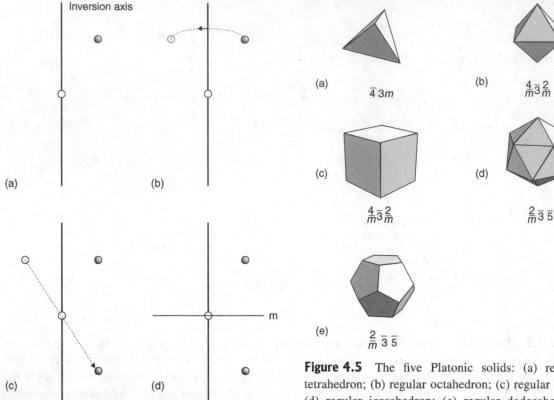

Inversion axis

(a)   (b)

(c)   (d)

● Atom   ○ Centre of symmetry

Figure 4.4 The operation of a two-fold improper rotation axis, $\bar{2}$: (a) the initial atom position; (b) rotation by 180° counter clockwise; invertion through the centre of symmetry; (d) the operation $\bar{2}$ is identical to that of a mirror plane, m

(a) $\bar{4}\,3m$

(b) $\frac{4}{m}\bar{3}\frac{2}{m}$

(c) $\frac{4}{m}\bar{3}\frac{2}{m}$

(d) $\frac{2}{m}\bar{3}\,\bar{5}$

(e) $\frac{2}{m}\,\bar{3}\,\bar{5}$

Figure 4.5 The five Platonic solids: (a) regular tetrahedron; (b) regular octahedron; (c) regular cube; (d) regular icosahedron; (e) regular dodecahedron. The point group symbol for each solid is given below each diagram

only one type of face, vertex and edge (Figure 4.5a–e). The regular tetrahedron, regular octahedron and regular icosahedron are made up from 4, 8 and 20 equilateral triangles respectively. The regular cube, or hexahedron, is composed of six squares, and the regular dodecahedron of 12 regular pentagons.

The tetrahedron illustrates the operation of a four-fold inversion axis, $\bar{4}$, (pronounced 'four bar'). A tetrahedron inscribed in a cube allows the three Cartesian axes to be defined (Figure 4.6a). With respect to these axes, the four-fold inversion axes lie parallel to the x-, y- and z-axes. One such axis is drawn in Figure 4.6b. The operation of this element is to move vertex A (Figure 4.6b) by a rotation of 90° in a counter-clockwise direction and then inversion through the centre of symmetry to generate vertex D. In subsequent application of the $\bar{4}$ operator, vertex D is transformed to vertex B, B to C and C back to A.

Apart from the $\bar{4}$ axis, rotation triad (3) axes pass through any vertex and the centre of the opposite triangular face, along the $\langle 111 \rangle$ directions, (Figure 4.6c). In addition, the tetrahedron has mirror symmetry. The mirror planes are {110} planes that contain two vertices

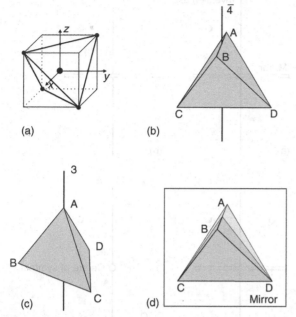

Figure 4.6 Symmetry elements present in a regular tetrahedron: (a) a tetrahedron inscribed in a cube, showing the three Cartesian axes; (b) four-fold inversion axes ($\bar{4}$) lie along x-, y- and z; (c) rotation triad (3) axes pass through a vertex and the centre of the opposite triangular face; (d) mirror planes contain two vertices and lie normal to the other two vertices

and lie normal to the other two vertices, (Figure 4.6d).

The regular octahedron can be used to describe the three-fold inversion axis, $\bar{3}$, (pronounced 'three bar'). As with the regular tetrahedron, it is convenient to refer to Cartesian axes to locate the symmetry operators, (Figure 4.7a). The highest order axis, however, is not $\bar{3}$, but a tetrad. These axes pass through a vertex and run along the x-, y- and z-axes. Each tetrad is accompanied by a mirror normal to it, (Figure 4.7b). This combination is written $4/m$ (pronounced '4 over em').

The three fold inversion axes relate the positions of the vertices to each other. A $\bar{3}$ inversion axis runs through the middle of each opposite

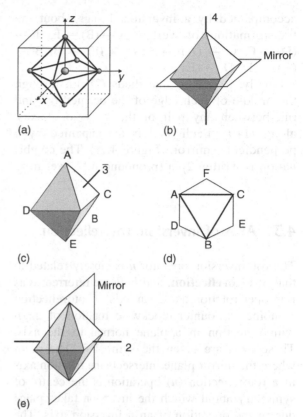

Figure 4.7 Symmetry elements present in a regular octahedron: (a) an octahedron in a cube, showing the three Cartesian axes; (b) each tetrad rotation axis (4) lies along either x-, y- or z and is normal to a mirror plane; (c) three-fold inversion axes ($\bar{3}$) pass through the centre of each triangular face; (d) a triangular face viewed from above; (d) diad axes through the centre of each edge lie normal to mirror planes

pair of triangular faces, along the cube $\langle 111 \rangle$ directions (Figure 4.7c). The operation of this symmetry element is to rotate a vertex such as A by 120° counter clockwise to B, and then invert it to generate the vertex F. The operation of the $\bar{3}$ axis is more readily seen from the view down the axis, (Figure 4.7d). Successive 120° rotations in a counter-clockwise direction, accompanied by a inversion, brings about the transformation of vertex $A \rightarrow (B) \rightarrow F$, $F \rightarrow (D) \rightarrow C$, $C \rightarrow (A) \rightarrow E$, $E \rightarrow (F) \rightarrow B$, $B \rightarrow (C) \rightarrow D$, $D \rightarrow (E) \rightarrow A$.

Finally, it is seen that diads (2) run through the middle of each edge of the octahedron, and run between any pair of the x-, y- or z-axes, along $\langle 110 \rangle$. Each diad is accompanied by a perpendicular mirror, (Figure 4.7e). The combination is written $2/m$ (pronounced '2 over m').

4.3   Axes of inversion: rotoreflection

The rotoinversion operator $\bar{n}$ is closely related to that of **rotoreflection**, and *both* are referred to as improper rotation axes. An axis of rotoreflection combines a counter clockwise rotation of $2\pi/n$ with reflection in a plane normal to the axis. These axes are given the symbol $\tilde{n}$. The point where the mirror plane intersects the rotation axis in a rotoreflection ($\tilde{n}$) operation is the centre of symmetry about which the inversion takes place during the operation of an $\bar{n}$ inversion axis. The operation of a rotoreflection axis has similar results to the operation of a rotoinversion axis, but the two sets of symbols *cannot* be directly interchanged. For example, the operation of a two-fold improper rotation axis $\tilde{2}$ is drawn in Figure 4.8. The initial atom position (Figure 4.8a), is rotated 180° counter clockwise (Figure 4.8b) then reflected across a mirror normal to the axis, (Figure 4.8c). It is seen that the operation is identical to that of a centre of symmetry, (Figure 4.8d), and this latter designation is used in preference to that of the improper axis. The operation of the improper axis $\bar{4}$ is equivalent to $\tilde{4}$, the only pair with direct correspondence.

The equivalence of the important rotoreflection axes with rotoinversion axes or other symmetry operators is given in Table 4.1. In crystallography the rotoinversion operation is always preferred,

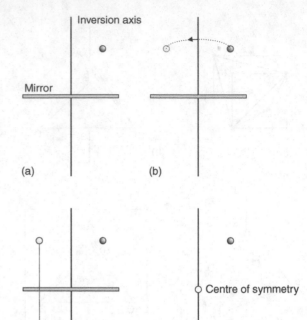

Figure 4.8   The operation of a two-fold rotoreflection improper rotation axis $\tilde{2}$: (a) the initial atom position; (b) rotation by 180° counter clockwise; (c) reflection across a mirror normal to the axis; (d) the operation of a centre of symmetry

Table 4.1   Correspondence of rotoreflection and rotoinversion axes

Axis of rotoreflection	Axis of rotoinversion
$\tilde{1}$	$\bar{2}$ (*m*)
$\tilde{2}$	$\bar{1}$
$\tilde{3}$	$\bar{6}$
$\tilde{4}$	$\bar{4}$
$\tilde{6}$	$\bar{3}$

but for molecular symmetry, which uses the Schoenflies system of symmetry nomenclature, the rotoreflection operations are chosen. Unfortunately, as mentioned above, *both* of these operations are often simply called *improper rotations*, without further specification, so care has to be exercised when comparing molecular symmetry descriptions (Schoenflies notation) with crystallographic descriptions (Hermann-Mauguin notation). A full correspondence between Schoenflies and Hermann-Mauguin nomenclature is given in Appendix 3.

4.4 The Hermann-Mauguin symbols for point groups

The symmetry operators that characterise a point group are collected together into the point group symbol. Crystallographers use the International or Hermann-Mauguin symbols. [The alternative Schoenflies symbols are given in Appendix 3.] As in two-dimensions, the order in which the symmetry operators are written is governed by specific rules, given in Table 4.2. The places in the symbol refer to **directions**. [When a number of directions are listed in Table 4.2 they are all equivalent, due to the symmetry of the object.] The first place or **primary** position is given to the most important or defining symmetry element of the group, which is often a symmetry axis. Symmetry axes are taken as parallel to the direction described. Mirror planes are taken to run **normal** to the direction indicated in Table 4.2. If a symmetry axis and the normal to a symmetry plane are parallel, the two symmetry characters are separated by a slash, /. Thus the symbol $2/m$ indicates that a diad axis runs parallel to the normal to a mirror plane, that is, the diad is normal to a mirror plane.

In the example of the regular tetrahedron, (Figures 4.5a, 4.6), the principle symmetry element is the $\bar{4}$ axis, which is put in the primary position in the point group symbol, and defines the [100] direction of the cubic axial set. The rotation triad (3) lies along the $\langle 111 \rangle$ directions

Table 4.2 The order of the Hermann-Mauguin symbols in point groups

Crystal system	Primary	Secondary	Tertiary
Triclinic	–	–	–
Monoclinic	[010], unique axis **b** [001], unique axis **c**	–	–
Orthorhombic	[100]	[010]	[001]
Tetragonal	[001]	[100], [010]	[1$\bar{1}$0], [110]
Trigonal, Rhombohedral axes	[111]	[1$\bar{1}$0], [01$\bar{1}$], [$\bar{1}$01]	
Trigonal, Hexagonal axes	[001]	[100], [010], [$\bar{1}\bar{1}$0]	
Hexagonal	[001]	[100], [010], [$\bar{1}\bar{1}$0]	[1$\bar{1}$0], [120], [$\bar{2}\bar{1}$0]
Cubic	[100], [010], [001]	[111], [1$\bar{1}\bar{1}$], [$\bar{1}$1$\bar{1}$], [$\bar{1}\bar{1}$1]	[1$\bar{1}$0], [110], [01$\bar{1}$], [011], [$\bar{1}$01], [101]

and so is given the secondary place in the symbol. In addition, the tetrahedron has mirror symmetry. The mirror planes are {110} planes that contain two vertices and lie normal to the other two vertices. These lie normal to some of the ⟨110⟩ directions, and are given the third place in the symbol, $\bar{4}3m$.

The same procedure can be used to collect the symmetry elements for a regular octahedron, (Figures 4.5b, 4.7), described above, into the point group symbol $4/m\,\bar{3}\,2/m$. The regular cube, (Figure 4.5c), can be described with the exactly the same point group, $4/m\,\bar{3}\,2/m$.

The two most complex Platonic solids, the icosahedron and dodecahedron, (Figure 4.5 d, e), both have $\bar{5}$ (five-fold inversion) axes. In these a vertex is rotated by 72°, (360/5)°, and then translated by the height of the solid. In the case of a dodecahedron a $\bar{5}$ axis passes through each vertex, while in the case of a regular icosahedron, a $\bar{5}$ axis passes through the centre of each face. The point group symbol for both of these solids is $2/m\,\bar{3}\,\bar{5}$, sometimes written as $\bar{5}\,\bar{3}\,2/m$.

To determine the point group of an object, it is only necessary to write down a list of all of the symmetry elements that it posses, (not always easy), and then order them following the rules set out above.

4.5   The symmetry of the Bravais lattices

The regular stacking of unit cells that can be translated but not rotated or reflected builds a crystal. This crystallographic limitation imposes a constraint upon the combinations of symmetry elements that are allowed in a lattice. In particular, the pentad axis and rotation axes higher than 6 are not allowed, for the reasons given in Chapter 3. It is also found that the pentad inversion axis, $\bar{5}$, and all inversion axes higher than $\bar{6}$ are forbidden. The only operators allowed within the Bravais

lattices are the centre of symmetry, $\bar{1}$, the mirror operator, m, the proper rotation axes 1, 2, 3, 4 and 6, and the improper rotation axes, $\bar{1}, \bar{2}, \bar{3}, \bar{4}$ and $\bar{6}$.

The operation of the allowed symmetry elements on the 14 Bravais lattices must leave each lattice point unchanged. The symmetry operators are thus representative of the point symmetry of the lattices. The most important lattice symmetry elements are given in Table 4.3. In all except the simplest case, two point group symbols are listed. The first is called the 'full' Hermann-Mauguin symbol, and contains the most complete description. The second is called the 'short' Hermann-Mauguin symbol, and is a condensed version of the full symbol. The order in which the operators within the symbol are written is given in Table 4.2.

The least symmetrical of the Bravais lattices is the triclinic primitive (aP) lattice. This is derived from the oblique primitive (mp) plane lattice by simply stacking other mp-like layers in the third direction, ensuring that the displacement of the second layer is not vertically above the first layer. Because the environment of each lattice point is identical with every other lattice point, a lattice point must lie at an inversion centre. This is the only symmetry element present, and the point group symbol for this lattice is $\bar{1}$.

The two monoclinic Bravais lattices, mP and mC, are generated in a similar way. Each layer of the lattice is similar to the plane mp lattice. In this case, the second and subsequent layers of the lattice are placed directly over the first layer, to generate the mP lattice, or displaced so that each lattice point in the second layer is vertically over the cell centre in the first layer to give the mC lattice. Because of this geometry, the diad axis of the plane lattice is preserved, and runs along the unique **b**- or **c**-axis of the lattice. Moreover, the fact that the layers are stacked vertically means that a mirror plane runs perpendicular to the diad, through

Table 4.3   The symmetries of the Bravais lattices

Crystal system	Bravais lattice	Point group symbol Full	Short[*]
Triclinic (anorthic)	aP	$\bar{1}$	
Monoclinic	mP	$\dfrac{2}{m}$	
	mC	$\dfrac{2}{m}$	
Orthorhombic	oP	$\dfrac{2}{m}\dfrac{2}{m}\dfrac{2}{m}$	mmm
	oC	$\dfrac{2}{m}\dfrac{2}{m}\dfrac{2}{m}$	mmm
	oI	$\dfrac{2}{m}\dfrac{2}{m}\dfrac{2}{m}$	mmm
	oF	$\dfrac{2}{m}\dfrac{2}{m}\dfrac{2}{m}$	mmm
Tetragonal	tP	$\dfrac{4}{m}\dfrac{2}{m}\dfrac{2}{m}$	$4/mmm$
	tI	$\dfrac{4}{m}\dfrac{2}{m}\dfrac{2}{m}$	$4/mmm$
Rhombohedral	hR	$\bar{3}\dfrac{2}{m}$	$\bar{3}m$
Hexagonal	hP	$\dfrac{6}{m}\dfrac{2}{m}\dfrac{2}{m}$	$6/mmm$
Cubic	cP	$\dfrac{4}{m}\bar{3}\dfrac{2}{m}$	$m\bar{3}m$
	cI	$\dfrac{4}{m}\bar{3}\dfrac{2}{m}$	$m\bar{3}m$
	cF	$\dfrac{4}{m}\bar{3}\dfrac{2}{m}$	$m\bar{3}m$

[*]Only given if different from the full symbol.

each plane of lattice points. No other symmetry elements are present. The point group symbol of both the mP and mC lattices is therefore $2/m$.

The rhombohedral primitive unit cell, often described in terms of a hexagonal unit cell, (see Section 1.1) can also be related to a cubic unit cell. If a cubic unit cell is compressed or stretched along one of the cube body diagonals it becomes rhombohedral, (Figure 4.9). The compression or stretching direction becomes the unique threefold axis that characterises the trigonal system (see Table 4.3). In the special case when the rhombohedral angle α_r is exactly equal to 60°,

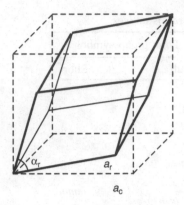

Figure 4.9 The relationship between a cubic unit cell and a rhombohedral unit cell. A cubic unit cell compressed or stretched along one of the cube body diagonals becomes rhombohedral

the points of the primitive rhombohedral lattice can be indexed in terms of a face-centred cubic lattice, where the lattice parameter of the cubic cell, a_c is given by:

$$a_c = (\sqrt{2})a_r$$

where a_r is the rhombohedral unit cell parameter.

The other point group symbols can be derived by similar considerations. It transpires that there are only seven point groups, corresponding to the seven crystal systems, although there are 14 lattices. They are given in Table 4.3.

4.6 The crystallographic point groups

A solid can belong to one of an infinite number of general three-dimensional point groups. However, if the rotation axes are restricted to those that are compatible with the translation properties of a lattice, a smaller number, the **crystallographic point groups**, are found. The operators allowed within the crystallographic

point groups are: the centre of symmetry, $\bar{1}$, the mirror operator, m, the proper rotation axes 1, 2, 3, 4 and 6, and the improper rotation axes, $\bar{1}$, $\bar{2}$, $\bar{3}$, $\bar{4}$ and $\bar{6}$. When these are combined together, 32 crystallographic point groups can be constructed, as listed in Table 4.4, of which 10 are centrosymmetric. The Schoenflies symbols for the crystallographic point groups are given in Appendix 3.

There is no significant symmetry present in the triclinic system, and the symmetry operator 1 or $\bar{1}$ is placed parallel to any axis, which is then designated as the **a**-axis.

In the monoclinic system, the point group symbols refer to the unique axis, conventionally taken as the **b**-axis. The point group symbol 2 means a diad axis operates parallel to the **b**-axis. The improper rotation axis $\bar{2}$, taken to run parallel to the unique **b**-axis, is the same as a mirror, m, perpendicular to this axis. The symbol $2/m$, indicates that a diad axis runs parallel to the unique **b**-axis and a mirror lies perpendicular to the axis. Because there are mirrors present in two of these point groups, the possibility arises that crystals with this symmetry combination will be enantiomorphous, and show optical activity, (see Sections 4.7, 4.8, below). Note that in some cases the unique monoclinic axis is specified as the **c**-axis. In this case the primary position refers to symmetry operators running normal to the direction [001], as noted in Table 4.3.

In the orthorhombic system, the three places refer to the symmetry elements associated with the **a**-, **b**- and **c**-axes. The most symmetrical group is $2/m\ 2/m\ 2/m$, which has diads along the three axes, and mirrors perpendicular to the diads. This is abbreviated to mmm because mirrors perpendicular to the three axes generates the diads automatically. In point group $2mm$, a diad runs along the **a**-axis, which is the intersection of two mirror planes. Group 222 has three diads along the three axes.

Table 4.4 Crystallographic point groups

Crystal system	Full[*]	Short[**]	Centrosymmetric groups
	Point group symbol		
Triclinic	1		
	$\bar{1}$		✓
Monoclinic	2		
	$m\ (\equiv \bar{2})$		
	$\dfrac{2}{m}\ (2/m)$		✓
Orthorhombic	222		
	$mm2$		
	$\dfrac{2}{m}\dfrac{2}{m}\dfrac{2}{m}\ (2/m\ 2/m\ 2/m)$	mmm	✓
Tetragonal	4		
	$\bar{4}$		
	$\dfrac{4}{m}\ (4/m)$		✓
	422		
	$4mm$		
	$\bar{4}2m$ or $\bar{4}m2$		
	$\dfrac{4}{m}\dfrac{2}{m}\dfrac{2}{m}\ (4/m\ 2/m\ 2/m)$	$4/mmm$	✓
Trigonal	3		
	$\bar{3}$		✓
	32 or 321 or 312		
	$3m$ or $3m1$ or $31m$		
	$\bar{3}\dfrac{2}{m}$ or $\bar{3}\dfrac{2}{m}1$ or $\bar{3}1\dfrac{2}{m}$	$\bar{3}m$ or $\bar{3}m1$ or $\bar{3}1m$	✓
	$(\bar{3}\ 2/m$ or $\bar{3}\ 2/m\ 1$ or $\bar{3}\ 1\ 2/m)$		
Hexagonal	6		
	$\bar{6}\left(\equiv \dfrac{3}{m}\right)$		
	$\dfrac{6}{m}\ (6/m)$		✓
	622		
	$6mm$		
	$\bar{6}2m$ or $\bar{6}m2$		
	$\dfrac{6}{m}\dfrac{2}{m}\dfrac{2}{m}\ (6/m\ 2/m\ 2/m)$	$6/mmm$	✓
Cubic	23		
	$\dfrac{2}{m}\bar{3}\ (2/m\ \bar{3})$	$m\bar{3}$	✓
	432		
	$\bar{4}3m$		
	$\dfrac{4}{m}\bar{3}\dfrac{2}{m}\ (4/m\ \bar{3}\ 2/m)$	$m\bar{3}m$	✓

[*]The symbols in parenthesis are simply a more compact way of writing the full symbol.
[**]Only given if different from the full symbol.

In the tetragonal system, the first place in the point group symbol refers to the unique **c**-axis. This must always be a proper or improper rotation tetrad axis. The second place is reserved for symmetry operators lying along the **a**-axis, and due to the tetrad, also along the **b**-axis. The third place refers to symmetry elements lying along the cell diagonals, ⟨110⟩. Thus the point group symbol $\bar{4}2m$ means that an inversion tetrad lies along the **c**-axis, a rotation diad along the **a**- and **b**-axes, and mirrors bisect the **a**- and **b**-axes.

In the trigonal groups, the first place is reserved for the defining three-fold symmetry element lying along the unique **c**-axis, (hexagonal axes), and the second position for symmetry elements lying along the **a**-axis, (hexagonal axes). Thus, the point group 32 has a rotation triad along the **c**-axis and a diad along the **a**-axis together with two other diads generated by the triad.

In the hexagonal groups, the first place is reserved for the defining six-fold symmetry element lying along the unique **c**-axis, sometimes with a perpendicular mirror, and the second position for symmetry elements lying along the **a**-axis, the [100] (or equivalent) direction. The third position is allocated to symmetry elements referred to a [120] (or equivalent) direction. This direction is at an angle of 30° to the **a**-axis, (Figure 4.10a). For example, the point group 6*mm* has a rotation hexad along the **c**-axis, a mirror with the mirror plane normal in a direction parallel to the **a**-axis, and a mirror with a normal parallel to the direction [120]. Operation of the hexad generates a set of mirrors, each at an angle of 30° to its neighbours, (Figure 4.10b). The point group $6/m\,2/m\,2/m$ has a hexad along the **c**-axis, a mirror plane normal parallel to the **c**-axis, (i.e. a mirror plane normal to the hexad), diads along [100] and [120], and mirror plane normals parallel to these directions, (Figure 4.10c).

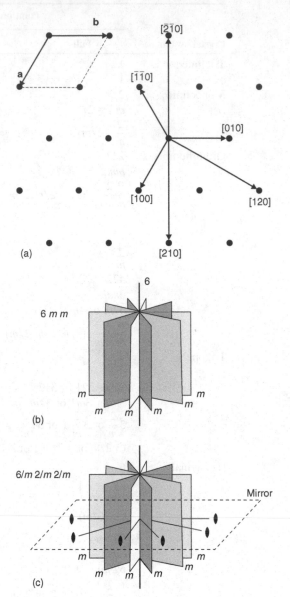

Figure 4.10 Symmetry elements present in a hexagonal crystal: (a) directions in a hexagonal lattice; (b) the point group 6*mm* has a rotation hexad along the **c**-axis which generates a set of mirrors, each at an angle of 30° to its neighbours; (c) the point group $6/m\,2/m\,2/m$ has a hexad along the **c**-axis, a mirror plane normal parallel to the **c**-axis, diads along [100] and [120], and mirror plane normals parallel to these directions

The cubic point groups show the greatest complexity in the arrangement of the symmetry elements. The first place in the symbol is reserved for symmetry elements associated with the **a**- (and hence **b**- and **c**-) axes. The second place refers to the type of triad lying along the cube body diagonal, $\langle 111 \rangle$. The third place refers to symmetry elements associated with the face diagonals, $\langle 110 \rangle$. Thus, the point group $4/m \; \bar{3} \; 2/m$ has four tetrad axes lying along the **a**- **b**- and **c**-axes, mirrors perpendicular to these axes, inversion triads along the cube body diagonals, diads along the face diagonals $\langle 110 \rangle$ with mirrors perpendicular to them. The 9 mirrors in this point group, three parallel to the cube faces and six parallel to the face diagonals, automatically generate the three tetrads and the six diads, so that the symbol is abbreviated to $m\bar{3}m$.

As remarked in Chapter 1, the external shape of a crystal can be classified into one of 32 crystal classes by making use of the symmetry elements that are present. These crystal classes correspond to the 32 crystallographic point groups. The two terms are often used completely interchangeably, and for practical purposes can be regarded as synonyms.

To determine the crystallographic point group of a crystal, write down a list of all of the symmetry elements that it possesses, order them following the rules set out above, and then compare them to the symmetry elements associated with the point groups listed in Table 4.4.

4.7 Point groups and physical properties

The physical properties of crystals reflect the crystal structure of the material. Although the magnitude of any property is related to the atom types that make up the solid, the existence or absence of a property is often controlled by the symmetry of the solid. This is set out in **Neumann's principle**, which states that:

"The symmetry elements of any physical property of a crystal must include the symmetry elements of the point group of the crystal."

Thus physical properties can be used as a probe of symmetry, and can reveal the crystallographic point group of the phase. Note that Neumann's principle states that the symmetry elements of a physical property must *include* those present in the point group, and not that the symmetry elements are identical with those of the point group. This means that a physical property may show more symmetry elements than the point group, and so not all properties are equally useful for revealing true point group symmetry. For example, the density of a crystal is controlled by the unit cell size and contents, but the symmetry of the material is irrelevant, (see Chapter 1). Properties similar to density, which do not reveal symmetry are called **non-directional**. **Directional** properties, on the other hand, may reveal symmetry.

Of the directional physical properties, some are able to disguise the lack of a centre of symmetry. For example, thermal expansion occurs equally in any $[uvw]$ and $[\bar{u}\bar{v}\bar{w}]$ direction, regardless of the symmetry of the crystal. Note that this does not mean that the thermal expansion is identical along all crystallographic directions simply that opposite directions behave in an identical fashion under the applied stimuli of heating and cooling. Such physical properties, called **centrosymmetric physical properties**, will only differentiate between 11 of the point groups. This reduced assembly are called **Laue classes**. The 11 Laue classes, each of which contains the centrosymmetric and non-centrosymmetric crystallographic point groups that behave in the same way when investigated by centrosymmetric physical properties, are listed in Table 4.5.

Table 4.5 The Laue classes

Crystal system	Laue class (centrosymmetric group)	Non-centrosymmetric groups included in the class
Triclinic	$\bar{1}$	1
Monoclinic	$2/m$	$2, m$
Orthorhombic	mmm	$222, mm2$
Tetragonal	$4/m$	$4, \bar{4}$
	$4/mmm$	$422, 4mm, \bar{4}2m$
Trigonal	$\bar{3}$	3
	$\bar{3}m$	$32, 3m$
Hexagonal	$6/m$	$6, \bar{6}$
	$6/mmm$	$622, 6mm, \bar{6}2m$
Cubic	$m\bar{3}$	23
	$m\bar{3}m$	$432, \bar{4}3m$

The symmetry of a material is also only partially revealed by the shape of etch pits on a crystal surface. These are created when a crystal begins to dissolve in a solvent. Initial attack is at a point of enhanced chemical reactivity, often where a dislocation reaches the surface. A pit forms as the crystal is corroded. The shapes of the pits, called **etch figures**, have a symmetry corresponding to one of the of 10 two-dimensional plane point groups. This will be the point group that corresponds with the symmetry of the face. An etch pit on a (100) face of a cubic crystal will be square, and on a (101) face of a tetragonal crystal will be rectangular.

4.8 Dielectric properties

Dielectrics are insulating materials, and the dielectric behaviour of a solid is its behaviour in an electric field. When ordinary insulators are exposed to an externally applied electric field the negative electrons and positive atomic nuclei are displaced in opposite directions, so that **electric dipoles** develop where none existed before. (An electric dipole occurs in a crystal when a region of positive charge is separated from a region of negative charge by a fixed distance r.) Electric dipoles are vectors, and need to be characterised by a description of both magnitude and a direction. The dipole, **p**, is represented by a vector pointing from the negative to the positive charge, (Figure 4.11a). A material that contains electric dipoles due to an applied electric field is said to become **polarised**.

In some dielectrics, such as pyroelectric and ferroelectric materials, **permanent electric dipoles** exist in the absence of an external electric field. In these latter types of materials, dielectric properties are related to the way in which the permanent electric dipoles, as well as the other electrons and atomic nuclei, respond to the applied electric field.

It is clear that if the effect of internal electric dipoles in a crystal is to be observed externally, as

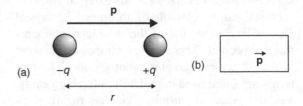

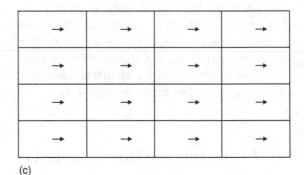

Figure 4.11 Pyroelectric crystals; (a) an electric dipole, **p**, represented by a vector pointing from the negative to the positive charge; (b) schematic representation of a unit cell containing a permanent electric dipole; (c) a pyroelectric crystal built of unit cells containing a permanent electric dipole

a change in physical property, the dipole arrangement within a unit cell must not cancel to zero. This constraint links the observable dielectric properties with the arrangement of atoms in a unit cell, and hence the symmetry of the crystal. The simplest starting point is to consider directions.

A **polar direction** in a crystal is a direction [uvw] that is not related by symmetry to the opposed direction [ūv̄w̄]. The two senses of the direction are physically different. No polar direc-

tions can therefore occur in a crystal belonging to any one of the 11 centrosymmetric point groups, listed in Table 4.5. Polar directions are found in the **polar point groups** or **polar crystal classes**. Unfortunately these terms are not exactly synonymous, and have slightly different meanings in the literature. In crystallographic usage, a polar class is simply one of the 21 non-centrosymmetric point groups, listed in Tables 4.5 and 4.6. In physics, a crystal class is

Table 4.6 Non-centrosymmetric crystal classes and polar directions

Crystal class	Polar directions	Unique polar axis	Pyroelectric effect	Enantiomorphous group	Optical activity possible
Triclinic					
1	All	None	✓	✓	✓
Monoclinic					
2	All not ⊥ 2	[010]	✓	✓	✓
m	All not ⊥ m	None	✓		✓
Orthorhombic					
mm2	All not ⊥ 2 or m	[001]	✓		✓
222	All not ⊥ 2	None		✓	✓
Tetragonal					
4	All not ⊥ 4	[001]	✓	✓	✓
4̄	All not ‖ or ⊥ 4̄	None			✓
422	All not ⊥ 4 or ⊥ 2	None		✓	✓
4mm	All not ⊥ 4	[001]	✓		
4̄2m	All not ⊥ 4̄ or ⊥ 2	None			✓
Trigonal		None			
3	All	[111]	✓	✓	✓
32	All not ⊥ 2	None		✓	✓
3m	All not ⊥ m	[111]	✓		
Hexagonal					
6	All not ⊥ 6	[001]	✓	✓	✓
6̄	All not ⊥ 6̄	None			
622	All not ⊥ 6 or 2	None		✓	✓
6mm	All not ⊥ 6	[001]	✓		
6̄m2	All not ⊥ 2	None			
Cubic					
23	All not ⊥ 2	None		✓	✓
432	All not ⊥ 4 or ⊥ 2	None		✓	✓
4̄3m	All not ⊥ 4̄	None			

⊥ = perpendicular to the axis following. ‖ = parallel to the axis following.

considered to be a polar class if it gives rise to the pyrolectric effect, (see below). This is a subset of the 21 non-centrosymmetric point groups, consisting of the 10 point groups 1, 2, 3, 4, 6, *m*, *mm*2, 3*m*, 4*mm*, and 6*mm*. They are listed in Table 4.6.

The difference between the two definitions rests with the fact that a pyroelectric crystal must possess an overall (observable) **permanent** electric dipole. Thus a pyroelectric crystal is built from unit cells, *each of which* must contain an overall electric dipole, (Figure 4.11b). The pyroelectric effect will only be observed, however, if all of these dipoles are aligned throughout the crystal, (Figure 4.11c). [Note that a ferroelectric crystal is defined in a similar way. The difference between a pyroelectric crystal and a ferroelectric crystal lies in the fact that the direction of each overall electric dipole in a ferroelectric crystal can be altered by an external electric field.]

Electric dipoles can arise in many ways in a crystal. One obvious way is for a crystal to contain **polar molecules**, which are molecules that carry a permanent electric dipole. For example, the charge on the N atom in the molecule nitric oxide, NO, is slightly positive, ($\delta+$), with respect to the O atom, ($\delta-$), and the molecule has a permanent electric dipole moment of 0.5×10^{-30} Cm, (Figure 4.12). Crystals containing NO molecules may therefore be polar themselves, depending upon the symmetry

of the molecular arrangement. An electric dipole can also arise simply as a consequence of the arrangement of the atoms in a non-molecular crystal. For example, a crystal such as ZnS contains electric dipoles because of the structure of the [ZnS$_4$] tetrahedra which make up the crystal. The centre of gravity of the negative charges located on the S atoms does not coincide with the positive charge located on the Zn atoms, so that each tetrahedron contains an internal electric dipole, pointing from the centre of gravity of the S atoms to the Zn atom (Figure 4.13a). Zinc sulphide crystallises in two main crystallographic forms, cubic zinc blende (sphalerite) or hexagonal wurtzite, (see also Chapter 8, Section 8.5). The arrangement of the tetrahedra in ZnS crystals with the hexagonal wurtzite structure is such that the dipoles all point in the same direction, leading to a pyroelectric structure, (Figure 4.13b). In the cubic zinc blende structure pairs of tetrahedra point in opposite directions, hence dipoles cancel out internally, and the material is not a pyroelectric material.

If a vector such as an electric dipole exists in a crystal, the operation of the symmetry

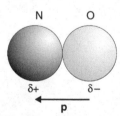

Figure 4.12 The NO molecule, which has a permanent dipole due to slight differences in charge $\delta+$, $\delta-$, on the component atoms.

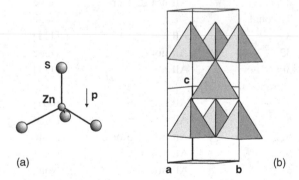

Figure 4.13 The wurtzite structure of zinc sulphide, ZnS: (a) each individual ZnS$_4$ tetrahedron has a permanent electric dipole **p**; (b) the crystal structure, which is built of tetrahedra all oriented in the same direction

elements present will give rise to a group of equivalent vectors, called a **form**. As has been hinted above, in a centrosymmetric crystal these vectors add to zero. In 11 of the non-centrosymmetric point groups, the vector sum of the form of vectors also adds to zero and the crystal will not show a pyroelectric effect. The 10 classes that do possess a resultant vector are the ones that show a pyroelectric effect. The resultant vector in these classes coincides with the solitary proper rotation axis, 2, 3, 4, or 6, which are referred to as the **unique polar axes**. They are listed in Table 4.6. As an example, ZnS crystals with the wurtzite structure belong to the point group 6*mm*, and the unique polar axis coincides with the crystallographic **c**-axis, that is, with the hexad axis.

4.9 Refractive index

Light can be regarded as a wave of wavelength λ with an electrical and magnetic component. The refractive index of a material is the physical property that is the external manifestation of the effect of the electric field component of a light wave on the electron distribution around each of the atoms in the structure. The refractive index of a material is sensitive to crystal structure. Atoms with easily polarisable, (i.e. easily displaced), electrons, give rise to a high refractive index, while those with tightly bound electrons give rise to a low refractive index. Thus the replacement of a closed shell ion such as Sr^{2+} by a lone pair ion of comparable size, such as Pb^{2+}, will greatly increase the refractive index of the material. This effect is used in the manufacture of high refractive index flint glasses, which contain appreciable amounts of lead.

Cubic crystals like halite (common salt or rock salt) have the same refractive index in all directions and are said to be optically **isotropic**. All other crystals are optically **anisotropic**. In tetragonal and hexagonal crystals the refractive indices along the **a**- and **b**-axes are the same and different from the refractive index along the **c**-axis. For example, the refractive index of the rutile form of TiO_2, which is tetragonal, is 2.609 along **a**- and **b**-, and 2.900 along the **c**-axis. In trigonal crystals, the refractive index along the three-fold rotation axis, [111] is different from that normal to this direction. In orthorhombic, monoclinic and triclinic crystals there are three refractive indices along mutually perpendicular axes.

This correlation is unsurprising. The refractive index depends upon the density of atoms in a crystal. In cubic crystals the atom density averages to be the same in all axial directions while in crystals of lower symmetry some axial directions contain more atoms than others.

The electrical and magnetic components of a light wave lie at right angles to one another, and each is most properly described as a vector. Only the electric vector is important for the optical properties described here. This vector is always perpendicular to the line of propagation of the light but can adopt any angle otherwise. The position of the electric vector defines the **polarisation** of the light wave. For ordinary light, such as that from the sun, the orientation of the electric vector changes in a random fashion every 10^{-8} seconds or so. Ordinary light is said to be **unpolarised**. Light is **linearly** or **plane polarised** when the electric vector is forced to vibrate in a single plane. The measured refractive index of a crystal depends upon the wavelength of the light and in many crystals upon the polarisation of the beam.

A beam of unpolarised light can be resolved into two linearly polarised components with vibration directions perpendicular to each other. For convenience these can be called the horizontally and vertically polarised components. When the beam enters a transparent medium, each of the two linearly polarised components experiences its own refractive index.

In the case of cubic, optically isotropic, crystals the refractive index is identical for both the horizontally and vertically polarised components, in any direction whatsoever. Hexagonal, trigonal and tetragonal crystals show two refractive indices, called the **principal indices**, written n_o and n_e. When a beam of light enters such a crystal along an arbitrary direction the refractive index differs for the two polarisation components. One experiences a refractive index n_o and the other a refractive index n_e', which has a magnitude between n_o and n_e, dependent upon the incident direction. However, if the beam of light is directed along the crystallographic **c**-axis in these crystals, each polarisation component experiences only one refractive index, equal to n_o. This direction is called the **optic axis**. There is only one such unique direction, and these materials are called **uniaxial crystals**. In the tetragonal, trigonal and hexagonal systems, the unique optic axis lies along the direction of highest symmetry, 4, 3 or 6. If the beam is directed perpendicular to the optic axis the two refractive indices observed are n_o and n_e.

Orthorhombic, monoclinic and triclinic crystals exhibit three principal refractive indices, n_α which has the smallest value, n_γ, which has the greatest value, and n_β which is between the other two. The horizontally and vertically polarised components of a beam of light entering such a crystal encounter different refractive indices, with magnitudes lying between the lowest, n_α and the highest, n_γ. Two optic axes exist in these crystals, and the refractive index encountered by both polarisation components of a light beam directed along either optic axis is n_β. These materials are known as **biaxial crystals**. Unlike uniaxial crystals, there is not an intuitive relationship between the optic axes and the crystallographic axes. However, one optic axis always lies along the direction of highest symmetry.

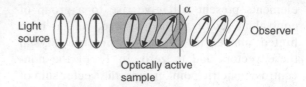

Figure 4.14   The rotation of the plane of polarisation of a beam of light by $\alpha°$ on passing through an optically active material

4.10   Optical activity

While the refractive index trends described in the previous section apply across all point groups, the property of optical activity is restricted to a smaller number. This physical property is the ability of a crystal (and also a melt or a solution) to rotate the plane of linearly polarised light to the right or the left (Figure 4.14). A material that rotates the plane of the light clockwise (when viewed into the beam of light) is called the **dextrorotatory**, *d*- or (+) form, while that which rotates the plane of the light counter clockwise is called the **laevorotatory**, *l*- or (−) form. The rotation observed, the **specific rotation** of the crystal, designated [α], is sensitive to both temperature and the wavelength of the light.

Optical activity was first understood with respect to naturally occurring tartaric acid crystals by Pasteur in 1848. At this stage, the puzzle was that tartaric acid crystals were optically active, while the chemically identical 'racemic acid' was not. The eventual results of the study revealed that 'racemic acid' crystallised to give two morphologically different crystals, which were mirror images of each other, that is, they existed as a left-handed and right-handed pair. Each of these crystal types was identical to tartaric acid in every way except that crystals of one hand would rotate the plane of polarised light one way, just as 'ordinary' tartaric acid, while those of the other hand rotated the plane of polarisation in the opposite direction. Such crystals, termed enantiomorphs, or an enantiomorphic pair, cannot be superimposed one upon the other, in the same way that a right-hand and left-hand glove cannot be superimposed one on the other.

The existence of left- and right-handed crystals implies an absence of a centre of symmetry, $\bar{1}$, in the crystal and optical activity is, in principle, confined to crystals which belong to one of the 21 non-centrosymmetric point groups listed in Tables 4.3 and 4.6. However, for a crystal to be enantiomorphic, the point group must also not contain a mirror, m, or a improper rotation axis $\bar{4}$. There are thus just 11 enantiomorphous point groups, 1, 2, 222, 4, 422, 3, 32, 6, 622, 23, 432, (Table 4.6), all of which can give crystals with a left- or right-handedness. From the other non-centrosymmetric point groups, optical activity can also occur in the four non-enantiomorphic classes m, $mm2$, $\bar{4}$ and $\bar{4}2m$, as directions of both left- and right-handed rotation occur in the crystal. There are thus 15 potentially optically active groups in all. In the remaining six non-centrosymmetric classes, optical activity is not possible because of the overall combination of symmetry elements present.

Optical activity is observable in any direction for crystals belonging to the two cubic enantiomorphic classes 23 and 432, but, in general, optical activity can only be observed in certain symmetry limited directions. For example, optical activity in the other enantiomorphic classes is only readily observed in a direction fairly close to an optic axis. This is because the different refractive indices that apply to light polarised vertically and horizontally masks the effect in directions further from an optic axis. In the non-enantiomorphic groups, no optical activity is found along an inversion axis or perpendicular to a mirror plane. Thus no optical activity occurs along the optic axis

in classes $\bar{4}$ or $\bar{4}2m$. In the remaining two groups, m and $mm2$, no optical activity is observed along the optic axes if these lie in a mirror plane.

4.11 Chiral molecules

The optical activity of many crystals is due to the presence of chiral molecules in the structure. A chiral molecule is one that cannot be superimposed upon its mirror image, and so exists in two mirror-image forms or enantiomers. Any such molecule is optically active. (See also section 4.1 and Figure 4.1). Although inorganic molecules with tetrahedral or octahedral bond geometry can form optically active pairs, optical activity has been explored in most detail with respect to organic molecules, where the mirror image molecules are known as **optical isomers**. Optical isomers occur in organic compounds whenever four *different* atoms or groups of atoms are attached to a tetrahedrally co-ordinated central carbon atom. Such a carbon atom is called a **chiral carbon atom** or **chiral centre**. Molecules with a single chiral centre exist as two enantiomers, one of which will rotate the plane of a linearly polarised light in one direction and the other in the opposite direction. The two enantiomers of the amino acid alanine, in which the chiral carbon atom is marked C^*, are shown in Figure 4.1. Only one, called the L-form, (see Chapter 5, Section 5.7), occurs naturally; the other can be made synthetically.

Pasteur's crystals of tartaric acid are more complex because the molecules contain two chiral carbon atoms. These can 'cancel out' internally in the molecule so that three molecular forms actually exist, the two optically active mirror image structures that cannot be superimposed on each other, as the laevorotatory (*l*-) and dextrorotatory (*d*-) forms, and the optically inactive (*meso-*) form, which can be superimposed on its mirror image[1]. These are drawn in Figure 4.15. Mixtures of enantiomers in equal proportions will not rotate the plane of linearly polarised light and are called **racemic** mixtures, after the 'racemic acid' of Pasteur. Note in the *meso-* form, the optical effects are internally cancelled, while in the racemic form they are externally compensated.

Although many optically active crystals contain optically active molecules, this is not mandatory. Crystals of quartz, one variety of silicon dioxide, SiO_2, occur in left or right-handed forms although no molecules are present. The structure of this material is composed of helices of corner-linked SiO_4 tetrahedra. The direction of the helix determines the left or right-hand nature of the crystals.

4.12   Second harmonic generation

The action of a light wave passing through a crystal causes a displacement of the electrons in the structure. In crystals without a centre of symmetry, (except for the cubic point group 432), this displacement generates light waves with higher frequencies, which are harmonics of the initial stimulating frequency. The most intense of these additional waves are those with double the frequency of the incident wave, and the production of these waves is called **second harmonic generation (SHG)**.

Generally, ordinary light sources are too weak to allow the effect to be observed. However laser light is of a sufficient intensity that the higher

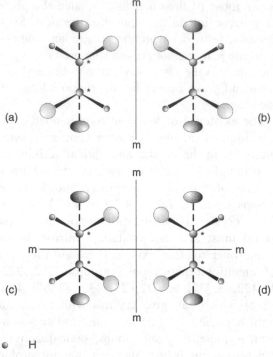

Figure 4.15   The structures of tartaric acid: (a) and (b) are enantiomorphs and optically active as the two structures cannot be superimposed upon each other and are mirror images; (a) (*d*)-, (+)- or (2*R*,3*R*)-tartaric acid; (b) (*l*)-, (−)- or (2*S*,3*S*)-tartaric acid; (c), (d), the *meso-* structure of tartaric acid; (c) (2*R*,3*S*)-tartaric acid; (d) (2*S*,3*R*)-tartaric acid. The chiral carbon atoms are marked[*]. The four bonds around each of these are arranged tetrahedrally; bonds in the plane of the page are drawn as full lines, those receding into the page as dashed lines, and those pointing out of the page as thin triangles

---

[1]The nomenclature of optically active organic molecules has been revised since their absolute configuration has been determinable via X-ray diffraction (Chapter 6). The equivalent terms are (*d*)-tartaric acid = (+)-tartaric acid = (2*R*,3*R*)-tartaric acid; (*l*)-tartaric acid = (−)-tartaric acid = (2*S*,3*S*)-tartaric acid; mesotartaric acid = (2*R*,3*S*)-tartaric acid = (2*S*,3*R*)-tartaric acid.

harmonics, especially the second harmonic, are quite easily observed. The presence of second harmonics is a very sensitive test for the absense of a centre of symmetry in a crystal. The situation in the anomalous point group 432 is that the symmetry elements present act so as to cancel the harmonics that are generated.

4.13 Magnetic point groups and colour symmetry

When the atoms that make up a material have unpaired electrons in the outer electron orbitals, the spin of these manifests itself externally as magnetism. Each atom can be imagined to have an associated elementary magnetic dipole attached to it. Paramagnetic compounds have the magnetic dipoles completely unaligned. These take random directions that are constantly changing in orientation. However, in some classes of magnetic materials, notably ferromagnetic, ferrimagnetic and antiferromagnetic solids, such as magnetic iron, Fe, or the ceramic magnet barium ferrite, $BaFe_{12}O_{19}$, the overall elementary magnetic dipoles on neighbouring atoms are aligned with each other (Figure 4.16).

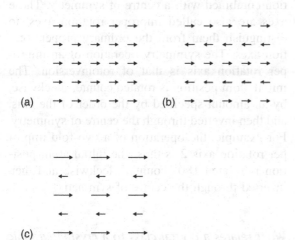

(a)

(b)

(c)

Figure 4.16 Magnetic ordering: (a) ferromagnetic; (h) antiferromagnetic; (c) ferrimagnetic. Magnetic dipoles on individual metal atoms are represented by arrows

From a macroscopic point of view, this is not apparent, because a crystal of a typical strongly magnetic material consists of a collection of domains, (small volumes), in which all the elementary magnetic dipoles are aligned. Adjacent domains have the magnetic dipoles aligned along different crystallographic directions, thus creating an essentially random array of magnetic dipoles. When searching for the point group of the crystal, this essentially random arrangement of dipoles can be safely ignored, as they can be in the case of paramagnetic solids. In both cases, the crystals can then be allocated to one of the 32 crystallographic point groups described above.

However, crystals can be obtained which consist of just a single magnetic domain. In such a case, the classical symmetry point group may no longer accurately reflect the overall symmetry of the crystal, because, in a derivation of the 32 (classical) point groups, the magnetic dipoles on the atoms were ignored. Clearly, in a single domain crystal of a magnetic compound, the magnetic dipoles ought to be taken into account when the symmetry operators described above are applied.

For example, the magnetic form of pure iron, α-iron, exists up to a temperature of 768°C. The non-magnetic form, once called β-iron, exists between the temperatures of 768°C and 912°C. Both of these forms of iron have the body-centred cubic structure, (Figure 4.17a). The point group of high temperature (β-iron), in which the magnetic dipoles are randomly oriented, or in multidomain samples of magnetic α-iron, is $m\bar{3}m$. If, however, a single domain crystal is prepared with all the magnetic moments aligned along [001], the **c**-axis is no longer identical to the **a**- or **b**-axes. The solid now belongs to the tetragonal system, (Figure 4.17b). When the spins are aligned along [110] the appropriate crystal system is orthorhombic, (Figure 4.17c), and when the spins are aligned along [111] the system is trigonal, (Figure 4.17d). In each of these cases

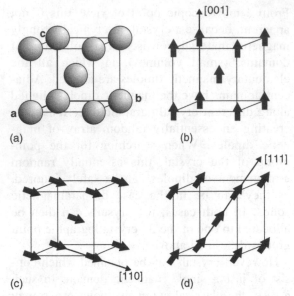

Figure 4.17 Magnetic structures of iron: (a) the structure of non-magnetic body-centred cubic iron; (b) magnetic spins, (arrows), aligned along the cubic [001] direction; (c) magnetic spins, (arrows), aligned along the cubic [110] direction; (d) magnetic spins, (arrows), aligned along the cubic [111] direction

new **magnetic point groups** describe the symmetry of the crystals. In all, there are 90 magnetic point groups.

The same increase in complexity is found when the point group of a ferroelectric crystal with aligned electric dipoles is considered, as, from a symmetry point of view the two cases are identical.

The 90 magnetic point groups are made up of the 32 classical point groups plus 58 point groups that specifically indicate the polarity of the atoms with magnetic or electric dipoles. These 58 non-classical point groups are called **antisymmetrical crystallographic point groups** or **'black and white'** point groups, where one colour, say black, is associated with one dipole orientation and the other, white, with the alterna-

tive dipole direction. The classical point groups are then termed the **neutral** groups.

The same considerations imply that the number of three-dimensional lattices should increase, and this is found to be so. There are a total of 36 magnetic lattices, made up of 22 **antisymmetry lattices**, together with the 14 neutral Bravais lattices.

More complex physical properties may require the specification of three or more 'colours'. In this case the general term **'colour symmetry'** is used, and the lattices and point groups so derived are the **colour lattices** and **colour point groups**.

Answers to introductory questions

What is a rotoinversion axis?

Apart from the rotation axes that occur in both two- and three-dimensional objects, an additional type of rotation axis occurs in a solid that is not found in planar shapes, the inversion axis, $\bar{n}$. The operation of an inversion axis consists of a rotation combined with a centre of symmetry. These axes are also called improper rotation axes, to distinguish them from the ordinary proper rotation axes. The symmetry operation of an improper rotation axis is that of rotoinversion. The initial atom position is rotated counter clockwise, by an amount specified by the order of the axis, and then inverted through the centre of symmetry. For example, the operation of a two-fold improper rotation axis $\bar{2}$ is thus: the initial atom position is rotated 180° counter clockwise and then inverted through the centre of symmetry.

What relates a crystal class to a crystallographic point group?

The external shape of a crystal can be classified into one of 32 crystal classes dependent upon the symmetry elements present. The collection of symmetry elements present in a solid is referred to as the general point group of the solid. A solid can belong to one of an infinite number of general three-dimensional point groups. However, if the rotation axes are restricted to those that are compatible with the translation properties of a lattice, a smaller number, the crystallographic point groups, are found. There are 32 crystallographic point groups. They are identical to the 32 crystal classes, and the terms are used interchangeably.

What are enantiomorphic pairs?

The property of optical activity is the ability of a crystal to rotate the plane of linearly polarised light to the right or the left. Enantiomorphs are the two forms of optically active crystals; one form rotating the plane of polarised light in one direction, and the other rotating the plane of polarised light in the opposite direction. Such crystals are found to exist as an enantiomorphic pair, consisting of right handed and left handed forms. These cannot be superimposed one upon the other, in the same way that a right-hand and left-hand glove cannot be superimposed one on the other.

The terms enantiomorphs and enaniomorphous pairs also apply to pairs of molecules that are optically active, and as such may endow crystals with optical activity, such as alanine, described in Chapter 5.

Problems and exercises

Quick quiz

1 The operation of an improper rotation $\bar{n}$ axis involves:
(a) A counter clockwise rotation plus reflection
(b) A counter clockwise rotation plus inversion through a centre of symmetry
(c) A counter clockwise rotation plus translation

2 The number of regular Platonic solids is:
(a) 5
(b) 7
(c) 9

3 The places in the Hermann-Mauguin symbol for a point group refer to:
(a) Planes
(b) Axes
(c) Directions

4 The point group symbol 4/m means:
(a) A mirror plane runs parallel to a four-fold rotation axis
(b) The normal to a mirror plane runs parallel to a four-fold rotation axis
(c) A mirror plane contains a four-fold rotation axis

5 The number of three-dimensional crystallographic point symmetry groups is:
(a) 14
(b) 23
(c) 32

6 Directional physical properties can give:
(a) A complete description of crystal symmetry
(b) A partial description of crystal symmetry
(c) No information about crystal symmetry

7 A polar direction can occur only in:
(a) A centrosymmetric point group

(b) Any non-centrosymmetric point group

(c) A subset of the non-centrosymmetric point groups

8 A monoclinic crystal will have:
(a) Two different principal refractive indices

(b) Three different principal refractive indices

(c) Four different principal refractive indices

9 An optically active crystal is one which is able to:
(a) Rotate the plane of polarised light as it traverses the crystal

(b) Split the light into two polarised components

(c) Generate harmonics of the fundamental frequency of polarised light

10 The number of magnetic point groups is:
(a) Less than the number of crystallographic point groups

(b) Equal to the number of crystallographic point groups

(c) Greater than the number of crystallographic point groups

Calculations and Questions

4.1 Write out the symmetry elements present and the corresponding point group for the molecules drawn in the figure below. The molecules are three-dimensional, with the following shapes: (a), SiCl$_4$, regular tetrahedron; (b), PCl$_5$, pentagonal bipyramid; (c), PCl$_3$, triangular pyramid; (d), benzene, hexagonal ring; (e), SF$_6$, regular octahedron. With reference to Appendix 3, write the Schoenflies symbol for each point group.

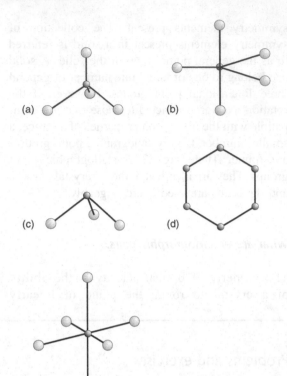

4.2 Determine the point group of each of the idealised 'crystals' in the figure below. (a) is a rhomb the shape of a monoclinic unit cell (see chapter 1) while (b) and (c) are similar, but with corners replaced by triangular faces. Presume that the triangular faces are identical. The axes apply to all objects with **b** perpendicular to the plane containing **a** and **c**.

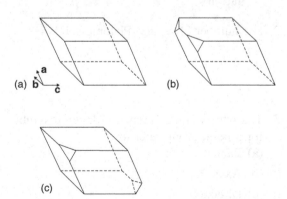

4.3 Determine the point group of each of the idealised 'crystals' in the figure below. (a) is a rhomb the shape of an orthorhombic unit cell (see chapter 1) while (b) and (c) are similar, but with corners replaced by triangular faces. Presume that the triangular faces are identical. The axes apply to all objects.

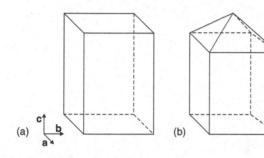

(a) (b)

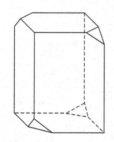

(c)

4.4 Write out the full Hermann-Mauguin symbols for the short symbols: (a), *mmm*; (b), 4/*mmm*; (c), $\bar{3}m$; (d), 6/*mmm*; (e), *m3m*.

4.5 (a) Sketch a regular cube (short symbol $m\bar{3}m$) and mark the symmetry axes and planes; (b) mark the cube faces so that the point group is changed to the cubic group $2/m\ \bar{3}$ (short symbol $m\bar{3}$, see Table 4.4).

4.6 Determine the point groups of the unit cells of the crystal structures of the metals Cu, Fe and Mg given in Chapter 1.

4.7 Chromium, Cr, has the A2 structure similar to that of iron, Figure(a) below but instead of having the magnetic moments aligned in a ferromagnetic order, they are aligned in an antiferromagnetic array (Figure 4.16b), so that the directions of the moments alternate from one layer to the next. Assuming the possible arrangements are as depicted in the figure below, what is the new crystal system applicable to single domain crystals, (b), (c) and (d)?

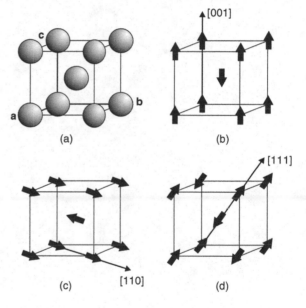

(a) (b)

(c) [110] (d)

5

Building crystal structures from lattices and space groups

- *What symmetry operation is associated with a screw axis?*

- *What is a crystallographic space group?*

- *What are Wyckoff letters?*

5.1 Symmetry of three-dimensional patterns: space groups

The seventeen plane groups, derived by combining the translations inherent in the five plane lattices with the symmetry elements present in the ten plane point groups, together with the glide operator, represent, in a compact way, all possible planar repeating patterns. In a similar way, a combination of the translations inherent in the 14 Bravais space lattices with the symmetry elements present in the 32 crystallographic point groups, together with a new symmetry element, the screw axis, described below, allows all possible three-dimensional repeating crystallographic patterns to be classified. The resulting 230 combinations are the **crystallographic space groups**.

There are parallels between the two-and three-dimensional cases. Naturally, mirror lines in two dimensions become **mirror planes**, and glide lines in two dimensions become **glide planes**. The glide translation vector, t, is constrained to be equal to half of the relevant lattice vector, T, for the same reason that the two-dimensional glide vector is half of a lattice translation (Chapter 3).

In addition, the combination of three-dimensional symmetry elements gives rise to a completely new symmetry operator, the **screw axis**. Screw axes are **rototranslational** symmetry elements, constituted by a combination of rotation and translation. A screw axis of order n operates on an object by (a) a rotation of $2\pi/n$ counter clockwise and then a translation by a vector t parallel to the axis, in a positive direction. The value of n is the **order** of the screw axis. For example, a screw axis running parallel to the c-axis in an orthorhombic crystal would entail a counter-clockwise rotation in the $a - b$ plane, (001), followed by a translation parallel to $+c$. This is a **right-handed screw rotation**. Now if the rotation component of the operator is applied n times, the total rotation is

---

Crystals and Crystal Structures.   Richard J. D. Tilley
© 2006 John Wiley & Sons, Ltd

equal to 2π. At the same time, the total displacement is represented by the vector $n\mathbf{t}$, running parallel to the rotation axis. In order to maintain the lattice repeat, it is necessary to write:

$$n\mathbf{t} = p\mathbf{T}$$

where p is an integer, and $\mathbf{T}$ is the lattice repeat in a direction parallel to the rotation axis. Thus:

$$\mathbf{t} = (p/n)\mathbf{T}$$

For example, the repeat translations for a three-fold screw axis are:

$$(\tfrac{0}{3})\,\mathbf{T}, (\tfrac{1}{3})\mathbf{T}, (\tfrac{2}{3})\mathbf{T}, (\tfrac{3}{3})\mathbf{T}, (\tfrac{4}{3})\mathbf{T}, \ldots.$$

Of these, only $(\tfrac{1}{3})\mathbf{T}$ and $(\tfrac{2}{3})\mathbf{T}$ are unique. The corresponding values of p are used in writing the three-fold screw axis as 3_1 (pronounced three sub one') or 3_2 (pronounced 'three sub two'). Similarly, the rotation diad 2 can only give rise to the single screw axis 2_1, (pronounced 'two sub 1'). The unique screw axes are listed in Table 5.1.

The operation of the screw axis 4_2 is described in Figure 5.1. The first action is a rotation counter clockwise by $2\pi/4$, i.e. 90°, (Figure 5.1a). The

Table 5.1   Rotation, inversion and screw axes allowed in crystals

Rotation axis, n	Inversion axis, $\bar{n}$	Screw axis, n_p			
1	$\bar{1}$ (centre of symmetry)				
2	$\bar{2}$ (m)	2_1			
3	$\bar{3}$	3_1	3_2		
4	$\bar{4}$	4_1	4_2	4_3	
6	$\bar{6}$	6_1	6_2	6_3	6_4	6_5

rotated atom is then translated by vector $\mathbf{t} = \mathbf{T}/2$, which is half of the lattice repeat distance parallel to the screw axis, (Figure 5.1b). This generates atom B from atom A. Repetition of this pair of operations generates atom C from atom B, (Figure 5.1c). The total distance displaced by the operations is equal to $2\mathbf{t} = 2 \times \mathbf{T}/2 = \mathbf{T}$, so that the atom C is also replicated at the origin of the screw vector, (Figure 5.1d). The operation of the screw axis on atom C generates atom D, (Figure 5.1e). The result of a 4_2 axis is to generate four atoms from the original one specified. This symmetry operation is often portrayed by a view along the axis (Figure 5.1f). In this figure, the motif is represented by a circle, (see Section 3.6), the $+$ means that the motif is situated above the plane of the paper and $\tfrac{1}{2} +$ indicates the position of a motif generated by screw operation. If the screw axis runs parallel to the **c**-axis, the heights could be written as $+z$ and $+z + \tfrac{1}{2}c$, where c is the lattice parameter.

5.2   The crystallographic space groups

The 230 crystallographic space groups summarise the total number of three-dimensional patterns that result from combining the 32 point groups with the 14 Bravais lattices and including the screw axes. Each space group is given a unique symbol and number, (Appendix 4). Naturally, just as the seventeen plane groups can give rise to an infinite number of different designs, simply by altering the motif, so the 230 crystallographic space groups can give rise to an infinite number of crystal structures simply by altering the atoms and their relative positions in the three-dimensional motif. However, each resultant structure must possess a unit cell that conforms to one of the 230 space groups.

Each space group has a space group symbol that summarises the important symmetry

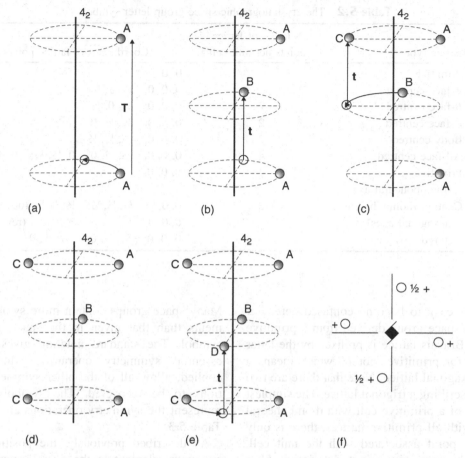

Figure 5.1 The operation of a 4_2 screw axis parallel to the z direction; (a) atom A at $z = 0$ is rotated counter clockwise by $90°$; (b) the atom is translated parallel to z by a distance of $\mathbf{t} = 2\mathbf{T}/4$, i.e. $\mathbf{T}/2$ to create atom B; (c) atom B is rotated counter clockwise by $90°$ and translated parallel to z by a distance of $\mathbf{t} = 2\mathbf{T}/4$, i.e. $\mathbf{T}/2$, to give atom C; (d) atom C is at $z = \mathbf{T}$, the lattice repeat, and so is repeated at $z = 0$; (e) repeat of the symmetry operation produces atom D at $z = \mathbf{T}/2$; (f) standard crystallographic depiction of a 4_2 screw axis viewed along the axis

elements present in the group. [As in previous chapters, the symbols used in crystallography, (International or Hermann-Mauguin symbols) are used throughout. The alternative Schoenflies symbols are listed in Appendix 4.] The space group symbols are written in two ways, a more explicit 'full' form, or an abbreviated 'short' form, both of which are given in Appendix 4. In

either case, the space group symbol consists of two parts: (i), a capital letter, indicating the lattice that underlies the structure, and (ii), a set of characters that represent the symmetry elements of the space group.

Table 5.2 lists the letter symbols used in the space groups to describe the underlying lattice, and the coordinates of the associated lattice

Table 5.2   The crystallographic space group letter symbols

Letter symbol	Lattice type	Number of lattice points per unit cell	Coordinates of lattice points
P	Primitive	1	0, 0, 0
A	A-face centred	2	0, 0, 0; 0, ½, ½
B	B-face centred	2	0, 0, 0; ½, 0, ½
C	C-face centred	2	0, 0, 0; ½, ½, 0
I	Body centred	2	0, 0, 0; ½, ½, ½
F	All-face centred	4	0, 0, 0; ½, ½, 0; 0, ½, ½; ½, 0, ½
R	Primitive (rhombohedral axes)	1	0, 0, 0
	Centred rhombohedral (hexagonal axes)	3	0, 0, 0; ⅔, ⅓, ⅓; ⅓, ⅔, ⅔ (obverse setting) 0, 0, 0; ⅓, ⅔, ⅓; ⅔, ⅓, ⅔ (reverse setting)
H	Centred hexagonal	3	0, 0, 0; ⅔, ⅓, 0; ⅓, ⅔, 0

points. It is easy to become confused here. For example, a space group derived from a primitive hexagonal Bravais lattice is prefixed by the letter symbol P, for primitive, not H, which means a centred hexagonal lattice. Note that there are two ways of describing a trigonal lattice. The simplest is in terms of a primitive cell with rhombohedral axes. As with all primitive lattices, there is only one lattice point associated with the unit cell. However, a more convenient description of a trigonal structure can often be made by using a hexagonal unit cell. When this option is chosen, the coordinates of the lattice points in the unit cell can be specified in two equivalent ways, the **obverse setting** or the **reverse setting**. The coordinates of the lattice points in each of these two arrangements are given in Table 5.2.

5.3   Space group symmetry symbols

The second part of a space group symbol consists of one, two or three entries after the initial letter symbol described above. At each position, the entry consists of one or two characters, describing a symmetry element, either an axis or a plane.

Many space groups contain more symmetry elements than that given in the space group full symbol. The standard notation gives only the essential symmetry operators, which, when applied, allow all of the other symmetry operations to be recovered. The symbols used to represent the symmetry operations are set out in Table 5.3.

As described previously, the position of the symmetry element in the overall symbol has a structural significance that varies from one crystal system to another as set out in Table 5.4.

Note that symmetry planes are always designated by their normals. If a symmetry axis and the normal to a symmetry plane are parallel, the two symbols are dived by a slash, as in the space group $P\,2/m$ (pronounced 'P two over m'). Full symbols carry place markers to make the position of the symmetry element in the structure unambiguous. For example, the unique axis in the monoclinic system is conventionally taken as the **b**-axis. The short symbol for the monoclinic space group number 3 is $P2$. Using the conventional axes, this means that the symmetry element of most significance is a diad along the **b**-axis. To make this clear, the full symbol is $P121$. If the

Table 5.3 Symmetry elements in space group symbols

Symbol	Symmetry operation	Comments
m	Mirror plane	reflection
a	Axial glide plane $\perp[010]$, $[001]$	Glide vector $\mathbf{a}/2$
b	Axial glide plane $\perp [001]$, $[100]$	Glide vector $\mathbf{b}/2$
c	Axial glide plane $\perp [100]$, $[010]$	Glide vector $\mathbf{c}/2$
	$\perp [1\bar{1}0]$, $[110]$	Glide vector $\mathbf{c}/2$
	$\perp [100]$, $[010]$, $[\bar{1}\bar{1}0]$	Glide vector $\mathbf{c}/2$, hexagonal axes
	$\perp [1\bar{1}0]$, $[120]$, $[\bar{2}\bar{1}0]$ $[\bar{1}\bar{1}0]$	Glide vector $\mathbf{c}/2$, hexagonal axes
n	Diagonal glide plane $\perp [001]$; $[100]$; $[010]$	Glide vector $\frac{1}{2}(\mathbf{a}+\mathbf{b})$; $\frac{1}{2}(\mathbf{b}+\mathbf{c})$; $\frac{1}{2}(\mathbf{a}+\mathbf{c})$
	Diagonal glide plane $\perp [1\bar{1}0]$; $[01\bar{1}]$; $[\bar{1}01]$	Glide vector $\frac{1}{2}(\mathbf{a}+\mathbf{b}+\mathbf{c})$
	Diagonal glide plane $\perp [110]$; $[011]$; $[101]$	Glide vector $\frac{1}{2}(-\mathbf{a}+\mathbf{b}+\mathbf{c})$; $\frac{1}{2}(\mathbf{a}-\mathbf{b}+\mathbf{c})$; $\frac{1}{2}(\mathbf{a}+\mathbf{b}-\mathbf{c})$
d	Diamond glide plane $\perp [001]$; $[100]$; $[010]$	Glide vector $\frac{1}{4}(\mathbf{a}\pm\mathbf{b})$; $\frac{1}{4}(\mathbf{b}\pm\mathbf{c})$; $\frac{1}{4}(\pm\mathbf{a}+\mathbf{c})$
	Diamond glide plane $\perp [1\bar{1}0]$; $[01\bar{1}]$; $[\bar{1}01]$	Glide vector $\frac{1}{4}(\mathbf{a}+\mathbf{b}\pm\mathbf{c})$; $\frac{1}{4}(\pm\mathbf{a}+\mathbf{b}+\mathbf{c})$; $\frac{1}{4}(\mathbf{a}\pm\mathbf{b}+\mathbf{c})$
	Diamond glide plane $\perp [110]$; $[011]$; $[101]$	Glide vector $\frac{1}{4}(-\mathbf{a}+\mathbf{b}\pm\mathbf{c})$; $\frac{1}{4}(\pm\mathbf{a}-\mathbf{b}+\mathbf{c})$; $\frac{1}{4}(\mathbf{a}\pm\mathbf{b}-\mathbf{c})$
1	None	–
2, 3, 4, 6	n-fold rotation axis	Rotation counter clockwise of $360°/n$
$\bar{1}$	Centre of symmetry	–
$\bar{2}(=m)$, $\bar{3}$, $\bar{4}$, $\bar{6}$	$\bar{n}$-fold inversion (rotoinversion) axis	$360°/n$ rotation counter clockwise followed by inversion
2_1, 3_1, 3_2, 4_1, 4_2, 4_3, 6_1, 6_2, 6_3, 6_4, 6_5	n-fold screw (rototranslation) axis, n_p	$360°/n$ right-handed screw rotation counter clockwise followed by translation by $(p/n)\mathbf{T}$

Table 5.4 Order of the Hermann-Mauguin symbols in space group symbols

Crystal system	Primary	Secondary	Tertiary
Triclinic	–	–	–
Monoclinic	$[010]$, unique axis $\mathbf{b}$ $[001]$, unique axis $\mathbf{c}$	–	–
Orthorhombic	$[100]$	$[010]$	$[001]$
Tetragonal	$[001]$	$[100]$, $[010]$	$[1\bar{1}0]$, $[110]$
Trigonal, Rhombohedral axes	$[111]$	$[1\bar{1}0]$, $[01\bar{1}]$, $[\bar{1}01]$	
Trigonal, Hexagonal axes	$[001]$	$[100]$, $[010]$, $[\bar{1}\bar{1}0]$	
Hexagonal	$[001]$	$[100]$, $[010]$, $[\bar{1}\bar{1}0]$	$[1\bar{1}0]$, $[120]$, $[\bar{2}\bar{1}0]$
Cubic	$[100]$, $[010]$, $[001]$	$[111]$, $[1\bar{1}\bar{1}]$, $[\bar{1}1\bar{1}]$, $[\bar{1}\bar{1}1]$	$[1\bar{1}0]$, $[110]$, $[01\bar{1}]$, $[011]$, $[\bar{1}01]$, $[101]$

unique axis is taken as the **c**-axis, the full symbol would be *P*112, but the short symbol remains *P*2. More examples from the monoclinic system are given in Table 5.5. The point group and space

Table 5.5 The meaning of some space group symbols

Space group number	Short symbol	Full symbol	Meaning
3	*P*2	*P*121	Diad along unique **b**-axis (conventional)
		*P*112	Diad along unique **c**-axis
6	*Pm*	*P*1*m*1	Mirror plane normal to the unique **b**-axis (conventional)
		*P*11*m*	Mirror plane normal to the unique **c**-axis
10	*P*2/*m*	*P*1 2/*m* 1	Diad and mirror plane normal along unique **b**-axis (conventional)
		*P*1 1 2/*m*	Diad and mirror plane normal along unique **c**-axis

Table 5.6 Point groups and space groups

Space group number	Short symbol	Full symbol	Crystallographic point group
3	*P*2	*P*121	2
6	*Pm*	*P*1*m*1	*m*
10	*P*2/*m*	*P*1 2/*m* 1	2/*m*
200	*Pm*$\bar{3}$	*P* 2/*m* $\bar{3}$	*m*$\bar{3}$
203	*Fd*$\bar{3}$	*F* 2/*d* $\bar{3}$	*m*$\bar{3}$
216	*F*$\bar{4}$3*m*	*F*$\bar{4}$3*m*	$\bar{4}$3*m*
219	*F*$\bar{4}$3*c*	*F*$\bar{4}$3*c*	$\bar{4}$3*m*

group are closely related. In order to determine the point group corresponding to a space group, the initial lattice letter of the space group is ignored, and the translation symmetry operators, *a*, *b*, *c*, *n*, *d*, are replaced by the mirror operator, *m*. Table 5.6 gives some examples.

5.4 The graphical representation of the space groups

A space group is generally represented by two figures, one that shows the effect of the operation of the symmetry elements present and one that shows the location of the various symmetry elements. In some space groups, it is possible to choose between several different origins for the diagrams, in which case both alternatives are presented.

Each diagram is a projection, down a unit cell axis, of a unit cell of the structure. The position of an atom or a motif is represented by a circle, and the origin of the cell is chosen with respect to the symmetry elements present. Symmetry elements are placed at the corners or along the edges of the unit cell whenever possible. The enantiomorphic form of a motif, generated by a mirror reflection from the original, is represented by a circle containing a comma. There are a set of standard graphical symbols used in these figures, the most important of which are set out in Table 5.7.

As an example, consider the diagrams appropriate to the space group number 75, *P*4, (Figure 5.2). The space group symbol reveals that the unit cell is primitive, and so contains only one lattice point. The principal symmetry element present is a tetrad axis parallel to the **c**-axis. The tetrad axis is always chosen to be parallel to the **c**-axis, and there is no need to specify the space group symbol as *P* 1 1 4. The origin of the unit cell coincides with the tetrad axis, and the projection is down this axis, and

Table 5.7 Graphical symbols used in space group diagrams

Symmetry axes perpendicular to plane of diagram

○	centre of symmetry
	Two-fold rotation axis
	Two-fold screw axis: 2_1
	Two-fold rotation axis with centre of symmetry
	Two-fold screw axis with centre of symmetry
▲	Three-fold rotation axis
△	Three-fold inversion axis: $\bar{3}$
	Three-fold screw axis: 3_1
	Three-fold screw axis: 3_2
◆	Four-fold rotation axis
	Four-fold inversion axis: $\bar{4}$
◆	Four-fold rotation axis with centre of symmetry
	Four-fold screw axis: 4_1
	Four-fold screw axis: 4_2
	Four-fold screw axis: 4_2 with centre of symmetry
	Four-fold screw axis: 4_3
	Six-fold rotation axis
	Six-fold inversion axis: $\bar{6}$
	Six-fold rotation axis with centre of symmetry
	Six-fold screw axis: 6_1

	Six-fold screw axis: 6_2
	Six-fold screw axis: 6_3
	Six-fold screw axis: 6_3 with centre of symmetry
	Six-fold screw axis: 6_4
	Six-fold screw axis: 6_5

Symmetry planes normal to plane of diagram

——	mirror plane
- - - -	axial glide plane: vector ½ **a**, **b** or **c** parallel to plane
··········	axial glide plane: vector ½ **a**, **b** or **c** normal to plane
— ·· — ··	diagonal glide plane: vector ½ **a**, **b** or **c** parallel to plane plus ½ **a**, **b** or **c** normal to plane

Symmetry planes parallel to plane of diagram

	mirror plane
	axial glide plane, vector ½ **a**, **b** or **c** in direction of the arrow
¼	diagonal glide plane, vector ½ (**a** + **b**), (**b** + **c**) or (**a** + **c**) in direction of the arrow

Equivalent position diagrams

○	motif in a general position in the plane of the diagram
○+	motif at arbitrary height x, y or z above the plane of the diagram
⊙	enantiomorph of motif in a general position in the plane of the diagram
⊙+	enantiomorph of motif at an arbitrary height x, y or z above the plane of the diagram

conventionally, the x-direction points from top to bottom and the y-direction from left to right.

The disposition of the symmetry elements is shown in Figure 5.2a. The diagrams reveal all of the symmetry elements present, including 'extra ones' generated by the principal symmetry elements described in the space group symbol. For example, the position of the tetrad axis at the unit cell corner also generates another at the cell centre and diad axes at the centre of each cell edge, (Figure 5.2a). The operation of the tetrad axis on a motif, (represented by a circle), generates four other copies in the unit cell, (Figure 5.2b). The + sign by the motif indicates that there is no change in the height of this unit due to operation of the tetrad axis.

The smallest part of the unit cell that will reproduce the whole cell when the symmetry

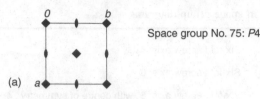

Space group No. 75: P4

◆   Four-fold rotation axis parallel to **c** (perpendicular to the plane of the diagram)

⬗   Two-fold rotation axis parallel to **c** (perpendicular to the plane of the diagram)

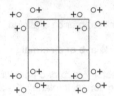

(b)   o+   motif at arbitrary height +z (0<z<c)

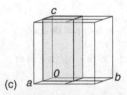

Figure 5.2 Space group diagrams for space group No 74, *P4*: (a) symmetry elements; (b) positions of motif generated by symmetry operations; (c) asymmetric unit

Table 5.8 Positions in the space group *P4*

Multiplicity	Wyckoff letter	Site symmetry	Coordinates of equivalent positions
4	d	1	(1) x, y, z (2) $\bar{x}, \bar{y}, z$ (3) $\bar{y}, x, z$ (4) $y, \bar{x}, z$
2	c	2	$0, \frac{1}{2}, z$ $\frac{1}{2}, 0, z$
1	b	4	$\frac{1}{2}, \frac{1}{2}, z$
1	a	4	$0, 0, z$

(x, y, z), is a general position in the unit cell, the positions generated by the symmetry operators present are called general equivalent positions. These are:

$$(1)\ x, y, z \quad (2)\ \bar{x}, \bar{y}, z \quad (3)\ \bar{y}, x, z \quad (4)\ y, \bar{x}, z$$

These are given the Wyckoff symbol *d*, and the multiplicity of the position is 4, as set out in Table 5.8.

If the atom position chosen initially falls upon a symmetry element, called a special position, the multiplicity will decrease. For example, an atom placed on the diad axis, will not be repeated by the operation of the diad. However, as there are two diad axes, an atom at one will necessarily imply an atom at the other. This position will then have a multiplicity of 2, whereas a general point has a multiplicity of 4. These special positions are found at:

$$0, \frac{1}{2}, z \quad \frac{1}{2}, 0, z$$

The multiplicity is two, and the Wyckoff symbol is *c*, (Table 5.8).

There are also special positions associated with the tetrad axes, at the cell origin, $(0, 0)$, and at the cell centre, $(\frac{1}{2}, \frac{1}{2})$, (Figure 5.2a). The multiplicity

operations are applied is called the **asymmetric unit**. This is not unique, and several alternative asymmetric units can be found for a unit cell, but it is clear that rotation or inversion axes must always lie at points on the boundary of the asymmetric unit. The conventional asymmetric unit for the space group *P4* is indicated in Figure 5.2c. The extension of the asymmetric unit is one unit cell in height, so that it is defined by the relations $0 \leq x \leq \frac{1}{2}, 0 \leq y \leq \frac{1}{2}, 0 \leq z \leq 1$.

Wyckoff letters, as described in Chapter 3 for the plane groups, label the different site symmetries that can be occupied by an atom in a unit cell. When the position chosen for the atom,

of an atom located on a tetrad axis will be 1. These two positions are:

$$\tfrac{1}{2}, \tfrac{1}{2}, z \quad \text{and} \quad 0, 0, z$$

The total number of positions in the unit cell is then as set out in Table 5.8.

5.5 Building a structure from a space group

As in two-dimensions, a crystal structure can be built from a motif plus a lattice. The motif is the minimal collection of atoms needed to generate the whole unit cell contents by application of the symmetry operators specified by the space group, or **generator**. The positions of the atoms in a structure can be organised in a compact way by using the symmetry properties of the space group that is appropriate to the crystal structure.

In this example, we will consider how to build a structure from the space group $P4_1$, (number 76). This is similar to the space group described in the previous section, $P4$, but here a four-fold screw axis along the **c**-axis replaces the ordinary tetrad. The space group diagrams are drawn in Figure 5.3. Note that the height of the motif is set out besides the symbol in Figure 5.3b. These heights can be obtained following the methods described in Figure 5.1 for the 4_2 screw axis. Figure 5.3 is superficially very similar to the diagrams applicable to the space group $P4$, (Figure 5.2). However, the screw axis does not generate diads, but twofold screw axes, 2_1. The only positions available are the equivalent general positions, set out in Table 5.9.

The structure of caesium phosphide, Cs_3P_7, belongs in space group $P4_1$, and has lattice parameters $a = b = 0.9046$ nm, $c = 1.6714$ nm. The structure is tetragonal, and the screw axis runs parallel to the **c**-axis. All of the atoms must be placed in equivalent general positions, the

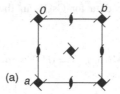

Space group No. 75: $P4_1$

↗ Four-fold screw axis: 4_1 parallel to **c** (perpendicular to the plane of the diagram)

♦ Two-fold screw axis: 2_1 parallel to **c** (perpendicular to the plane of the diagram)

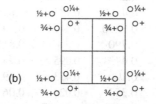

O+   motif at arbitrary height $+z$ $(0 < z < c)$
O¼+   motif at arbitrary height $z + c/4$
O½+   motif at arbitrary height $z + c/2$
O¾+   motif at arbitrary height $z + 3c/4$

Figure 5.3 Space group diagrams for space group No 75, $P4_1$: (a) symmetry elements; (b) positions of motif generated by symmetry operations

coordinates of which are set out in Table 5.10, making Z, the number of Cs_3P_7 units in the unit cell equal to four.

Note that there are three different Cs atoms specified, numbered Cs1, Cs2 and Cs3. The unit cell with just these three atoms present is drawn,

Table 5.9 Positions in the space group $P4_1$

Multiplicity	Wyckoff letter	Site symmetry	Coordinates of equivalent positions
4	a	1	(1) x, y, z
			(2) $\bar{x}, \bar{y}, z + \tfrac{1}{2}$
			(3) $\bar{y}, x, z + \tfrac{1}{4}$
			(4) $y, \bar{x}, z + \tfrac{3}{4}$

Table 5.10 Crystallographic data for Cs_3P_7 in the space group $P4_1$

Crystal system: tetragonal

Lattice parameters: $a = 0.9046$ nm, $b = 0.9046$ nm, $c = 1.6714$ nm, $\alpha = \beta = \gamma = 90°$

$Z = 4$.

Space group, $P4_1$, Number 75

Atom	Multiplicity and Wyckoff letter	Coordinates of atoms		
		x	y	z
Cs1	$4a$	−0.2565	0.3852	0
Cs2	$4a$	0.4169	0.7330	0.8359
Cs3	$4a$	0.0260	0.8404	0.9914
P1	$4a$	0.790	0.600	0.811
P2	$4a$	0.443	0.095	0.947
P3	$4a$	0.106	0.473	0.893
P4	$4a$	0.357	0.024	0.061
P5	$4a$	0.629	0.794	0.032
P6	$4a$	0.998	0.341	0.705
P7	$4a$	0.011	0.290	0.840

Data taken from T. Meyer, W. Hoenle and H.G. von Schnering, Z. anorg. allgem. Chemie, **552**, 69–80 (1987), provided by the EPSRC's Chemical Database Service at Daresbury (see Bibliography)

projected down the **c**-axis, (Figure 5.4a), down the **a**-axis, (Figure 5.4b), and in perspective, (Figure 5.4c). Note that the atom at a height $z = 0$ is repeated at the top and bottom of the unit cell. Each of these atoms will be replicated four times, by the action of the symmetry axis, so that the unit cell will contain 12 Cs atoms. The positions of these atoms are given in Table 5.11, listed as Cs1, Cs1_2, Cs1_3 etc, to show which atoms are derived from those in the original specification. The structure, with all the Cs atoms shown, projected down the **c**-axis, is drawn in Figure 5.4d.

Similarly, there are seven different P atoms listed, giving 28 P in the unit cell. The positions of these atoms are obtained in the same way as those for the Cs atoms.

The use of the cell symmetry allows one to specify just 10 atoms instead of the total cell contents of 40 atoms. The projection of the complete structure down the **c**-axis is drawn in Figure 5.4e.

Note that even in a relatively simple structure such as this one, it is difficult to obtain an idea of the disposition of the atoms in the unit cell, or how the crystal is built up. For this, computer

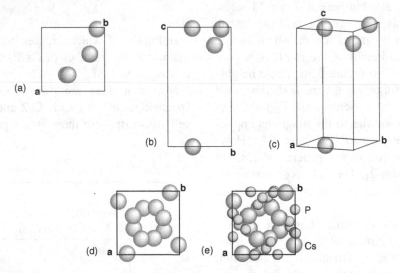

Figure 5.4 The structure of Cs_3P_7: (a) unit cell projected down the **c**-axis, atoms Cs1, Cs2, Cs3 only; (b) unit cell projected down the **a**-axis, atoms Cs1, Cs2, Cs3 only; (c) perspective view of unit cell, atoms Cs1, Cs2, Cs3 only; (d) unit cell projected down the **c**-axis, with all 12 Cs atoms present; (e) unit cell projected down the **c**-axis, with all atoms present

Table 5.11 The positions of the Cs atoms in a unit cell of Cs$_3$P$_7$

Atom	Coordinates of atoms		
	x	y	z
Cs1	0.7435	0.3852	0
Cs1_2	0.6148	0.7435	¼
Cs1_3	0.2565	0.6148	½
Cs1_4	0.3852	0.2565	¾
Cs2	0.4169	0.7330	0.8359
Cs2_2	0.2670	0.4169	0.0859
Cs2_3	0.5831	0.2670	0.3359
Cs2_4	0.7330	0.5831	0.5859
Cs3	0.0260	0.8404	0.9914
Cs3_2	0.1596	0.0260	0.2414
Cs3_3	0.9740	0.1569	0.4914
Cs3_4	0.8404	0.9740	0.7414

Table 5.12 Crystallographic data for diopside, CaMgSi$_2$O$_6$

Crystal system: monoclinic
Lattice parameters: $a = 0.95848$ nm, $b = 0.86365$ nm, $c = 0.51355$ nm, $\alpha = 90°$, $\beta = 103.98°$, $\gamma = 90°$. $Z = 4$.
Space group, C 1 2/c 1, Number 15

Atom	Multiplicity and Wyckoff letter	Coordinates of atoms		
		x	y	z
Ca1	4e	0	0.3069	¼
Mg1	4e	0	0.9065	0.464
Si1	8f	0.284	0.0983	0.2317
O1	8f	0.1135	0.0962	0.1426
O2	8f	0.3594	0.2558	0.3297
O3	8f	0.3571	0.0175	0.9982

Data taken from M. T. Dove, American Mineralogist, **74**, 774–779 (1989), provided by the EPSRC's Chemical Database Service at Daresbury (see Bibliography).

graphics are necessary (see Bibliography for further details). Alternatively, a description in terms of polyhedra, as discussed in Chapter 7, may be helpful.

5.6 The structure of diopside, CaMgSi$_2$O$_6$

As a further example of a structure building exercise, consider the mineral diopside, CaMgSi$_2$O$_6$, which was one of the first crystal structures to be determined, during the early years of X-ray crystallography. The crystallographic data is given in Table 5.12.

The space group diagrams are given in Figure 5.5a, b. The convention for monoclinic crystals is to project the cell down the unique **b**-axis, (Figure 5.5a). The origin of the unit cell is chosen at a centre of symmetry, the **a**-axis points to the right and the **c**-axis down the page. The **b**-axis is then perpendicular to the plane of the figure. The space group symbol indicates that the most important symmetry element is a diad, which runs parallel to the **b**-axis. Perpendicular

to this is a glide plane, the glide vector of which is parallel to the **c**-axis, and of value $c/2$, which generate a set of 2_1 screw axes. Heights by some symbols represent the height of the operator above the plane of the figure as fractions of the cell parameter b, (Figure 5.5a). The action of the symmetry operators on a motif at a general position shows that this is sometimes transformed into its mirror image, shown by the presence of a comma within the circle that represents the motif. A motif above the plane of the figure is denoted by a + sign, and those below the plane by a − sign. Those designated ½+ and ½− are translated by a distance of ½ of the cell parameter b with respect to those unlabelled, (Figure 5.5b). The equivalent positions in the cell are given in Table 5.13. When these are applied to the atoms listed in Table 5.12 the value of the space group approach to crystal structure building is apparent, because only six atoms are listed in Table 5.12, although the unit cell contains 40 atoms. The structure is projected down the unique **b**-axis, (Figure 5.6a),

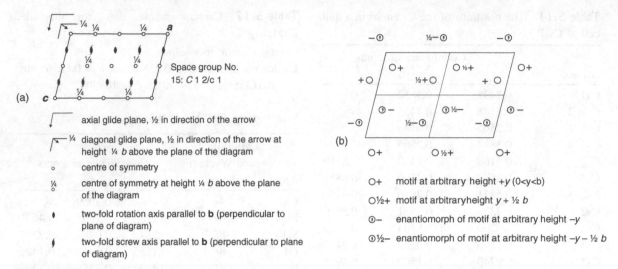

(a)

axial glide plane, ½ in direction of the arrow

diagonal glide plane, ½ in direction of the arrow at height ¼ *b* above the plane of the diagram

o   centre of symmetry

¼   centre of symmetry at height ¼ *b* above the plane of the diagram

•   two-fold rotation axis parallel to **b** (perpendicular to plane of diagram)

two-fold screw axis parallel to **b** (perpendicular to plane of diagram)

(b)

O+   motif at arbitrary height +*y* (0<*y*<*b*)

O½+   motif at arbitraryheight *y* + ½ *b*

⊙−   enantiomorph of motif at arbitrary height −*y*

⊙½−   enantiomorph of motif at arbitrary height −*y* − ½ *b*

Figure 5.5   Space group diagrams for space group No 15, *C* 1 2/*c* 1: (a) symmetry elements; (b) positions of motif generated by symmetry operations

and down the **c**-axis, (Figure 5.6b). Once again, notice that the structure is not easily visualised in this atomic representation. (See Chapter 7, Figure 7.17 for an alternative representation of the structure in terms of polyhedra).

Table 5.13   Positions in the space group C 1 2/*c* 1

Multiplicity	Wyckoff letter	Site symmetry	Coordinates of equivalent positions
8	*f*	1	(1) x, y, z
			(2) $\bar{x}, y, \bar{z} + \frac{1}{2}$
			(3) $\bar{x}, \bar{y}, \bar{z}$
			(4) $x, \bar{y}\ z + \frac{1}{2}$
			(5) $x + \frac{1}{2}, y + \frac{1}{2}, z$
			(6) $\bar{x} + \frac{1}{2}, y + \frac{1}{2}, \bar{z} + \frac{1}{2}$
			(7) $\bar{x} + \frac{1}{2}, \bar{y} + \frac{1}{2}, \bar{z}$
			(8) $x + \frac{1}{2}, \bar{y} + \frac{1}{2}, z$
4	*e*	2	$0, y, \frac{1}{4};\ \ 0, \bar{y}, \frac{3}{4}$
4	*d*	$\bar{1}$	$\frac{1}{4}, \frac{1}{4}, \frac{1}{2};\ \ \frac{3}{4}, \frac{1}{4}, 0$
4	*c*	$\bar{1}$	$\frac{1}{4}, \frac{1}{4}, 0;\ \ \frac{3}{4}\ \frac{1}{4}, \frac{1}{2}$
4	*b*	$\bar{1}$	$0, \frac{1}{2}, 0;\ \ 0, \frac{1}{2}, \frac{1}{2}$
4	*a*	$\bar{1}$	$0, 0, 0;\ \ 0, 0, \frac{1}{2}$

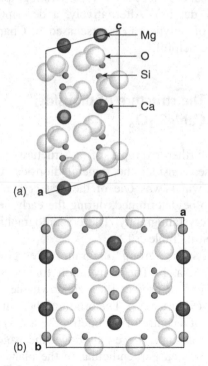

(a)

(b)

Figure 5.6   The structure of diopside, CaMgSi$_2$O$_6$: (a) unit cell projected down the **b**-axis; (b) unit cell projected down the **c**-axis

5.7 The structure of alanine, $C_3H_7NO_2$

Alanine, 2-aminopropionic acid, $C_3H_7NO_2$, is an amino acid, and one of the naturally occurring building blocks of proteins. Naturally occurring amino acid molecules occur in only one configuration, the 'left-handed' form. Thus, the naturally occurring modification of alanine was described in older literature as L-alanine, or, L-α-aminopropionic acid, where the letter L- describes the configuration of the molecule. This nomenclature is confusing, as the prefix, L- is similar to the prefix, l-, meaning laevorotatory. In fact the natural molecule is dextrorotatory, and can be correctly written d-alanine or (+)-alanine. The use of the letter symbol L- to designate the configuration has now been superseded, and a modern description of the molecule is (S)-(+)-2-aminopropionic acid, often shortened to (S)-alanine. The letter symbol S describes the configuration of the molecule and the symbol (+) records that it is dextrorotatory. The laevorotatory molecule is written as (R)-alanine or (R)-(−)-2-aminopropionic acid. Normal chemical synthesis gives an equal mixture of the (R)- and (S)- molecules. Such a mixture does not rotate the plane of linearly polarised light, and is designated (R, S)-alanine. In older literature, this is also called DL-alanine.

Of central interest is the arrangement in space of the four groups surrounding the chiral carbon atom, the molecular configuration. Organic chemists have known for many years that optical activity arises when four different atoms or groups of atoms are attached to a central carbon atom by tetrahedrally oriented bonds. However, which of the two possible mirror images, or configurations, caused light to rotate to the left as against to the right was not known. Careful diffraction studies of molecular crystals are able to solve this problem, (see Chapter 6).

The naturally occurring optically active molecules, (S)-alanine, and the laboratory synthesised optically inactive mixture (R, S)-alanine, form different crystals. To determine the absolute configuration of the molecule it is simplest to start with crystals of the optically active (S)-alanine, [L-alanine, (S)-(+)-2-aminopropionic acid]. The presence of a single enantiomer in a crystal has implications for the possible symmetry of the crystal. For example, the point group of a crystal containing only one enantiomer of an optically active molecule must be one of the 11 that do not contain an inversion axis. In addition, all enantiomorphous biological molecules crystallise in one of the space groups that lack inversion centres or mirror planes. There are only 65 appropriate space groups. The crystallographic data for naturally occurring (S)-alanine is given in Table 5.14, where it is seen

Table 5.14 Crystallographic data for (S)-alanine [L-alanine, (S)-(+)-2-aminopropionic acid]

Crystal system: orthorhombic
Lattice parameters: $a = 0.5928$ nm, $b = 1.2260$ nm, $c = 0.5794$ nm, $\alpha = 90°$, $\beta = 90°$, $\gamma = 90°$. Z = 4.
Space group, $P 2_1 2_1 2_1$, Number 19

Atom	Multiplicity and Wyckoff letter	Coordinates of atoms x	y	z
C1	4a	0.55409	0.14081	0.59983
C2	4a	0.46633	0.16105	0.35424
C3	4a	0.25989	0.09069	0.30332
N1	4a	0.64709	0.08377	0.62438
O1	4a	0.72683	0.25580	0.32970
O2	4a	0.44086	0.18403	0.76120
H1	4a	0.70360	0.05720	0.19790
H2	4a	0.77830	0.18970	0.20860
H3	4a	0.58070	0.14920	0.01100
H4	4a	0.42130	0.24980	0.33590
H5	4a	0.19360	0.11090	0.13400
H6	4a	0.12610	0.10560	0.43570
H7	4a	0.30410	0.00510	0.30560

Data taken from R. Destro, R. E. Marsh and R. Bianchi, J. Phys. Chem., **92**, 966–973 (1989).

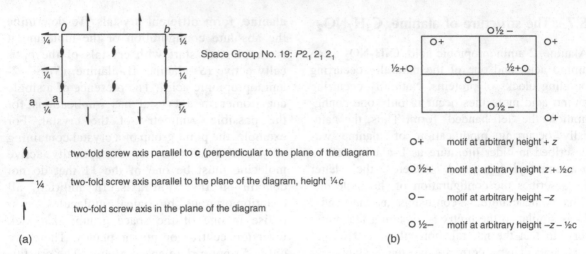

Space Group No. 19: $P2_1 2_1 2_1$

⚍   two-fold screw axis parallel to **c** (perpendicular to the plane of the diagram

——¼   two-fold screw axis parallel to the plane of the diagram, height ¼**c**

↑   two-fold screw axis in the plane of the diagram

O+   motif at arbitrary height + z

O ½+   motif at arbitrary height $z + \frac{1}{2}c$

O–   motif at arbitrary height $-z$

O ½–   motif at arbitrary height $-z - \frac{1}{2}c$

(a)   (b)

Figure 5.7   Space group diagrams for space group No 19, $P2_1 2_1 2_1$: (a) symmetry elements; (b) positions of motif generated by symmetry operations

that the space group of the crystals is $P2_1 2_1 2_1$, Number 19.

The space group diagrams are given in Figure 5.7. The space group symbol indicates that the most important symmetry elements are two-fold screw axes running throughout the unit cell parallel to the three axes. These symmetry operators turn a motif above the plane of the diagram into one below the plane, shown by the presence of $+$, $\frac{1}{2}+$, $-$ and $\frac{1}{2}-$, where the fractions represent half of the appropriate unit cell edge. The equivalent positions in the cell are given in Table 5.15. It is seen that only one general position occurs, with multiplicity of four, generating four molecules of alanine in the unit cell.

The structure of the optically active (S)-alanine consists of layers of alanine molecules, all of the same enantiomer, (Figure 5.8). The structure displayed as atom packing is not helpful, (see i.e. urea, in Chapter 1), and it is preferable to delineate molecules using the familiar ball and stick approach. This allows the molecular geometry and packing to be visualised. Even so, the structure of a molecule is not easy to see from a two-dimensional figure, but in fact, the hydrogen atom is below the central carbon atom, the NH_2

group lies in front and to the right of it, the CH_3 group lies to the left and the COOH group lies to the rear. The configuration is identical to that given in Figure 5.9(a). (To confirm the correspondence, it is useful to make a small model using dough and matchsticks or similar. Ideally, view the molecules using computer software, see Bibliography.)

A comparison of the optically inactive crystals of synthetic (R, S)-alanine with the optically active natural crystals of is of interest. The crystallographic data for (R, S)-alanine are

Table 5.15   Positions in the space group $P2_1 2_1 2_1$

Multiplicity	Wyckoff letter	Site symmetry	Coordinates of equivalent positions
4	a	1	(1) x, y, z (2) $\bar{x} + \frac{1}{2}, \bar{y}, z + \frac{1}{2}$ (3) $\bar{x}, y + \frac{1}{2}, \bar{z} + \frac{1}{2}$ (4) $x + \frac{1}{2}, \bar{y} + \frac{1}{2}, \bar{z}$

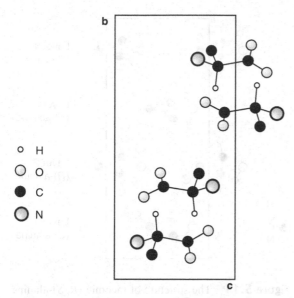

Figure 5.8 The structure of optically active (*S*)-alanine projected down the **a**-axis. Most of the hydrogen atoms have been omitted for clarity, and only that attached to the chiral carbon atom is included

given in Tables 5.16 and 5.17. Because mirror image forms of the alanine molecule pack into the unit cell, the space group should contain mirror or glide planes. The space group diagrams, (Figure 5.10), confirm this supposi-

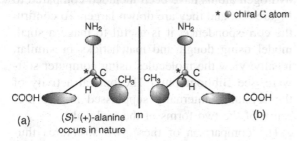

Figure 5.9 'Ball and stick' representation of the enantiomorphic (mirror image) forms of the molecule alanine, $CH_3CH(NH_2)COOH$: (a) the naturally occurring form, (*S*) (+) alanine; (b) the synthetic form (*R*)-(−)-alanine. The chiral carbon atom is marked

Table 5.16 Crystallographic data for (*R*, *S*)-alanine [DL-alanine, (*R*, *S*)-2-aminopropionic acid]

Crystal system: orthorhombic
Lattice parameters: $a = 1.2026\,nm$, $b = 0.6032\,nm$, $c = 0.5829\,nm$, $\alpha = 90°$, $\beta = 90°$, $\gamma = 90°$. Z = 4.
Space group, $Pna2_1$, Number 33

Atom	Multiplicity and Wyckoff letter	Coordinates of atoms		
		x	y	z
C1	4*a*	0.66363	0.22310	0.33800
C2	4*a*	0.64429	0.31330	0.58180
C3	4*a*	0.59233	0.02160	0.29160
N1	4*a*	0.63948	0.39760	0.16590
O1	4*a*	0.58960	0.48620	0.60450
O2	4*a*	0.68614	0.20090	0.74090
H1	4*a*	0.74190	0.18040	0.32370
H2	4*a*	0.60600	−0.03080	0.13870
H3	4*a*	0.51530	0.06050	0.30690
H4	4*a*	0.61040	−0.09300	0.29970
H5	4*a*	0.68150	0.51600	0.19420
H6	4*a*	0.56800	0.43510	0.17410
H7	4*a*	0.65440	0.34630	0.02610

Data taken from M. Subha Nandhini, R. V. Krishnakumar and S. Natarajan, Acta Crystallogr., C**57**, 614–615 (2001).

tion. The symmetry elements consist of a two-fold screw axis along the **c**-axis, an axial glide plane perpendicular to the **b**-axis, and a diagonal glide plane perpendicular to the **a**-axis,

Table 5.17 Positions in the space group $Pna2_1$, No. 33

Multiplicity	Wyckoff letter	Site symmetry	Coordinates of equivalent positions
4	*a*	1	(1) x, y, z
			(2) $\bar{x}, \bar{y}, z + \frac{1}{2}$
			(3) $x + \frac{1}{2}, \bar{y} + \frac{1}{2}, z$
			(4) $\bar{x} + \frac{1}{2}, y + \frac{1}{2}, z + \frac{1}{2}$

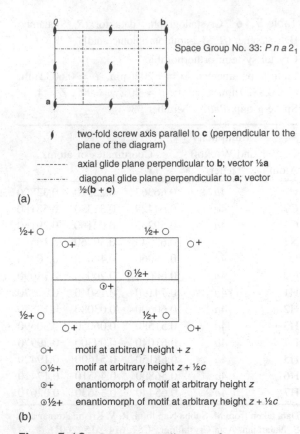

two-fold screw axis parallel to **c** (perpendicular to the plane of the diagram)

-------- axial glide plane perpendicular to **b**; vector ½**a**

—·—·— diagonal glide plane perpendicular to **a**; vector ½(**b** + **c**)

(a)

○+   motif at arbitrary height + z

○½+   motif at arbitrary height z + ½c

⊙+   enantiomorph of motif at arbitrary height z

⊙½+   enantiomorph of motif at arbitrary height z + ½c

(b)

Figure 5.10 Space group diagrams for space group No 33, *Pna2₁*: (a) symmetry elements; (b) positions of motif generated by symmetry operations

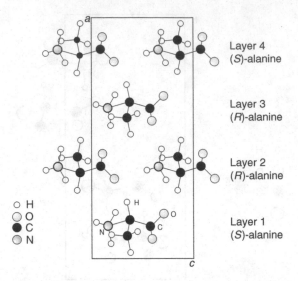

Figure 5.11 The structure of racemic (*R*, *S*)-alanine projected down the **b**-axis. All hydrogen atoms are included

(Figure 5.10a). The axial glide plane has the effect of moving the motif by a vector ½ **a**, parallel to the **a**-axis, and the diagonal glide moves the motif by a vector ½ **b** parallel to the **b**-axis and ½ **c** parallel to the **c**-axis, (Figure 5.10b). The effect of these symmetry operations is to create enantiomorphic copies of any motif in the unit cell. If the motif is taken as an alanine molecule, the symmetry elements present in the space group will generate a structure that contains layers of alanine molecules of one type interleaved with layers of the other enantiomorph.

The crystal structure confirms this arrangement, (Figure 5.11). As with (*S*)-alanine, the structure contains layers of alanine molecules. Each layer contains of just one of the enantiomers. The molecules in layer 1 are identical to those in layer 4, and both consist of the naturally occurring configuration, (*S*)-alanine. Enantiomers with the opposite configuration, (*R*)-alanine, form layers 2 and 3. (To show this difference, more hydrogen atoms have been included compared to Figure 5.8, and they are drawn larger. To confirm the correspondence, it is useful to make a small model using dough and matchsticks or similar. Ideally, view the molecules using computer software, see Bibliography.) The optical activity of the crystal is internally suppressed by the presence of the two forms of the molecule.

The comparison of these two structures thus provides a good insight into the relationship between structure, symmetry and physical properties of crystals. Be careful to note, though, that the unit cell and symmetry properties of the crystal are a consequence of the way in which the molecules pack into a minimum energy arrangement and not *vice versa*.

Answers to introductory questions

What symmetry operation is associated with a screw axis?

Screw axes are rototranslational symmetry elements. A screw axis of order n operates on an object by (a) a rotation of $2\pi/n$ counter clockwise and then a translation by a distance $\mathbf{t}$ parallel to the axis, in a positive direction. The value of n is the order of the screw axis. For example, a screw axis running parallel to the $\mathbf{c}$-axis in an orthorhombic crystal would entail a counter-clockwise rotation in the $\mathbf{a}$-$\mathbf{b}$-plane, (001), followed by a translation parallel to $+\mathbf{c}$. This is a right-handed screw rotation. If the rotation component of the operator is applied n times, the total rotation is equal to 2π. The value of n is the principle symbol used for a screw axis, together with a subscript giving the fraction of the repeat distance moved in the translation. Thus, a three-fold screw axis 3_1 involves a counter clockwise rotation of $2\pi/3$, followed by a translation of a distance equal to ⅓ the lattice repeat in a direction parallel to the axis. The three-fold screw axis 3_2 involves a counter clockwise rotation of $2\pi/3$, followed by a translation of a distance equal to ⅔ the lattice repeat in a direction parallel to the axis.

What is a crystallographic space group?

A crystallographic space group describes the symmetry of a three-dimensional repeating pattern, such as found in a crystal. Each space group is thus a collection of symmetry operators. There are 230 space groups, which can be derived by combining the 32 crystallographic point groups with the 14 Bravais lattices. Any three-dimensional crystal unit cell must have an internal symmetry that corresponds to one of the 230 crystallographic space groups. Obviously, by varying the motif, an infinite number of crystal structures can be created, but they will all possess a combination of symmetry elements corresponding to one of the crystallographic space groups.

What are Wyckoff letters?

If an atom is placed in the unit cell of a crystal at a position, or site, (x, y, z), identical copies of the atom must, by operation of the symmetry elements present, lie at other sites, (x_1, y_1, z_1), (x_2, y_2, z_2) and so on. A Wyckoff letter is a site label, and each different type of site in the unit cell is given a different Wyckoff letter.

The maximum number of copies of a site is generated when the initial atom is at a general position in the unit cell, and the resulting set are called general equivalent positions. If, however, position of the atom coincides with a symmetry element, a special position, the symmetry operators will generate fewer copies. The number of copies of an atom that are generated for each distinguishable site in the unit cell is called the site multiplicity. The sites are listed in order of decreasing multiplicity, and Wyckoff letters are added, starting at the lowest position in the list, with letter a and then proceeding alphabetically to the equivalent general positions, which are labelled last, as b, c, d etc, depending upon the number of different sites in the unit cell.

Problems and exercises

Quick quiz

1 The translation vector associated with a glide plane is:
   (a) A quarter of a lattice translation vector
   (b) A half of a lattice translation vector
   (c) A full lattice translation vector

2 The translation vector associated with a 3_2 axis is:
   (a) A two thirds of a lattice translation vector

(b) A half of a lattice translation vector

(c) Two lattice translation vectors

3 The number of crystallographic space groups is:
(a) 210
(b) 230
(c) 250

4 The first place in the Hermann-Mauguin symbol for a space group refers to:
(a) The crystal system, (cubic etc.)
(b) The point group
(c) The Bravais lattice

5 The last three places in the Hermann-Mauguin symbol for a space group refer to:
(a) Axes
(b) Planes
(c) Directions

6 The letter d in a space group symbol means:
(a) A diamond glide plane
(b) A diagonal glide plane
(c) A double glide plane

7 A rototranslation axis in a space group symbol is represented by:
(a) $n(n = 2, 3, 4, 6)$
(b) $\bar{n}(\bar{n} = \bar{3}, \bar{4}, \bar{6})$
(c) $n_p(n = 2, 3, 4, 6, p = 1, 2, 3, 4, 5)$

8 The symbol $2/m$ in a space group symbol means:
(a) A mirror runs parallel to a diad axis
(b) The normal to a mirror plane runs parallel to a diad axis
(c) A mirror plane contains a diad axis

9 The smallest part of the unit cell that will reproduce the whole cell when the symmetry operations are applied is:
(a) The motif
(b) The asymmetric unit
(c) The point group

10 The Wyckoff letter a is given to:
(a) The equivalent general sites
(b) The equivalent sites with the highest multiplicity
(c) The equivalent sites with the lowest multiplicity

Calculations and Questions

5.1 Draw diagrams similar to Figure 5.1f, for the axes $3, \bar{3}, 3_1, 3_2$.

5.2 Write out the full Hermann-Mauguin space group symbols for the short symbols:

(a), monoclinic, No. 7, Pc; (b), orthorhombic, No. 47, $Pmmm$; (c), tetragonal, No. 133, $P4_2/nbc$; (d), hexagonal, No. 193, $P6_3/mcm$; (e), cubic, No. 225, $Fm\bar{3}m$.

5.3 Tetragonal space group No. 79 has space group symbol $I4$.
(a) What symmetry elements are specified by the space group symbol?
(b) Draw diagrams equivalent to Figure 5.2a, b, for the space group.
(c) What new symmetry elements appear that are not specified in the space group symbol?

5.4 Determine the general equivalent positions for a motif at x, y, z in the tetragonal space group No. 79, $I4$, (see Q5.3).

5.5 Structural data for PuS_2 are given in the following Table:

Crystal system: tetragonal
Lattice parameters: $a = b = 0.3943$ nm, $c = 0.7962$ nm
Space group, $P4mm$, Number 99

Atom	Multiplicity and Wyckoff letter	Coordinates of atoms		
		x	y	z
Pu1	$1a$	0	0	0
Pu2	$1b$	½	½	0.464
S1	$1a$	0	0	0.367
S2	$1b$	½	½	0.097
S3	$2c$	½	0	0.732

Data taken from J. P Marcon and R. Pascard, Comptes Rendues Acad. Sci. Paris, Serie C, **262**, 1679–1681 (1966), provided by the EPSRC's Chemical Database Service at Daresbury (see Bibliography).

The equivalent positions in this space group are:

Positions in the space group No. 99, $P4mm$

Multiplicity	Wyckoff letter	Site symmetry	Coordinates of equivalent positions
8	g	1	(1) x, y, z
			(2) $\bar{x}, \bar{y}, z$
			(3) $\bar{y}, x, z$
			(4) $y, \bar{x},\ z$
			(5) $x, \bar{y}, z$
			(6) $\bar{x}, y, z$
			(7) $\bar{y}, x, z$
			(8) y, x, z
4	f	$.m.$	$x, ½, z;\ \bar{x}, ½, z;$ $½, x, z;\ ½, \bar{x}, z$
4	e	$.m.$	$x, 0, z;\ \bar{x}, 0, z;$ $0, x, z; 0, \bar{x}, z$
4	d	$..m$	$x, x, z;\ \bar{x}, \bar{x}, z;$ $\bar{x}, x, z; x, \bar{x}, z$
2	c	$2mm$	$½, 0, z;\ 0, ½, z$
1	b	$4mm$	$½, ½, z$
1	a	$4mm$	$0, 0, z$

The space group diagrams are given in the following figure

Space Group No. 99: $P4mm$

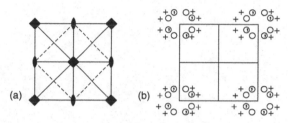

(a) (b)

Space group diagrams for space group No 99, $P4mm$: (a) symmetry elements; (b) positions of motif generated by symmetry operations

(a) What symmetry elements are specified in the space group symbol?

(b) What additional symmetry elements arise due to the operation of those in (a)?

(c) How many Pu atoms are there in the unit cell?

(d) What are the positions of the Pu atoms?

(e) How many S atoms are there in the unit cell?

(f) What are the positions of the S atoms?

6

Diffraction and crystal structures

- *What crystallographic information does Bragg's law give?*

- *What is an atomic scattering factor?*

- *What are the advantages of neutron diffraction over X-ray diffraction?*

Radiation incident upon a crystal is scattered in a variety of ways. When the wavelength of the radiation is similar to that of the atom spacing in a crystal the scattering, which is termed **diffraction**, gives rise to a set of well defined beams arranged with a characteristic geometry, to form a **diffraction pattern**. The positions and intensities of the diffracted beams are a function of the arrangements of the atoms in space and some other atomic properties, especially, in the case of X-rays, the atomic

number of the atoms. Thus, if the positions and the intensities of the diffracted beams[1] are recorded, it is possible to deduce the arrangement of the atoms in the crystal and their chemical nature.

This bland statement does not do justice to the many years of skilful experimental and theoretical work that underpins the determination of crystal structures, structural crystallography. The first crystal structure reported was that of NaCl, (by W. H and W. L. Bragg) in 1913. By 1957 the subject had advanced sufficiently for the structure of penicillin to be published, and in 1958 the first three-dimensional model of a protein, myoglobin, was determined, (see Bibliography). Currently the structures of many complex proteins have been published, allowing for great advances in the understanding of the biological function of these vital molecules.

Crystal structures are mainly determined using the diffraction of X-rays, supplemented by neutron and electron diffraction, which give information that X-ray diffraction cannot supply. The topic falls into three easily separated sections. Initially one can use the positions of the diffracted beams to give information about the size of the unit cell of the material. A second stage is to calculate the intensities of the diffracted beams and to relate these data to the crystal structure.

---

[1] Diffracted beams are frequently referred to as 'reflections', 'spots' or 'lines'. The use of the word reflection stems from the geometry of the diffraction (see section 6.1, Bragg's Law). The use of the terms 'spots' and 'lines' arose because X-ray diffraction patterns were once recorded on photographic film. Single crystals gave rise to a set of spots, while polycrystalline samples gave a series or rings or lines on the film. The terms 'reflections', 'lines' or 'spots' will be used here as synonyms for beams.

Crystals and Crystal Structures.   Richard J. D. Tilley
© 2006 John Wiley & Sons, Ltd

Finally it is necessary to recreate an image of the crystal, the crystal structure itself, using the information contained in the diffraction pattern. Aspects of these steps are considered in this Chapter, but a detailed account of the ways in which crystal structures are derived is beyond the scope of this book, for which the Bibliography should be consulted[2].

6.1   The position of diffracted beams: Bragg's law

A beam of radiation will only be diffracted when it impinges upon a set of planes in a crystal, defined by the Miller indices (hkl), if the geometry of the situation fulfils quite specific conditions, defined by **Bragg's law**:

$$n\lambda = 2d_{hkl}\sin\theta \qquad (6.1a)$$

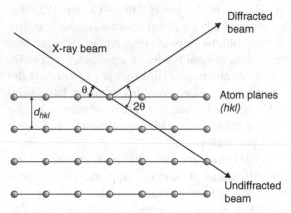

Figure 6.1   The geometry of Bragg's law for the diffraction of X-rays from a set of crystal planes, (hkl), with interplanar spacing d_{hkl}

---

[2]Note that it is crystallographic convention to use the Ångström unit, Å, as the unit of length. The conversion factor is $1\,\text{Å} = 10^{-10}\,\text{m}$; $1\,\text{nm} = 10\,\text{Å}$. Here the SI unit, (nm), will be mainly used, but Å may be employed from time to time, to avoid unnecessary confusion with crystallographic literature.

where n is an integer, λ is the wavelength of the radiation, d_{hkl} is the interplanar spacing (the perpendicular separation) of the (hkl) planes and θ is the diffraction angle or **Bragg angle** (Figure 6.1). Note that the angle between the direction of the incident and diffracted beam is equal to 2θ. Bragg's law defines the conditions under which diffraction occurs, and gives the **position** of a diffracted beam, without any reference to its intensity.

There are a number of features of Bragg's law that need to be emphasised. Although the geometry of Figure 6.1 is identical to that of reflection, the physical process occurring is diffraction, and the angles of incidence and reflection conventionally used to describe the reflection of light are not the same as those used in the Bragg equation. Moreover, there is no constraint on the angle at which a mirror will reflect light. However, planes of atoms will diffract radiation only when illuminated at the angle $\sin^{-1}(n\lambda/2d_{hkl})$.

Equation 6.1a includes an integer n, which is the **order** of the diffracted beam. Crystallographers take account of the different orders of diffraction by changing the values of (h k l) to (nh nk nl). Thus the first order reflection from, for example, (111) planes, occurs at an angle given by:

$$\sin\theta\,(1^{\text{st}}\,\text{order}\,111) = 1\lambda/2d_{111}$$

The second order reflection from the same set of planes then occurs at an angle:

$$\sin\theta\,(2^{\text{nd}}\,\text{order}\,111) = 2\lambda/2d_{111}$$

but is always referred to as the first order reflection from (222) planes, i.e.

$$\sin\theta\,(1^{\text{st}}\,\text{order}\,222) = 1\lambda/2d_{222}$$

or more simply, the (222) reflection:

$$\sin \theta \, (222) = 1\lambda/2d_{222}$$

Similarly, the third order reflection from (111) planes is at an angle:

$$\sin \theta \, (3^{\mathrm{rd}} \text{ order } 111) = 3\lambda/2d_{111}$$

but is always referred to as the (333) reflection, i.e.

$$\sin \theta \, (333) = 1\lambda/2d_{333}$$

Note that as $d_{111} = 2d_{222} = 3d_{333}$ the equations are formally identical. The same is true for all planes, so that the Bragg equation used in crystallography is simply:

$$\lambda = 2d_{hkl} \sin \theta \qquad (6.1b)$$

Within a crystal there are an infinite number of sets of atom planes, and Bragg's law applies to all these. Thus if a crystal in a beam of radiation is rotated, each set of planes will, in its turn, diffract the radiation when the value of $\sin \theta$ becomes appropriate. This is the principle by which diffraction data is collected for the whole of the crystal. The arrangement of the diffracted beams, when taken together, is the diffraction pattern of the crystal.

The d_{hkl} values for any crystal can be calculated from knowledge of the lattice parameters (see Chapter 2). The Bragg equation, applied to diffraction data, results in a list of d_{hkl} values for a compound. It is possible, by putting these to data sets together, to determine the size of the unit cell of the material producing the diffraction pattern. In effect, this means allocating a value hkl to each diffracted beam, a process called **indexing** the diffraction pattern.

This procedure, although simple in principle, requires skill and ingenuity. It is now carried out automatically by computers, which can test large numbers of alternative unit cells in minimal time.

6.2 The geometry of the diffraction pattern

As described above, when a beam of suitable radiation is incident on a rotating crystal, the beam will generally pass directly through the crystal, but if it is at the Bragg angle for a particular set of (hkl) planes some of the energy of the beam will pass into a diffracted beam. It is difficult to gain a complete idea of what happens when radiation is incident on a real crystal, because there are many sets of planes to be considered simultaneously, all of them capable of diffracting the radiation. A simple graphical construction first described by Ewald, allows this to be achieved.

To visualise the diffraction pattern of a crystal:

(i) draw the reciprocal lattice of the crystal, in an orientation equivalent to that of the crystal being irradiated (Figure 6.2a).

(ii) draw a vector of length $1/\lambda$ from the origin of the reciprocal lattice in a direction parallel, and in an opposite direction, to the beam of radiation. Here λ is the wavelength of the radiation, (Figure 6.2b).

(iii) from the end of the vector, construct a sphere, called the **Ewald sphere**, (shown as a circle in cross-section), of radius $1/\lambda$. Each reciprocal lattice point that the sphere (circle) touches (or nearly touches) will give a diffracted beam (Figure 6.2c, d).

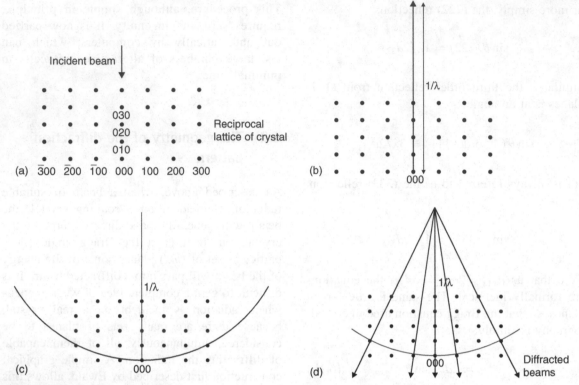

Figure 6.2   The Ewald construction: (a) the reciprocal lattice; (b) a vector of length $1/\lambda$, dawn parallel to the beam direction; (c) a sphere passing through the 000 reflection, drawn using the vector in (b) as radius; (d) the positions of the diffracted beams

Note that this construction is equivalent to the Bragg law. The distance of the hkl reciprocal lattice point from the origin of the reciprocal lattice is given by $1/d_{hkl}$, where d_{hkl} is the interplanar spacing of the (hkl) planes, (Figure 6.3). The diffraction angle, θ, then conforms to the geometry:

$$\sin\theta = [1/2d_{hkl}]/[1/\lambda] = \lambda/2d_{hkl}$$

Rearrangement leads to the Bragg equation:

$$2d_{hkl}\sin\theta = \lambda$$

The utility of the Ewald method of visualising a diffraction pattern is best understood via the formation of an electron diffraction pattern. This is because the Ewald sphere is much larger than the spacing between the reciprocal lattice points. For electrons accelerated through 100 kV, which is typical for an electron microscope, the electron wavelength is 0.00370 nm, giving the Ewald sphere a radius of 270.27 nm^{-1}. The lattice parameter of copper (see Chapter 1), which can be taken as illustrative of many inorganic compounds, is 0.360 nm, giving a reciprocal lattice spacing a^* of 2.78 nm^{-1}. That is, for a reciprocal lattice spacing of 2.78 cm, the Ewald sphere has a

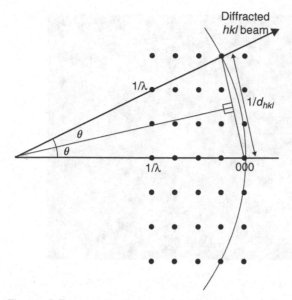

Figure 6.3 The geometry of the Ewald construction, showing it to be identical to Bragg's law

radius of nearly 3 metres, (for a reciprocal lattice spacing of about 1 inch, the sphere radius is approximately 9 feet). Because of this disparity in size, the surface of the Ewald sphere is nearly flat and lattice points in the plane which is perpendicular to the electron beam and which passes through the origin will touch the Ewald sphere. Hence all of the diffracted beams corresponding to these points will be produced simultaneously. If the diffracted beams are intercepted by a screen and displayed visually the arrangement of diffraction spots will resemble a plane section through the reciprocal lattice, at a scale dependent upon the geometry of the arrangement. In essence, an electron diffraction pattern is thus a projection of a plane section through the reciprocal lattice, (Figure 6.4a, b), although the way it is produced, by a series of electron lenses, is more complex, (see Section 6.11, below, for more information). Tilting the crystal will allow other sections of the reciprocal

lattice to be visualised, and so the whole reciprocal lattice can be built up.

If the material examined consists of a powder of randomly arranged small crystallites, each will produce its own pattern of reflections. When large numbers of crystallites are present all reciprocal lattice sections will be present. In this case, a pattern of spotty rings will be seen, (Figure 6.4c). The definition of the rings improves as the number of crystallites increases, although if the crystal size becomes very small, the rings start to loose sharpness and become diffuse, as explained in the following section.

Although the diffraction patterns obtained from an electron microscope are easy to understand in terms of the reciprocal lattice, it is still necessary to allocate the appropriate *hkl* value to each spot in order to obtain crystallographic information. This is called indexing the pattern. A comparison of the **real space** situation, in an electron microscope, with the **reciprocal space** equivalent, the Ewald sphere construction, (Figure 6.5), shows that the relationship between the d_{hkl} values of the spots on the recorded diffraction pattern is given by the simple relationship:

$$r/l = (1/d_{hkl})/(1/\lambda)$$
$$d_{hkl} = \lambda l/r$$

where r is the distance from the origin of the diffraction pattern, 000, to the *hkl* spot and l is the distance over which the diffraction pattern is projected. In reality, as the pattern is produced by the action of lenses, the value of l is not readily accessible, and the pair of constants λl, called the **camera constant** for the equipment, is given by the manufacturer. Alternatively the camera constant can be determined experimentally by the observation of a ring pattern from a known material. A polycrystalline evaporated thin film of TlCl, which adopts

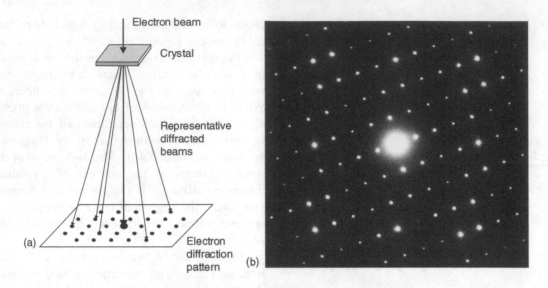

Electron beam

Crystal

Representative
diffracted
beams

(a)

Electron
diffraction
pattern

(b)

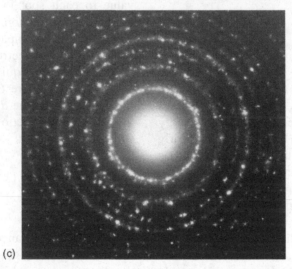

(c)

Figure 6.4   Electron diffraction patterns: (a) schematic diagram showing the formation of an electron diffraction pattern in an electron microscope; (b) an electron diffraction pattern from $WNb_{12}O_{33}$, resembling a plane section through the reciprocal lattice; (c) a 'spotty' electron diffraction ring pattern from a polycrystalline TiO_2 photocatalyst

the CsCl structure, $a = 0.3834$ nm, was often used for this purpose. The successive diffraction rings are indexed as 100, 110, 111, 200, 210, 211 ... with the smallest first. The

d-values of each plane are easily calculated using the formula in Chapter 2,

$$1/(d_{hkl})^2 = [h^2 + k^2 + l^2]/a^2$$

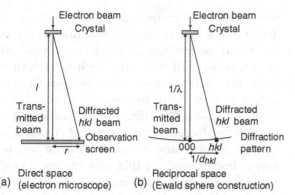

(a) Direct space (electron microscope) (b) Reciprocal space (Ewald sphere construction)

Figure 6.5 Comparison of real space and reciprocal space formation of diffraction patterns: (a) schematic formation of a diffraction pattern in an electron microscope; (b) the Ewald construction of a diffraction pattern

A measurement of the ring diameters allows a value of λl to be calculated.

To index an electron diffraction pattern, that is, to give each spot appropriate hkl values is straightforward when a value of the camera constant, λl is known. The distance from the centre of the pattern to the reflection in question is measured, to give a value of r, and this is converted into a value of d using

$$d = \lambda l / r$$

A list of the d_{hkl}-values for the material under investigation is calculated using the equations given in Chapter 2 and the measured value of d is identified as d_{hkl} using the list. When two or more diffraction spots have been satisfactorily identified the whole pattern can be indexed. Naturally ambiguities arise because of errors in the assessment of the d-values from the diffraction pattern, but they can generally be removed by calculating the angles between the possible planes.

In the case of X-ray diffraction, the radius of the Ewald sphere is similar to that of the reciprocal lattice spacing. For example, the wavelength of CuKα radiation, typically used for X-ray investigations, is 0.15418 nm, giving an Ewald sphere radius of 6.486 nm^{-1}, which is comparable to the reciprocal lattice spacing of copper, (2.78 nm^{-1}), and other inorganic materials. This means that far fewer spots are intercepted, and fewer diffracted beams are produced. The appearance of an X-ray diffraction pattern is then not so easy to visualise as that of an electron diffraction pattern, and rather complex recording geometry must be used to obtain diffraction data from the whole of the reciprocal lattice.

Whatever the way in which the reciprocal lattice is derived, it is often the key to the determination of the correct unit cell of a phase.

6.3 Particle size

Although Bragg's law gives a precise value for the diffraction angle, θ, diffraction actually occurs over a small range of angles close to the ideal value because of crystal imperfections. More perfect crystals diffract over a very narrow range of angles while very disordered crystals diffract over a wide range. The *approximate* range of angles over which diffraction occurs, $\delta\theta$, centred upon the exact Bragg angle, θ, is given by:

$$\delta\theta \approx \lambda / (D_{hkl} \cos\theta)$$

where D_{hkl} is the crystal thickness in the direction normal to the (hkl) planes diffracting the radiation and $\delta\theta$ is measured in radians. Thus,

$$D_{hkl} = m \, d_{hkl}$$

where m is an integer. The *apparent* crystal size in a direction perpendicular to the (hkl) planes is then given by:

$$m\,d_{hkl} \approx \lambda / (\delta\theta\cos\theta)$$

Thus the size and sharpness of the diffracted beams provide a measure of the perfection of the crystal; a sharp spot indicating a well ordered crystal and a fuzzy spot an extremely disordered or very small crystallite. Amorphous materials give only a few broad ill-defined rings if anything at all. Before the advent of electron microscopy, (see Section 6.13), the extent of broadening of X-ray reflections was one of the main methods used to estimate particle sizes, although a far more sophisticated analysis of the data than that given above needs to be employed for accurate work.

These size effects can easily be pictured in terms of the reciprocal lattice. Ideally, each reciprocal lattice point is sharp and diffraction will only occur when the Ewald sphere intersects a particular point. The effect of finite crystal dimensions is to modify the shape of each reciprocal lattice point by drawing the point out in reciprocal space in a direction normal to the small dimension in real space. The following descriptions are a *first approximation only*. (More accurate depictions of the change in shape of a reciprocal lattice reflection can be determined by applying diffraction theory.) A square crystal of side w will have each reciprocal lattice point modified to a cross with each arm of approximate length $1/w$, extended along a direction perpendicular to the square faces; a cubic crystal will be similar, with each arm pulled out along directions normal to the cube faces. A spherical crystal of diameter w will have each reciprocal lattice point broadened into a sphere of approximately $1/w$ diameter; a needle of diameter w

will have each reciprocal lattice point broadened into a disc of approximate diameter $1/w$ normal to the needle axis and a thin crystal, of thickness w, will have the reciprocal lattice point drawn out into a spike of approximate length $1/w$ in a direction normal to the sheet, (Figure 6.6a–d). The effect of this broadening is to greatly increase the areas of reciprocal lattice sections sampled in an electron microscope. For example, a thin film has reciprocal lattice points drawn out into spikes normal to the film, allowing the Ewald sphere to intersect considerably more points than in the case of a thick film, (Figure 6.7).

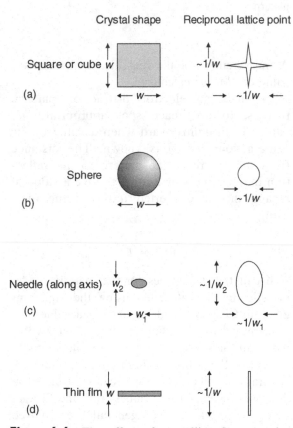

Figure 6.6 The effect of crystallite shape on the form of a diffraction spot: (a) square or cube; (b) sphere; (c) elliptical needle; (d) thin film

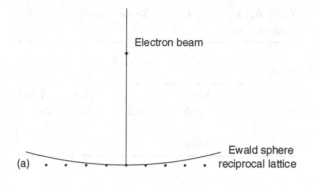

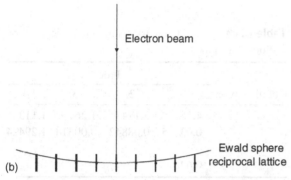

Figure 6.7 The Ewald sphere intersects fewer reciprocal lattice points in a thick film, (a) than a thin film, (b) due to the elongation of the reciprocal lattice points into spikes

6.4 The intensities of diffracted beams

The experimental techniques outlined in the previous sections allow the lattice parameters of a crystal to be determined. However, the determination of the appropriate crystal lattice, face-centred cubic as against body-centred, for example, requires information on the intensities of the diffracted beams. More importantly, in order to proceed with a determination of the complete crystal structure, it is vital to understand the relationship between the intensity of a beam diffracted from a set of (hkl) planes and the atoms that make up the planes themselves.

The intensities of diffracted beams vary from one radiation type to another, and are found to depend upon the following factors:

(i) the nature of the radiation.

(ii) the Bragg angle of the diffracted beam.

(iii) the diffracting power of the atoms present – the atomic scattering factor.

(iv) the arrangement of the atoms in the crystal – the structure factor.

(v) the thermal vibrations of the atoms – the temperature factor.

(vi) the polarisation of the beam of radiation.

(vii) the thickness, shape and perfection of the crystal – the form factor. This aspect has already been touched upon in Section 6.3, and is mentioned again in Section 8.6.

(viii) for diffraction from a powder rather than a single crystal, the number of equivalent (hkl) planes present – the multiplicity.

To explain these factors it is convenient to describe the determination of intensities with respect to X-ray diffraction. The differences that arise with other types of radiation are outlined below (Sections 6.12, 6.14, 6.17).

In the following four sections the important contribution made by the arrangement and types of atoms in crystals is considered. The other factors, which can be regarded as correction terms, are described later in this Chapter. In the case of X-rays, the initial beam is almost undiminished on passing through a small crystal, and the intensities of the diffracted beams are a very small percentage of the incident beam intensity. For this reason, it is reasonable to assume that each diffracted

X-ray photon is scattered only once. Scattering of diffracted beams back into the incident beam direction is ignored. This reasonable approximation is the basis of the **kinematical theory** of diffraction, which underpins the theoretical calculation of the intensity of a diffracted beam of X-rays.

6.5 The atomic scattering factor

X-rays are diffracted by the electrons on each atom. The scattering of the X-ray beam increases as the number of electrons, or equally, the atomic number (proton number), of the atom, Z, increases. Thus heavy metals such as lead, Pb, $Z = 82$, scatter X-rays far more strongly than light atoms such as carbon, C, $Z = 6$. Neighbouring atoms such as cobalt, Co, $Z = 27$, and nickel, Ni, $Z = 28$, scatter X-rays almost identically. The scattering power of an atom for a beam of X-rays is called the **atomic scattering factor**, f_a.

Atomic scattering factors were originally determined experimentally, but now can be calculated using quantum mechanics. The atomic scattering factors, derived from quantum mechanical calculations of the electron density around an atom, are (approximately) given by the equation:

$$f_a = \sum_{i=1}^{4} a_i \cdot \exp\left[-b_i\left(\frac{\sin\vartheta}{\lambda}\right)^2\right] + c \quad (6.2)$$

The nine constants, a_i, b_i and c_i, called the **Cromer-Mann coefficients**, vary for each atom or ion. The Cromer-Mann coefficients for sodium, Na, with an atomic number, $Z = 11$, for use in equation (6.2) are given in Table 6.1. The units of f_a are electrons, and the wavelength of the radiation is in Å. To use the SI unit of nm, $(10\ \text{Å} = 1\ \text{nm})$, for the wavelength, use the same values of a_i and c, but divide the values of b_i by 100.

Table 6.1a The Cromer-Mann coefficients for sodium, λ in Å

Coefficient	Index			
	1	2	3	4
a	4.763	3.174	1.267	1.113
b	3.285	8.842	0.314	129.424
c	0.676			

Table 6.1b The Cromer-Mann coefficients for sodium, λ in nm

Coefficient	Index			
	1	2	3	4
a	4.763	3.174	1.267	1.113
b	0.03285	0.08842	0.00314	1.29424
c	0.676			

These and other values used in this Chapter are taken from http://www-structure.llnl.gov.

It is found that the scattering is strongly angle dependent, this being expressed as a function of $(\sin\theta)/\lambda$, which, by using the Bragg law, equation 6.1, is equal to $1/2d_{hkl}$. The scattering factor curves for all atoms have a similar form, (Figure 6.8). At $(\sin\theta)/\lambda = \text{zero}$ the scattering factor is equal to the atomic number, Z, of the element in question. The element sodium, with $Z = 11$, has a value of the scattering factor at $(\sin\theta)/\lambda = \text{zero}$ of 11. For ions, the scattering factor is similarly dependent upon the number of electrons present. For example, the scattering factor for the sodium ion Na^+ at $(\sin\theta)/\lambda = \text{zero}$ is equal to $Z-1$, i.e. 10, rather than 11. Similarly, the atomic scattering factors of K^+ $(Z = 18)$ and Cl^- $(Z = 18)$ are identical.

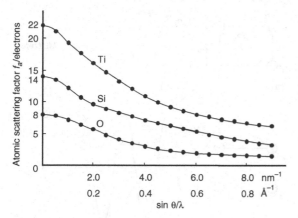

Figure 6.8 Atomic scattering factors for titanium, Ti, silicon, Si, and oxygen, O, as a function of $\sin \theta / \lambda$

Electrons also scatter radiation by another effect, the Compton effect. Compton scattering adds to the general background of scattered X-rays, and is usually ignored in structure determination.

6.6 The structure factor

To obtain the total intensity of radiation scattered by a unit cell, the scattering of all of the atoms in the unit cell must be combined. This is carried out by adding together the waves scattered from each set of (hkl) planes independently, to obtain a value called the **structure factor**, **F**(hkl), for each hkl plane. It is calculated in the following way.

The *amplitude* (i.e. the 'strength') of the diffracted radiation scattered by an atom in a plane (hkl) will be given by the value of f_a appropriate to the correct value of $\sin\theta/\lambda$ $(=1/2d_{hkl})$ for the (hkl) plane. However, because the scattering atoms are at various locations in the unit cell, the waves scattered by each atom are out of step with each other as they leave the unit cell. The difference by which the waves are out of step is called the **phase difference** between the waves.

The way in which the scattered radiation adds together is governed by the relative phases of each scattered wave. The phase difference between two scattered waves, the amount by which they are out of step, is represented by a 'phase angle' which is measured in radians. Waves that are in step have a phase angle of 0, or an even multiple of 2π. Waves that are completely out of step have a phase difference of π, or an odd multiple of π. If two waves with a phase difference of 2π are added the result is a wave with *double the amplitude* of the original, while if the phase difference is π, the result is *zero*. Intermediate values of the relative phases give intermediate values for the amplitude of the resultant wave, (Figure 6.9).

The phase difference between the waves scattered by two atoms will depend upon their relative positions in the unit cell and the directions along which the waves are superimposed. The directions of importance are those specified by the Bragg equation, equation (6.1), which, for simplicity, are denoted by the indices of the (hkl) planes involved in the scattering, rather than the angle itself. The phase of the wave (in radians) scattered from an atom A at a position x_A, y_A, z_A, into the (hkl) reflected beam is:

$$2\pi(hx_A + ky_A + lz_A) = \phi_A$$

where x_A, y_A, z_A are the fractional coordinates with respect to the unit cell edges, as normally used.

To illustrate this, suppose that the unit cell contains just two identical atoms, M1, at the cell origin (0, 0, 0), and M2 at the cell centre, (½, ½, ½), (Figure 6.10a). The reflection from (100) planes will be non-existent, because the waves scattered by the M1 atoms at the cell origin, which have a phase:

$$M1 \quad 2\pi(1 \times 0) = 0$$

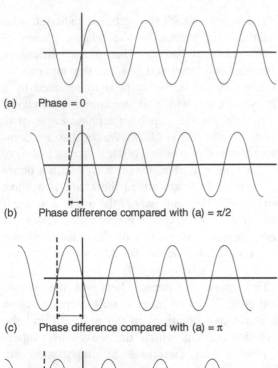

(a)   Phase = 0

(b)   Phase difference compared with (a) = π/2

(c)   Phase difference compared with (a) = π

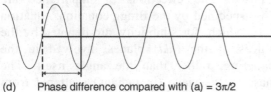

(d)   Phase difference compared with (a) = 3π/2

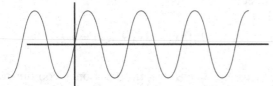

(e)   Phase difference compared with (a) = 0 or 2π

Figure 6.9   The phase difference between waves: (a) 0; (b) π/2; (c) π; (d) 3π/2; (e) 2π

are combined with a similar wave, arising from the M2 atoms at the cell centre, with phase:

$$M2 \quad 2\pi(1 \times \tfrac{1}{2}) = \pi$$

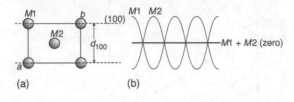

(a)   (b)

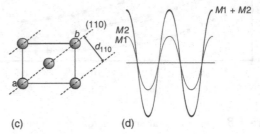

(c)   (d)

Figure 6.10   zThe scattering of waves from a unit cell: (a) (100) planes, resulting in zero amplitude; (b) (110) planes, resulting in doubled amplitude

to give a zero resultant amplitude, (Figure 6.10b). In contrast, the amplitude of the reflection from (110), (Figure 6.10c), will be high, because this time the waves from both M1 and M2 are in step, with relative phases:

$$M1 \quad 2\pi(1 \times 0 + 1 \times 0) = 0$$
$$M2 \quad 2\pi(1 \times \tfrac{1}{2} + 1 \times \tfrac{1}{2}) = 2\pi$$

(Figure 6.10d). Thus, even in this simple case, the amplitude of the resultant scattered wave can vary between the limits of zero to $2 \times f_M$, depending upon the planes that are scattering the waves.

When the scattered waves from all of the atoms in a unit cell are added, the form of the resultant wave will depend, therefore, upon both the scattering power of the atoms involved and the individual phases of all of the separate waves. A clear picture of the relative importance of the scattering from each of the atoms in a unit cell, including the phase information, and how they add together to give a final value of $\mathbf{F}(hkl)$ can be obtained by adding the scattering from each atom using vector addition (Appendix 1). The scattering from an atom A is represented by a vector $\mathbf{f}_A$ that has a length equal to the scattered amplitude (i.e. the strength of the scattering), of f_A and a phase angle ϕ_A. To draw the diagram, a vector, $\mathbf{f}_A$, of length f_A, is drawn at an angle ϕ_A in an anticlockwise direction to the horizontal (x-) axis, (Figure 6.11a). In order to picture the scattering from a complete unit cell, the scattering from each of the N atoms in the cell is represented by vectors $\mathbf{f}_n$, of length f_n drawn at an angle ϕ_n, to the horizontal, (Figure 8.11b), where n runs from 1 to N. Note that the scattering from an atom at the origin of the unit cell, position 000, will be represented by a vector along the horizontal, as the phase angle will be zero, (Figure 6.11c). Each successive vector is added, head to tail, to the preceding one, with ϕ_n always being drawn in an anticlockwise direction with respect to the horizontal. The resultant of the addition of all of the $\mathbf{f}_n$ vectors, using the 'head to tail' method of graphical vector addition, (see Appendix 1), gives a vector $\mathbf{F}(hkl)$ with a phase angle ϕ_{hkl}, (Figure 6.11d). The numerical length of the vector $\mathbf{F}(hkl)$ is written $F(hkl)$. It is seen that the value of ϕ_{hkl} is equal to the sum of all of the phase angles, ϕ_n, of the scattered waves from the n atoms.

6.7 Structure factors and intensities

The graphical method gives a lucid picture of scattering from a unit cell, but is impractical as a method for calculation of the intensities of diffracted beams. The pictorial summation must be expressed algebraically for this purpose. The simplest way of carrying this out is to express the scattered wave as a **complex amplitude**, (see Appendices 5 and 6):

$$\mathbf{f}_A = f_A \exp[2\pi i(hx_A + ky_A + lz_A)] = f_A e^{i\phi_A}$$

$$(6.3)$$

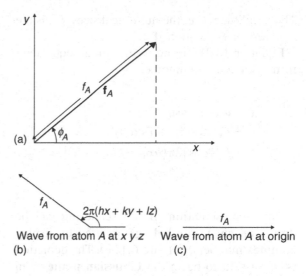

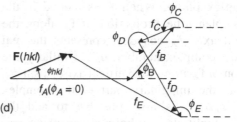

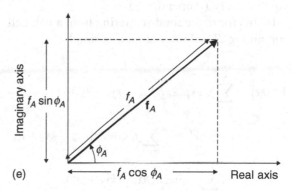

Figure 6.11   The representation of scattered waves as vectors: (a) a scattered wave vector in the x, y plane; (b) the wave scattered by an atom A at (x, y, z); (c) the wave scattered by atom A at the origin, (000); (d) the addition of waves scattered by five atoms, A, B, C, D, and E; (e) represention of $\mathbf{f}_A$ as a complex number on an Argand diagram

The amplitude, (i.e. the numerical size), f_A, is the *modulus* of $\mathbf{f}_A$, written $|\mathbf{f}_A|$.

Equation (6.3) can be written in an equivalent form as a complex number:

$$
\begin{aligned}
\mathbf{f}_A &= a_{hkl} + \mathrm{i}\, b_{hkl} \\
&= f_A \{\cos \phi_A + \mathrm{i} \sin \phi_A\} \\
&= f_A \{\cos [2\pi(hx_A + ky_A + lz_A)] \\
&\quad + \mathrm{i} \sin [2\pi(hx_A + ky_A + lz_A)]\}
\end{aligned}
$$

This representation of the scattering can be drawn on an **Argand diagram** used to display complex numbers, (Figure 6.11 e). The depiction is often said to be in the **Gaussian plane** or in the **complex plane**. When $\mathbf{f}_A$ is plotted in the Gaussian plane the projection of $\mathbf{f}_A$ along the horizontal axis, $f_A \cos \phi_A$, represents the real part of the complex number, a_{hkl}. Similarly, the projection of $\mathbf{f}_A$ along the vertical axis, $f_A \sin \phi_A$, represents the imaginary part of the complex number, b_{hkl}. It is now possible to add the scattered amplitudes in the complex plane graphically, as in Figure 6.11 d, or better still, algebraically (Appendix 6).

In this form, the total scattering from a unit cell containing N different atoms, is simply:

$$
\begin{aligned}
\mathbf{F}(hkl) &= \sum_{n=1}^{N} f_n \exp[2\pi i(hx_n + ky_n + lz_n)] \\
&= F(hkl)e^{i\phi_{hkl}} = \sum_{n=1}^{N} f_n \cos 2\pi(hx_n + ky_n + lz_n) \\
&\quad + \mathrm{i} \sum_{n+1}^{N} f_n \sin 2\pi(hx_n + ky_n + lz_n) \\
&= A_{hkl} + \mathrm{i} B_{hkl}
\end{aligned}
$$

where hkl are the Miller indices of the diffracting plane, and the summation is carried out over all N atoms in the unit cell, each of which has an atomic scattering factor, f_n, appropriate to the

(hkl) plane being considered. The magnitude of the scattering is $F(hkl)$, the modulus of $\mathbf{F}(hkl)$, written $|\mathbf{F}(hkl)|$, and ϕ_{hkl} is the phase of the scattering factor, which is the sum of all of the phase contributions from the various atoms present in the unit cell.

The **intensity** scattered into the hkl beam by all of the atoms in the unit cell, $I_0(hkl)$, is given by $|\mathbf{F}(hkl)|^2$, the modulus of $\mathbf{F}_{hkl}$ squared, (see Appendix 6):

$$
\begin{aligned}
I_0(hkl) &= |\mathbf{F}(hkl)|^2 = F(hkl)^2 \\
&= \left\{ \sum_{n=1}^{N} f_n \exp[2\pi i(hx_n + ky_n + lz_n)] \right\}^2 \\
&= \left\{ \sum_{n=1}^{N} f_n \cos 2\pi(hx_n + ky_n + lz_n) \right. \\
&\quad \left. + \mathrm{i} \sum_{n+1}^{N} f_n \sin 2\pi(hx_n + ky_n + lz_n) \right\}^2 \\
&= A_{hkl}^2 + B_{hkl}^2
\end{aligned}
$$

If the unit cell has a centre of symmetry, the sine terms, B_{hkl}^2, can be omitted, as they sum to zero.

The calculation of the intensity reflected from a plane of atoms in a known structure is straightforward, although tedious without a computer. Because of the instrumental and other factors that affect measured intensities, calculated intensities, which ignore these factors, are generally listed as a fraction or percentage of the strongest reflection.

6.8 Numerical evaluation of structure factors

Although structure factor calculations are carried out by computer, it is valuable to evaluate a few examples by hand, to gain an appreciation of the steps involved. As an example, the calculation of the structure factor and intensity for the 200 reflection from a crystal of the rutile form of TiO$_2$, (see Chapter 1), is reproduced here. The details of the tetragonal unit cell are:

Lattice parameters,

a(=b) = 0.4594 nm, c = 0.2959 nm.

Atom positions: Ti: 000; ½ ½ ½

O : ³⁄₁₀ ³⁄₁₀ 0; ⁴⁄₅ ¹⁄₅ ½; ⁷⁄₁₀ ⁷⁄₁₀ 0; ¹⁄₅ ⁴⁄₅ ½

There are two Ti atoms and four oxygen atoms in the unit cell.

The calculation proceeds in the following way.

(i) Estimate the value of $(\sin\theta/\lambda)$ appropriate to (200):

$$(\sin\theta/\lambda) = 1/(2d_{200})$$
$$= 2.177\,\text{nm}^{-1}(0.2177\,\text{Å}^{-1})$$

(ii) Determine the atomic scattering factors for Ti and O at this value from tables of data or equation (6.2), using the appropriate Cromer-Mann coefficients. (Approximate values can found from Figure 6.7). The precise values are: f_{Ti}, 15.513, f_O, 5.326.

(iii) Calculate the phase angles of the waves scattered by each of the atoms in the unit cell. Note that the unit cell is centrosymmetric, and so the sine terms, (B_{hkl}), can be omitted. In this case, the phase angles are simply the cosine, (A_{hkl}), terms, $[\cos 2\pi\,(hx_n + ky_n + lz_n)]$, which in this case simplifies to $[\cos 2\pi\,(2x_n)]$. The results are given in Table 6.2.

(iv) $F(200)$ is given by adding the values in the final column of Table 6.2.

$$F(200) = 2(15.513) - 4(4.3097) = 13.79.$$

Table 6.2 Calculation of $F(200)$ for TiO$_2$, rutile

Atom	Phase angle, ϕ/radians	Phase angle, $\phi/°$	Cos ϕ	$f_a\cos\phi = A_{200}$
Ti(1)	0	0	1.0	15.513
Ti(2)	2π	360	1.0	15.513
O(1)	1.2π	216	−0.8090	−4.3087
O(2)	3.2π	576 (= 216)	−0.8090	−4.3087
O(3)	2.8π	504 (= 144)	−0.8090	−4.3087
O(4)	0.8π	144	−0.8090	−4.3087

(v) The phase angle ϕ_{200}, associated with $F(200)$, is the sum of all of the individual phase angles in Table 6.2. It is seen that $\phi_{200} = 10\pi\,(1800°) = 2\pi\,(360° = 0°)$.

(vi) Calculate the intensity, $I_0(200)$, of the 360°= diffracted beam, equal to $|\mathbf{F}(hkl)|^2$:

$$I_0(200) = |\mathbf{F}(200)|^2 = F(200)^2 = 186.16$$

This procedure is repeated for all of the other hkl reflections from the unit cell.

To relate this numerical assessment to the vector addition described above, the values given in Table 6.2, are displayed graphically in Figure 6.12. The value obtained for $F(200)$ from this figure is 13.8, compared with the arithmetical value of 13.6. The phase angle ϕ_{hkl} is zero.

An examination of Table 6.2 and Figure 6.12 makes clear the importance of the phases of the waves. To a large extent these control the intensity. If the rutile calculation is repeated for the

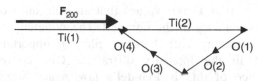

Figure 6.12 The vector addition of scattered waves contributing to the 200 reflection of rutile, TiO$_2$

Table 6.3 Calculation of $F(100)$ for TiO_2, rutile

Atom	Phase angle, ϕ/radians	Phase angle, ϕ/°	$f_a \cos \phi = A_{100}$
Ti(1)	$2\pi\,(1 \times 0)$	0	$f(Ti) \times 1$
Ti(2)	$2\pi\,(1 \times \frac{1}{2})$	180	$f(Ti) \times -1$
O(1)	$2\pi\,(1 \times \frac{3}{10})$	108	$f(O) \times -0.3090$
O(2)	$2\pi\,(1 \times \frac{4}{5})$	288	$f(O) \times 0.3090$
O(3)	$2\pi\,(1 \times \frac{7}{10})$	252	$f(O) \times -0.3090$
O(94)	$2\pi\,(1 \times \frac{1}{5})$	72	$f(O) \times 0.3090$

(100) plane, the phase angles for the two Ti atoms are:

$$Ti(1), \quad \cos 2\pi(1 \times 0) = 1$$
$$Ti(2), \quad \cos 2\pi(1 \times \tfrac{1}{2}) = -1$$

That is, the waves scattered by the two Ti atoms are exactly out of step and will cancel. It is immediately apparent that the (100) reflection will be weak, as it will depend only upon the oxygen atoms. In fact these cancel too, so that the intensity turns out to be zero, (Table 6.3). The phase angle, ϕ_{100}, equal to the sum of all of the separate phase angles, is equal to 5π (900°) $= \pi$ (180°), (Table 6.3).

6.9 Symmetry and reflection intensities

The atoms in the unit cell will determine the intensity of a diffracted beam of radiation via the structure factor, as described above. This means that there will be relationships between the various $\mathbf{F}(hkl)$ values that arise because of the symmetry of the unit cell. The symmetry of the structure will therefore play an important part in the intensity diffracted. One consequence of this is **Friedel's law**, which states that the structure factors, $\mathbf{F}$, of the pair of reflections hkl and $\bar{h}\,\bar{k}\,\bar{l}$ are **equal in magnitude but have opposite phases**. That is:

$$F(hkl) = F(\bar{h}\,\bar{k}\,\bar{l}); \quad \phi_{hkl} = -\phi_{\bar{h}\,\bar{k}\,\bar{l}}$$

Such pairs of reflections are called **Friedel pairs**. Because of this, the intensities can be expressed as:

$$F(hkl)^2 = I_0(hkl) = F(\bar{h}\,\bar{k}\,\bar{l})^2 = I_0(\bar{h}\,\bar{k}\,\bar{l})$$

The intensities of Friedel pairs will be equal. This will cause the diffraction pattern from a crystal to appear centrosymmetric even for crystals that lack a centre of symmetry. Diffraction is thus a centrosymmetric physical property, which means that the point symmetry of any diffraction pattern will belong to one of the 11 Laue classes, (see Section 4.7).

The most important consequence the symmetry elements present in a crystal is that some (hkl) planes have $F(hkl)$ = zero, and so will never give rise to a diffracted beam, irrespective of the atoms present. Such 'missing' diffracted beams are called **systematic absences**. This can most easily be understood in terms of the vector representation described above. Suppose that a crystal is derived from a body centred lattice. In the simplest case, the motif is one atom per lattice point, and the unit cell contains atoms at 000 and ½ ½ ½, (Figure 6.13a). [Note the cell can have any symmetry]. The structure factor for each (hkl) set is given by:

$$F(hkl) = f_a \exp 2\pi i(h \times 0 + k \times 0 + l \times 0)$$
$$+ f_a \exp 2\pi i(h \times \tfrac{1}{2} + k \times \tfrac{1}{2} + l \times \tfrac{1}{2})$$

If $h + k + l$ is *even*, $F(hkl)$ becomes

$$F(hkl) = f_a \exp 2\pi i(0) + f_a \exp 2\pi i(2n/2)$$

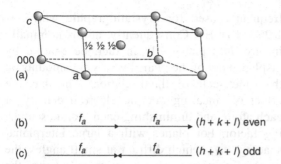

Figure 6.13 Reflections from a body-centred unit cell: (a) a body-centred unit cell; (b) vector addition of waves for reflections $h + k + l$ even, gives rise to a scattered amplitude; (c) vector addition of waves for reflections $h + k + l$ odd, gives rise to zero amplitude

Table 6.4 Reflection conditions for Bravais lattices

Lattice type	Reflection conditions[*]
Primitive (P)	None
A-face centred (A)	$k + l = 2n$
B-face centred (B)	$h + l = 2n$
C-face centred (C)	$h + k = 2n$
Body centred (I)	$h + k + l = 2n$
All-face centred (F)	$h + k, h + l, k + l = 2n,$
	(h, k, l, all odd or all even)

[*]n is an integer.

where n is an integer. The Gaussian plane representation (Figure 6.13b) shows that the vectors are parallel and add together to give a positive value for $F(hkl)$ for any value of n.

If $h + k + l$ is *odd*, $F(hkl)$ becomes:

$$F(hkl) = f_a \exp 2\pi i(0) + f_a \exp 2\pi i(n/2)$$

where n takes odd integer values, 1, 3 etc. The Gaussian plane diagram (Figure 6.13c) now shows that the vectors are opposed, and $F(hkl)$ is 0, irrespective of the value of n. Thus all planes with $h + k + l$ even will give rise to a reflection while all planes with $h + k + l$ odd will not.

All crystallographic unit cells derived from a body-centred lattice give rise to the same systematic absences. Similar considerations apply to the other Bravais lattices. The conditions that apply for diffraction *to occur* from (hkl) planes in the Bravais lattices, called **reflection conditions**, are listed in Table 6.4.

Apart from the Bravais lattice, other crystallographic features give rise to systematic absences. The symmetry elements that are responsible for systematic absences are (i), a centre of symmetry,

(ii), screw axes and (iii), glide planes. The systematic absences that occur on a diffraction pattern thus give detailed information about the symmetry elements present in a unit cell and the lattice type upon which it is based.

Systematic absences arise from symmetry considerations and always have $F(hkl)$ equal to zero. They are quite different from **structural absences**, which arise because the scattering factors of the atoms combine so as to give a value of $F(hkl) =$ zero for other reasons. For example, the (100) diffraction spots in NaCl and KCl are systematically absent, as the crystals adopt the *halite* structure, which is derived from an all-face centred (F) lattice, (see Table 6.4). On the other hand, the (111) reflection is present in NaCl, but is (virtually) absent in KCl for structural reasons – the atomic scattering factor for K^+ is virtually equal to that of Cl^-, as the number of electrons on both ions is 18.

The reciprocal lattice described in Chapter 2 consists of an array of points. Because a diffraction pattern is a direct representation of the reciprocal lattice, it is often useful to draw it as a **weighted reciprocal lattice**, in which the area allocated to each node is proportional to the structure factor $F(hkl)$ of each reflection, where

$$F(hkl) = \sum_{n=1}^{N} f_n \exp[2\pi i(hx_n + ky_n + lz_n)]$$

The weighted reciprocal lattice omits all reflections that are systematically absent, and so gives a clearer impression of the appearance of a diffraction pattern from a crystal (Figure 6.14).

6.10   The temperature factor

The most important correction term applied to intensity calculations is the **temperature factor**. The calculations described above assume that the atoms in a crystal are stationary. This is not so, and in molecular crystals and some inorganic crystals the vibrations can be considerable, even at room temperature. This vibration has a considerable effect upon the intensity of a diffracted beam, as can be seen from the following example. The vibration frequency of an atom in a crystal is often taken to be approximately 10^{13} Hz at room temperature. The frequency of an X-ray beam of wavelength 0.154 nm, (the wavelength of copper Kα radiation, which is

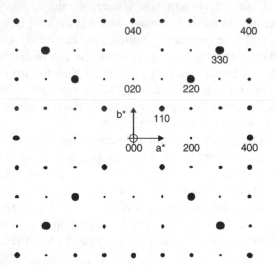

Figure 6.14   The $hk0$ section of the weighted reciprocal lattice of rutile, TiO_2. The lattice points are drawn as circles with radii proportional to the structure factor of the reflection

frequently used for crystallographic work), is 1.95×10^{18} Hz. Consequently, atoms nominally in any (hkl) plane will, in fact, be continually displaced out of the plane by varying amounts on the time scale of the radiation. This has the effect of smearing out the electron density of each atom, and diminishing each atomic scattering factor. For planes with a large interplanar spacing, d_{hkl}, which diffract at small angles, the displacements are only a fraction of d_{hkl}, and the effect is small. However, the atomic displacements in planes with a small interplanar spacing, d_{hkl}, may be equal or greater than d_{hkl} itself, causing considerable loss in diffracted intensity. It is for this reason that X-ray diffraction patterns from organic molecular crystals, which usually have low melting points and large thermal vibrations, are very weak at high Bragg angles, while high melting point inorganic crystals usually give sharp diffraction spots at high Bragg angles. Diffraction patterns from crystals in which atoms display large degrees of thermal vibration are obtained from crystals maintained at low temperatures, mounted on special cryogenic holders, to offset this problem. [The data for (S)-alanine given in Chapter 5, Section 5.7, were obtained from a crystal maintained at 23 K.]

Although chemical bonds link the atomic vibrations throughout the crystal, the thermal motion of any atom in a crystal is generally assumed to be independent of the vibration of the others. Under this approximation, the atomic scattering factor for a thermally vibrating atom, f_{th}, is given by:

$$f_{th} = f_a \exp\left[-B\left(\frac{\sin\vartheta}{\lambda}\right)^2\right] \quad (6.4)$$

Where f_a is the atomic scattering factor defined for stationary atoms, B is the **atomic temperature factor**, (also called the **Debye-Waller** factor or **B-factor**), θ is the Bragg angle and λ the wavelength of the radiation. The effect of the temperature factor is to reduce the value of the atomic scattering factor considerably at higher values of $\sin\theta/\lambda$, (Figure 6.15). The structure factor for a diffracted (hkl) beam is then:

$$F(hkl) = \sum_{n=1}^{N} f_n \exp[2\pi i(hx_n + ky_n + lz_n)]$$

$$\times \exp\left[-B\left(\frac{\sin\vartheta}{\lambda}\right)^2\right]$$

$$F(hkl)^2 = \left\{ \sum_{n=1}^{N} f_n \exp[2\pi i(hx_n + ky_n + lz_n)] \right.$$

$$\left. \times \exp\left[-B\left(\frac{\sin\vartheta}{\lambda}\right)^2\right] \right\}^2 = I(hkl)$$

and the corrected diffracted intensity, $I(hkl)$ is given by:

$$I(hkl) = I_0(hkl) \exp\left[-2B\left(\frac{\sin\vartheta}{\lambda}\right)^2\right]$$

The atomic temperature factor is related to the magnitude of the vibration of the atom concerned by the equation:

$$B = 8\pi^2 U$$

where U is the **isotropic temperature factor**, which is equal to the square of the mean displacement of the atom, $\langle \bar{r}^2 \rangle$, from the normal equilibrium position, r_0. Because crystallographers record the atomic dimensions in Å (10^{-10} m) B is quoted in units of Å^2. The value of B can

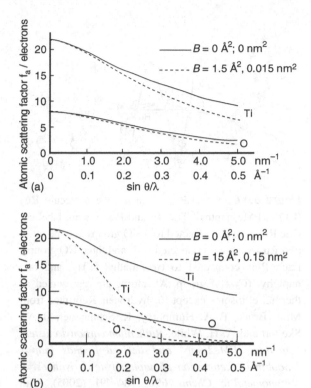

Figure 6.15 The effect of the temperature factor on the scattering power of titanium, Ti, and oxygen, O, atoms: (a) $B = 0$ and 0.015 nm^2 (1.5 Å^2); (b) $B = 0$ and 0.15 nm^2 (15 Å^2)

range from $1 - 100$ Å^2, ($0.1 - 10$ nm^2), although values of the order of $1 - 10$ Å^2 ($0.1 - 1$ nm^2), are normal. To use the SI unit of nm for the wavelength of the radiation in equation 6.4, the value of B in Å^2 must be divided by 100. In crystal structure drawings, the atom positions are often drawn as spheres with a radius proportional to B.

Although the value of U gives a good idea of the overall magnitude of the thermal vibrations of an atom, these are not usually isotropic (the same in all directions). To display this, the atom positions in a crystal structure are not indicated by spheres of a radius proportional to

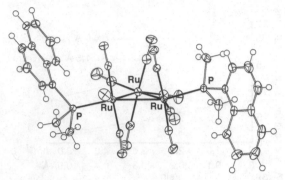

Figure 6.16 ORTEP diagram of the molecule Ru_3 $(CO)_{10}(PMe_2napth)_2$. The Ru and P atoms are labelled. One Ru atom is connected to 4 CO groups and the other two Ru atoms are connected to P and three CO groups. Each P is connected to two methyl (CH_3) and one naphthyl ($C_{10}H_7$) group. All atoms are represented as thermal ellipsoids except for hydrogen. Reprinted from M. I. Bruce, P. A. Humphrey, R. Schmutzler, B. W. Skelton and A. H. White, *Ruthenium carbonyl clusters containing PMe2(nap) and derived ligands (nap = 1-naphthyl): generation of naphthalyne derivatives. J. Organometallic Chem.*, **690**, 784–791 (2005), with permission from Elsevier

B, but by ellipsoids with axes proportional to the mean displacement of the atoms in three principal directions. The isotropic temperature factor, *U*, is replaced by the **anisotropic temperature factors** U_{11}, U_{22}, U_{33}, U_{12}, U_{13}, U_{23}. These define the size and orientation of the thermal ellipsoid with respect to the crystallographic axes. Diagrams of molecules in particular, are displayed in this way, using a representation of the structure called an ORTEP diagram, which is an acronym for Oak Ridge Thermal Ellipsoid Program. This pictorial presentation gives an idea of the three-dimensional molecular shape by drawing the bond distances and angles and representing the atoms by ellipsoids. Figure 6.16 shows an ORTEP diagram of the molecule $Ru_3(CO)_{10}$ $(PMe_2napth)_2$.

6.11   Powder X-ray diffraction

Although powder diffraction may not be the first choice for structure determination, powder X-ray diffraction is used routinely for the identification of solids, especially in mixtures, in a wide range of sciences from geology to forensics[3]. Two main sample geometries are used in X-ray powder diffraction experiments. In the first, a finely powdered sample of the material to be investigated is formed into a cylinder by introducing it into a hollow glass fibre, or by binding it with a gum. In the second, the powder is formed into a flat plate, by applying it to sticky tape or a similar adhesive. In either case, the powder is irradiated with a beam of X-rays.

Each crystallite will have its own reciprocal lattice, and if the crystals are randomly orientated each reciprocal lattice will be randomly orientated. Because of this, the overall reciprocal lattice appropriate to the powder will consist of a series of concentric spherical *hkl* shells rather than discrete spots. In such cases the diffraction pattern from a powder placed in the path of an X-ray beam gives rise to a series of cones rather than spots, (Figure 6.17a). The positions and intensities of the diffracted beams are recorded along a narrow strip (Figure 6.17b) to yield a characteristic pattern of diffracted 'lines' or peaks, (Figure 6.17c). The position of a line, (not the intensity), is found to depend only upon the spacing of the crystal planes involved in the diffraction and the wavelength of the X-rays used, via the Bragg law. The angle between the transmitted or 'straight through' beam and the

[3]In the very first Sherlock Holmes story (A Study in Scarlet, first published in *Beeton's Christmas Annual*, 1887), emphasis is placed upon Holmes' ability to tell different soil types 'at a glance', and this important ability features in a number of stories. Of course, Holmes had to make do with optical means to characterise the differing soils. The routine use of powder X-ray diffraction to quantify different soil types has made this an vastly more powerful tool, which is widely in use today, both geologically and in forensic science.

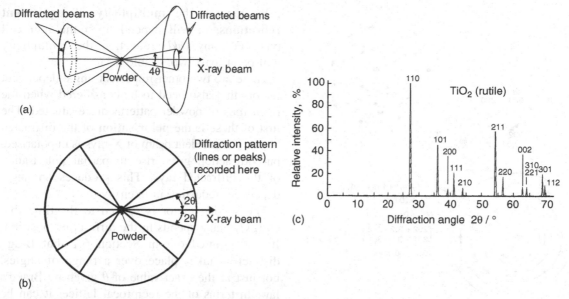

Figure 6.17 Powder X-ray diffraction: (a) a beam of X-rays incident upon a powder is diffracted into a series of cones; (b) the diffracted beams are recorded along a circle, to give the diffraction pattern; (c) the diffraction pattern from powdered rutile, TiO_2

diffracted beam is then 2θ, so the angle at the apex of a cone of diffracted rays is 4θ, where θ is the Bragg angle. In essence, the positions of the lines on a powder diffraction pattern simply depend upon the unit cell dimensions of the phase.

The intensities of the diffraction lines depend upon the factors already mentioned, but further corrections must be included if the calculated intensities are to correspond accurately with the observed intensities. An important factor is the **multiplicity** of the reflections in the powder pattern, p. This correction term can be easily understood by deriving the powder pattern from the diffraction 'spot' pattern of a single crystal (Figure 6.18a). The intensity of each spot is well defined, and the intensity of say, the 200 spot, $I(200)$, can be measured independently of all other reflections. If an X ray beam is incident upon a random array of crystals, instead of a single crystal, each diffraction spot will lie at a

random angle. With a small number of crystals, a 'spotty ring pattern' will form. Figure 6.18b shows the pattern generated from the single crystal pattern in Figure 6.18a if there are four other identical crystals present, rotated by (i) 25° (ii) 39° (iii) 43° and (iv) 55°. In principle it is still possible to measure unique intensities, such as $I(200)$, but it is becoming difficult. If large numbers of crystallites are present a 'ring pattern' will form, (Figure 6.18c). It is now impossible to measure a value for $I(200)$. The ring arising from a (200) reflection will consist of the superposition of both (200) and $(\bar{2}00)$ intensities. The best that can now be obtained is $I(200)$ plus $I(\bar{2}00)$, and the intensity is double that expected from a single (200) diffracted beam. Hence the intensities measured on a powder pattern are greater than that of a single reflection by a reflection multiplicity, p. For the (200) reflection described, the multiplicity is 2. The multiplicities of the other rings are noted on Figure 6.18c.

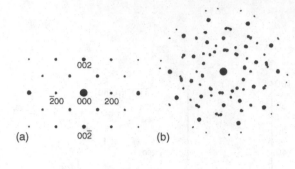

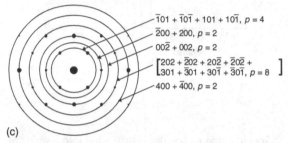

(c)

Figure 6.18 Powder diffraction patterns: (a) a single crystal diffraction pattern; (b) a 'spotty ring pattern' created by five single crystals identical to that in (a) rotated from the original by 0°, 25°, 39°, 43° and 32°. (c). The ring pattern generated by a large number of randomly oriented crystallites, superimposed upon the single crystal diffraction pattern. The intensity of each ring is the sum of the intensities of more than one single crystal spot. The multiplicities of the equivalent reflections that make up the rings are shown as p

In general, the **multiplicity of equivalent reflections**, p, will depend upon the unit cell type. For any (hkl) reflection the multiplicity will be at least two, (Table 6.5).

There are two other important angle dependent factors that also need to be considered when the intensities of powder patterns are evaluated. The first of these is the **polarisation** of the diffracted beam. The incident beam of X-rays is unpolarised, but diffraction gives rise to partial polarisation of the diffracted rays. This produces an angle dependent reduction in intensity.

The second factor is related to the time that an (hkl) plane spends in the diffracting position. It was pointed out in Section 6.1 that Bragg diffraction takes place over a range of angles, not just at the exact value of θ given by Bragg's law. In terms of the reciprocal lattice, it can be likened to each reciprocal lattice point having a volume, (Figure 6.6). The time that each diffraction spot spends near enough to the Ewald sphere to give rise to a diffracted beam is then found to be angle dependent. This is termed the **Lorentz factor**.

The Lorentz and polarisation factors are usually combined, for powder X-ray diffraction, into a single correction term:

$$(1 + \cos^2 2\theta)/(\sin^2 \theta \cos \theta)$$

Table 6.5 Multiplicity of equivalent reflections for powder diffraction patterns

Crystal Class	Diffracting plane and multiplicity[*]						
Triclinic	all, 2						
Monoclinic	$0k0$, 2	$h0l$, 2	hkl, 4				
Orthorhombic	$h00$, 2	$0k0$, 2	$00l$, 2	$hk0$, 4	$0kl$, 4	$h0l$, 4	hkl, 8
Tetragonal	$00l$, 2	$h00$, 4	$hh0$, 4	$hk0$, 8[*]	$0kl$, 8	hhl, 8	hkl, 16[*]
Trigonal	$00l$, 2	$h00$, 6	$hh0$, 6	$hk0$, 12[*]	$0kl$, 12[*]	hhl, 12[*]	hkl, 24[*]
Hexagonal	$00l$, 2	$h00$, 6	$hh0$, 6	$hk0$, 12[*]	$0kl$, 12[*]	hhl, 12[*]	hkl, 24[*]
Cubic	$h00$, 6	$hh0$, 12	$hk0$, 24[*]	hhh, 8	hhl, 24	hkl, 48[*]	

[*]In some point groups the multiplicity is made up from contributions of two different sets of diffracted beams at the same angle but with different intensities, (see, i.e. Figure 6.18c)

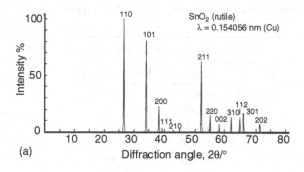

(a)

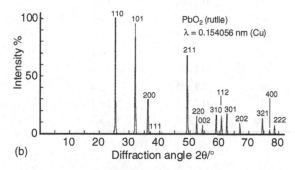

(b)

Figure 6.19 Powder diffraction patterns of materials with the *rutile* structure: (a) tin dioxide, cassiterite, SnO_2; (b) lead dioxide, PbO_2

The intensity of a powder diffraction ring is then written as:

$$I(hkl) = I_0(hkl) \exp\left[-2B\left(\frac{\sin\vartheta}{\lambda}\right)^2\right]\left(\frac{1+\cos^2 2\vartheta}{\sin^2\vartheta\cos\vartheta}\right)\cdot p$$

Another correction factor arises because the X-ray beam is absorbed as it passes through the sample. The effect of this is to reduce the intensities of reflections that pass through large volumes of sample compared to those that pass through small volumes. Correction factors will depend upon sample shape and size, and a general formula applicable to all sample geometries can

not be given. However, this term must be included in careful work.

In essence, a powder diffractogram contains as much information as a single crystal experiment. When the intensity and the positions of the diffraction pattern are taken into account, the pattern is unique for a single substance. The X-ray diffraction pattern of a substance can be likened to a fingerprint. In effect, the pattern of lines on the powder diffraction pattern of a single phase is virtually unique, and mixtures of different crystals can be analysed if a reference set of patterns is consulted.

To illustrate this, the powder diffraction patterns of two materials with closely similar structures and unit cells are shown in Figure 6.19. These are the *rutile* form of PbO_2, with a tetragonal unit cell, $a = 0.4946$ nm, $c = 0.3379$ nm and SnO_2, which also adopts the *rutile* structure with $a = 0.4737$ nm, $c = 0.3186$ nm. The data for these structures is given in Table 6.6. It is seen that although the two powder patterns show a family similarity they are readily differentiated from each other and also from rutile (TiO_2) itself (Figure 6.17c). A mixture of the two phases also gives a pattern that is readily interpreted, and if the intensities of the two sets of lines from the two phases are compared a quantitative assessment of the relative amounts of the two materials present can be made.

6.12 Electron microscopy and structure images

It is apparent that, given a crystal structure, it is quite an easy matter to compute both the positions, the intensities and the phases of the diffracted beams that arise when the crystal is irradiated. The main task that faces the crystallographer, however, is to derive an unknown crystal structure from experimentally collected diffraction data. To understand the physical

Table 6.6 Powder diffraction data for the rutile forms of SnO_2 and PbO_2, for copper radiation, $\lambda = 0.1540562$ nm

SnO₂ a = 0.4737 nm, c = 0.3186 nm			PbO₂ a = 0.4946 nm, c = 0.3379 nm		
hkl	*d*/nm	Relative intensity	*hkl*	*d*/nm	Relative intensity
110	0.3350	100	110	0.3497	100
101	0.2644	81	101	0.2790	95.5
200	0.2368	22.4	200	0.2473	29.5
111	0.2309	3.8	111	0.2430	1.5
210	0.2118	1.3	211	0.1851	67.4
211	0.1764	61.7	220	0.1749	15.1
220	0.1675	14.6	002	0.1689	7.4
002	0.1593	7.0	310	0.1564	16.1
310	0.1498	13.2	112	0.1521	16.1
112	0.1439	13.2	301	0.1482	17.4
301	0.1415	16.5	202	0.1395	9.4
202	0.1322	6.3	321	0.1271	13.7
321	0.1215	9.7	400	0.1237	3.8

process involved, it is helpful to initially consider electron microscopy[4].

Electrons are charged, and, as a consequence, they interact much more strongly with the outer electron cloud of atoms than X-rays. Because of this, electron scattering is far more intense than X-ray scattering, and an electron beam accelerated by 100 kV is only able to penetrate about 10 nm into a crystal before being either absorbed or entirely diffracted into other directions. Thus, although the condition for the diffraction of electrons is given by Bragg's law, a more complex theory is needed to calculate the intensity of the diffracted beams. In the case of X-rays it is reasonable to assume

that each diffracted X-ray photon is scattered only once, and the kinematical theory of diffraction is adequate for X-ray diffraction, (see Section 6.4). Electrons are generally diffracted frequently on passing through a crystal, and each electron is scattered many times, even when traversing a slice of crystal only 1 or 2 nanometres in thickness. This implies that the intensity of the incident beam not only falls as electrons are diffracted, but also increases as electrons are re-diffracted back into the incident beam direction. The theory that is needed to account for this is called the **dynamical theory** of diffraction. The calculation of the diffraction pattern intensities is therefore not as easy as in the case of X-rays. However, the major advantage of electrons, which quite offsets this disadvantage, is that, because they are charged, they can be focussed by magnetic lenses, to yield an image of the structure. This reciprocal relationship between the formation of the diffraction pattern and the image is at the heart of structure determination, irrespec-

---

[4]There are a number of different types of electron microscope, but for the present purposes they can be divided into two categories, scanning and transmission. Scanning electron microscopy, broadly speaking, is used to reveal surface topography, while transmission electron microscopy is of more relevance to crystal structure determination. This is the only type of electron microscopy considered here.

tive of the technique used. It is displayed most clearly, however, in high resolution electron microscopy.

The steps by which an electron (or any) microscope forms an image is as follows. An objective lens, which sits close to the sample, takes each of the separate sets of beams that have been diffracted by the crystal and focuses them to a spot in the back focal plane of the lens, (Figure 6.20). This set of spots makes up the diffraction pattern. As described above, the electron diffraction pattern is a good approximation to a plane section through the weighted reciprocal lattice of the material. If desired, lenses below the objective lens, called intermediate and projector lenses, are then focussed so as to form a magnified image of the diffraction pattern on the viewing screen, (Figure 6.21), which can then be recorded. [Earlier in this Chapter, an electron diffraction pattern was simply described as a projection of the reciprocal lattice onto an observing screen, (see Figure 6.4). This does not introduce errors of interpretation, but omits the complexity of the instrument.]

However, if the diffracted beams are allowed to continue, they create an image of the object in the image plane of the objective lens, some way beyond the diffraction pattern. This image comes about by the *recombination* of the diffracted beams. The focal length of the intermediate and projector lenses can be altered to focus either upon the diffraction pattern (Figure 6.22a) or the image (Figure 6.22b). Thus either can be viewed at high magnification. More than this, because the focal lengths can be varied continuously, the various intensity patterns that form on the viewing screen between, on the one hand, the diffraction pattern, and on the other the image, can be observed. The way in which the diffracted beams recombine to form an image is therefore lucidly demonstrated.

Thus it is apparent that the image of the structure is formed by the recombination of the beams in the diffraction pattern. Now the detail in the structure will depend critically upon how many diffracted beams contribute to the image.

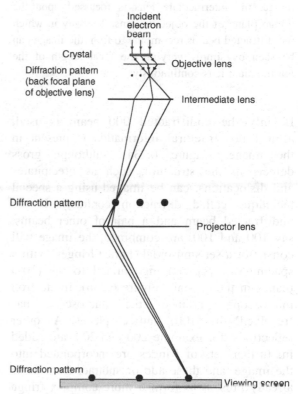

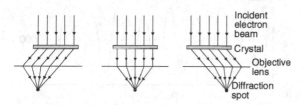

Figure 6.20 The formation of an electron diffraction pattern in a transmission electron microscope. All the diffracted beams come to a focus in the back focal plane of the objective lens, to form the diffraction pattern. [Only three diffracted beams are drawn for clarity]

Figure 6.21 The formation of a magnified diffraction pattern in an electron microscope, (schematic)

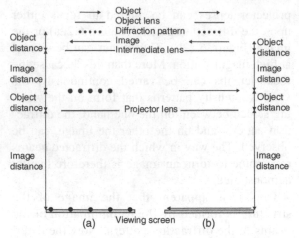

Figure 6.22 The formation of images in an electron microscope (schematic): (a) to view a diffraction pattern, the intermediate lens is focussed upon the back focal plane of the objective lens; (b) to view an image, the intermediate lens is focussed upon the image plane of the objective lens. The way in which the diffracted beams recombine to form the image can be seen by simply varying the focal length of the intermediate lens continuously

If only the undiffracted 000 beam is used, almost no structural information is present in the image, (Figure 6.23a), (although gross defects in the structure, such as precipitates and dislocations, can be imaged using a special technique called diffraction contrast). If the undiffracted beam and a pair of other beams, say 100 and $\bar{1}00$, are combined, the image will consist of a set sinusoidal 'lattice fringes' with a spacing of d_{100}, running parallel to the (100) planes in the crystal. (Figure 6.23b). In electron microscope parlance, the microscope has 'resolved' the (100) 'lattice' planes. As other reflections, for example 200 and $\bar{2}00$, are added in, further sets of fringes are incorporated into the image, and these add or subtract from those already present to form a more complex fringe pattern, but still parallel to the original (100) set, and with a spacing of d_{200}, so that the micro-

scope has 'resolved' the (200) 'lattice' planes, (Figure 6.23c). (The fringe profiles in Figure 6.23c are idealised. In reality a more complex wave profile is found, dependent upon the intensities of the 100 and 200 beams, but the overall fringe repeat remains equal to d_{200}).

As more and more pairs of $h00$ and $\bar{h}00$ beams are added in, the fringe profile becomes more and more complex, but the result can always be viewed in simple terms as set of fringes parallel to (100), with spacing equal to d_{h00}. Should this degree of detail be imaged by the instrument, these planes, 300, 400 and higher, have been 'resolved'. If reflections corresponding to $0k0$ diffraction spots are considered, the fringes in the image will run parallel to the (010) planes in

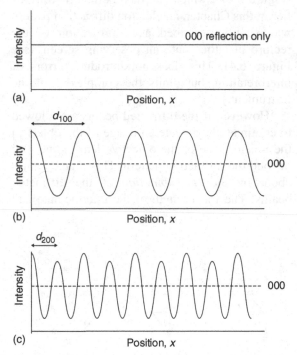

Figure 6.23 The formation of an image in an electron microscope: (a) 000 beam only, no contrast; (b) 000, 100 and $\bar{1}00$ beams, giving 100 'lattice fringes'; (c) 000, 100, $\bar{1}00$, 200 and $\bar{2}00$ beams, giving 200 'lattice fringes'

the crystal. When these are added to the fringes representing ($h00$) planes, a more complex contrast again results, a sort of 'tweed' or cross-hatched effect, which is a much better representation of the structure than the one-dimensional fringes. More detail is added by including fringes at an angle, such as the members of the $hk0$ set. The same is true for all series of hkl and $\bar{h}kl$ reflections. Each extra pair of reflections produces a set of sinusoidal fringes, of spacing, d_{hkl} running parallel to the (hkl) set of planes in the crystal, which are added into the resultant image contrast in proportion to the intensity of the reflections. With each step, the image contrast resembles the object more and more.

There is another significant feature of these images. Because electrons interact with the electron clouds around the atomic nuclei, the electron microscope image contrast is, in fact, a map of the electron density in the crystal, projected in a direction parallel to the electron beam. However, as greatest electron density usually occurs close to atomic nuclei, the image can also be interpreted in terms of projected atoms without significant loss of precision.

There are two important points to note. Firstly, the amount of information in the image, its **resolution**, is dependent upon the *number* of diffracted beams that are included in the image formation process. Secondly, **all of the information in each diffracted beam**, (including spurious information introduced by lens defects), contributes to image formation. Thus an electron micrograph contains information about non-periodic structures, such as defects, that may be present. As electron microscopes are able to image at atomic resolution, the **defect structure** of complex materials can be understood in a degree of detail not available to X-ray diffraction. For example, Figure 6.24a is a micrograph of a flake of niobium pentoxide, H-Nb_2O_5. The structure is composed of columns of corner-linked NbO_6 octahedra, which in projection, look like blocks or tiles. The blocks are of two

sizes, approximately 1×1.4 nm and 1×1.8 nm. The regular way in which the blocks tile the surface is clearly revealed on the right of Figure 6.24a, but careful scrutiny of the upper left side

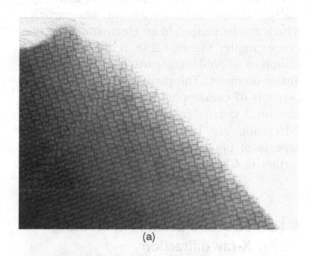

(a)

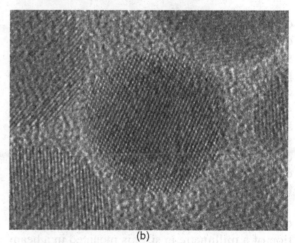

(b)

Figure 6.24 Electron micrographs: (a) a crystal of H–Nb_2O_5. In projection the structure is built of interlocking rectangular 'tiles'. Irregular tile sizes are visible; (b) A cadmium sulphide particle of approximately 8 nm diameter. The particle is revealed by the two sets of overlapping 'lattice' fringes. The uneven background contrast arises from the amorphous carbon film used to support the particles. Courtesy Dr. J. H. Warner and Dr. R. D. Tilley, Victoria University of Wellington, New Zealand

reveals the presence of blocks of different sizes, and a lamella of structure in which the blocks, although of normal sizes, are arranged differently. These variants would not be revealed by X-ray diffraction. The technique is also especially valuable in the study of nanoparticles, which can be imaged in an electron microscope. For example, Figure 6.24b is the image of a cadmium sulphide nanoparticle approximately 8 nm in diameter. The particle is revealed by the two sets of overlapping 'lattice' fringes. Microstructural details for such particles via X-ray diffraction are impossible to obtain. Related aspects of crystalline structures are considered further in Chapter 8.

6.13   Structure determination using X-ray diffraction

Crystal structures are most often determined by diffraction of X-rays from single crystals of the sample, although structure determination from polycrystalline powders is also important, especially via neutron diffraction, as described below. As noted in the previous section, in order to generate a structure, (i.e. an image), it is necessary to recombine the beams that make up the diffraction pattern. Unfortunately, this cannot be carried out by using lenses, and the process must be done mathematically.

The technique is simple in principle. A small crystal of the material, of the order of a fraction of a millimetre in size, is mounted in a beam of X-rays. Each plane of atoms in the crystal gives rise to a narrow diffracted beam. The position of each beam is recorded, along with the intensity of each spot. Each member of the resulting data set comprises a position, intensity and hkl index.

The problem is then to mathematically convert the data set into an electron density map. The way in which this is done is similar to that described for the formation of an electron microscope image. The contributions from all the sets of hkl reflections are added together. Consider a one-dimensional crystal, composed of a line of atoms of various types. The contribution of the undiffracted beam is $F(000)$, the structure factor for the 000 reflection. To this is added (mathematically) the contributions of, say, the 100 and $\bar{1}00$ reflections, $F(100)$ and $F(\bar{1}00)$ to give a low resolution 'image'. More pairs of reflections are then added in order to improve the resolution and obtain a more realistic 'image' of the atom chain. The image contrast can be represented by:

$$\text{contrast} = \mathbf{F}(000) + \mathbf{F}(100) + \mathbf{F}(\bar{1}00) + \ldots$$

Unfortunately values of $\mathbf{F}(hkl)$ are not available in the X-ray data set. Recall that

$$\mathbf{F}(hkl) = F(hkl)e^{i\phi_{hkl}}$$

The value $F(hkl)$ is easily obtained as it is equal to the square root of the measured intensity. However, the phase angle, ϕ_{hkl}, for the reflection cannot be recovered from the intensity. This fact, which has been the continued bane of structural crystallographers, is known as the **phase problem**. The equation for the contrast must be rewritten in terms of the known experimentally determined $F(h00)$ values and the unknown phases, ϕ_{100}, ϕ_{200}, ϕ_{300} etc. The equation for a one-dimensional chain is:

$$\begin{aligned}\text{contrast} &= F(000) + F(100)\cos 2\pi(x - \phi_{100}) \\ &+ F(200)\cos 2\pi(2x - \phi_{200}) \\ &+ F(300)\cos 2\pi(3x - \phi_{300}) + \ldots \\ &= F_{000} + \sum_{h=-\infty}^{+\infty} F_h \cos 2\pi(hx - \phi_h)\end{aligned}$$

where the summation is taken over all of the $h00$ and $\bar{h}00$ reflections. The units of the various scattering factors, $F(h00)$, are electrons. The electron density of the crystal in the x direction, given in electrons/unit distance, is then given by $\rho(x)$, where:

$$\rho(x) = (1/d_{100})\left[F_{000} + \sum_{h=-\infty}^{+\infty} F_h \cos 2\pi(hx - \phi_h)\right]$$

The one-dimensional equation can be generalised to three dimensions and the electron density of a crystal at any point in the unit cell x, y, z, is given by:

$$\rho(x, y, z) = \frac{1}{V} \sum_{h=-\infty}^{+\infty} \sum_{k=-\infty}^{+\infty} \sum_{l=-\infty}^{+\infty} F(hkl)$$
$$\times \exp[-2\pi i(hx + ky + lz - \phi_{hkl})]$$

where V is the volume of the unit cell, the indices h, k and l run from $-\infty$ to $+\infty$. Because values of $F(hkl)$ are available experimentally, the problem in computing the electron density or, equivalently, determining the crystal structure, reduces to how to obtain the phase, $\phi_{(hkl)}$ for each reflection, viz. how to solve the phase problem.

Clearly the vast numbers of structures that have been solved is a testament to the fact that the phase problem has been solved. A brief introduction to some techniques that achieve this, especially those of relevance to protein crystallography, is given in Section 6.16.

The procedure of structure determination from a single crystal X-ray diffraction experiment can be summarised in the steps below.

(i) Obtain an accurate set of intensity values. Obviously, this is essential as everything else depends upon this raw data. The cor-

rection factors mentioned above, as well instrumental factors associated with the geometry of the diffractometer, must be taken into account. [In the early days of X-ray crystallography this in itself was no easy task.]

(ii) Determine the unit cell and index the diffracted reflections in terms of hkl. Determine the point group and space group of the crystal, making use of systematic absences.

(iii) Construct a possible model of the crystal, using physical or chemical intuition, or established techniques for solving the phase problem (see Section 6.16).

(iv) Compare the intensities, or, more usually, the structure factors, expected of the model structure with those obtained experimentally.

(v) Adjust the atom positions in the model repeatedly to obtain improved agreement between the observed and calculated values. This process is called **refinement**. The crystal structure is generally regarded as satisfactory when a low value of the **reliability factor**, **residual**, or **crystallographic R factor**, is obtained. There are a number of ways of assessing the R factor, each of which gives a different number, and often more than one value is cited. The most usual expression used is:

$$R = \frac{\sum_{hkl} ||F_{obs}| - |F_{calc}||}{\sum_{hkl} |F_{obs}|}$$

where F_{obs} and F_{calc} are the observed and calculated structure factors, and the modulus values, written $|F_{obs}|$, etc, means that the absolute value of the structure factor is taken, and negative signs are ignored. Structure determinations for well crystallised inorganic or metallic compounds have R values of the order of 0.03. The values of R for organic compounds or complex molecules such as proteins are generally higher.

When a powder is examined, many diffracted beams overlap, (see Section 6.11), so that the procedure of structure determination is more difficult. In particular this makes space group determination less straightforward. Nevertheless, powder diffraction data is now used routinely to determine the structures of new materials. An important technique used to solve structures from powder diffraction data is that of **Rietveld refinement**. In this method, the exact shape of each diffraction line, called the **profile**, is calculated and matched with the experimental data. Difficulties arise not only because of overlapping reflections, but also because instrumental factors add significantly to the profile of a diffracted beam. Nevertheless, Rietveld refinement of powder diffraction patterns is routinely used to determine the structures of materials that cannot readily be prepared in a form suitable for single crystal X-ray study.

6.14 Neutron diffraction

Neutrons, like X-rays, are not charged, but unlike X-rays, do not interact significantly with the electron cloud around an atom, only with the massive nucleus. Because of this, neutrons penetrate considerable distances into a solid. Neutrons are diffracted following Bragg's law, and the intensities of diffracted beams can be calculated in a similar way to that used for X-ray beams. Naturally, neutron atomic scattering factors, which differ considerably from X-ray atomic scattering factors, need to be used. The technique is quite different from X-ray diffraction in practice, not least because neutrons need to be generated in a nuclear reactor and need to be slowed to an energy that is suitable for diffraction experiments. Despite this difficulty,

neutron diffraction has a number of significant strong points, and is used routinely for structure determination. Neutron diffraction is usually carried out on powder samples, and structure refinements are carried out via the Rietveld method, often in conjunction with X-ray data for the same material.

An advantage of neutron diffraction is that it is often able to distinguish between atoms in a crystal that are difficult to separate with X-rays. This is because X-ray scattering factors are a function of the atomic number of the elements, but this is not true for neutrons. Of particular importance is the fact that the neutron scattering factors of light atoms, such as hydrogen, H, carbon, C, nitrogen, N, and oxygen, O, are similar to, or even greater than, those of transition metals and heavy metal atoms. This makes it easier to determine their positions in the crystal, compared to X-ray diffraction and neutron diffraction is the method of choice for the precision determination of the positions of light atoms in a structure.

Another advantage of neutron diffraction is that neutrons have a spin and so interact with the magnetic structure of the solid, which arises from the alignment of unpaired electron spins in the structure. The magnetic ordering in a crystal is quite invisible to X-rays, but is revealed in neutron diffraction as extra diffracted beams and hence a change in unit cell.

6.15 Protein crystallography

In principle, protein crystallography, or indeed, the crystallography of any large organic molecule, is identical to that of the crystallography of inorganic solids. However, structural studies of proteins and other large organic molecules are difficult and success has hinged upon the development of a number of important crystallographic techniques. In addition, proteins are involved in all aspects of life, and this gives the study of protein structures a special place in crystallography.

The most significant difference between proteins and the other structures considered earlier is the enormous increase in the complexity between the two. For example, one of the smaller proteins, insulin, has a molar mass of approximately $5,700 \, g \, mol^{-1}$, and a run of the mill (in size terms) protein, haemoglobin, has a molar mass of approximately $64,500 \, g \, mol^{-1}$. When it is remembered that proteins are mostly made up of light atoms, H, C, N and O, the atomic complexity can be appreciated. This complexity is reflected in the difference between the rate of development of inorganic crystallography compared with protein crystallography. The first inorganic crystal structures were solved by W. H. and W. L. Bragg in the early years of the 20th century, but it was not until the middle of the same century, a lapse of 50 years or so, before the structures of macromolecules and proteins began to be resolved.

Proteins are polymers of amino acids. Amino acids, typified by alanine, (see Section 5.15), have an acidic –COOH group and a basic –NH$_2$ group on the same molecule. The naturally occurring amino acids are often called α-amino acids, this meaning that the –NH$_2$ group is bonded to the carbon atom, (the α carbon), next to the –COOH group. Naturally occurring proteins are built from 20 natural α-amino acids, and in all but one, glycine, (NH$_2$CH$_2$COOH) the α carbon atom is a chiral centre, so that the molecules exist as enantiomers (see chapter 4). However, only one enantiomer is used to construct proteins in cells, a configuration often called the L-form. Thus the naturally occurring form of the α-amino acid alanine is often called L-alanine, (see Section 5.15).

Amino acids are linked together to form polymers via the reaction of the acidic –COOH group

on one molecule with the basic –NH$_2$ group on another. The resulting link is called a **peptide bond**, (Scheme 6.1). If there are fewer than about 50 amino acids in the chain, the molecules are referred to as **polypeptides**, while if there are more than this number, they are usually called **proteins**. For convenience, naturally occurring proteins are often divided into two categories. **Fibrous proteins**, are those in which the polymer chains are arranged so as to form long fibres. These are common constituents of muscle, tendons, hair and silk, and are both strong and flexible. **Globular proteins** have the polymer chains coiled into a compact, roughly spherical shape. These are vital molecules in the management of life processes, and include most enzymes and hormones, including such important molecules as haemoglobin, that controls oxygen transport in the blood, and insulin, which mediates glucose metabolism.

The first step in the determination of the structure of a protein is to discover the sequence of amino acids present along the protein chain. This is known as the **primary structure** of the molecule, (Scheme 6.2). The primary structure can be determined by chemical methods, but gives little information upon the structure adopted by the molecule in living organisms. The task of protein crystallographers is to determine the three dimensional arrangement of the primary structure.

For a satisfactory crystallographic study, it is vital to start with a high quality crystal. This is not easy for proteins. Crystallisation of such large molecular complexes frequently leads to trapped solvent, and a reasonable protein crystal may still contain between 30–80 % solvent. Because the

H$_2$N - R - COOH + H$_2$N - R' - COOH $\rightarrow$ H$_2$N - R - CO - NH - R' - COOH + H$_2$O

Amino acid 1   Amino acid 2   Peptide bond

R and R' represent the rest of the amino acid molecules.

Scheme 6.1 The formation of a peptide bond

(a) E - K - F - D - K - F - L - T - A - M - K

(b) Glu - Lys - Phe - Asp - Lys - Phe - Leu - Thr - Ala - Met - Lys -

(c) Glutamic acid - Lysine - Phenylalanine - Aspartic acid - Lysine
    - Phenylalanine - Leucine - Threonine - Alanine - Methionine - Lysine -

Scheme 6.2   A primary sequence of amino acids in a protein: (a) single-letter abbreviations for the amino acids; (b) three-letter abbreviations for the amino acids; (c) full names

solvent is disordered, it is not registered as such by diffraction data, but large quantities of solvent do degrade the quality of the data and limit the resolution of the final structure.

At the outset, crystallographers determine the unit cell dimensions of the protein and its density. The density can yield a value for the molar mass of the protein, provided that the included solvent is taken into account. The space group of the crystal is also important. Protein molecules are enantiomorphous (see Chapter 4), and as a consequence, will only crystallise in space groups with no inversion centres or mirror planes. There are just 65 possibilities from the 230 space groups.

The major experimental difficulty in protein crystallography is the solution of the phase problem (Section 6.13). There are a number of ways in which this can be tackled; the first to be perfected, and the method which is still a cornerstone of protein structural work, is that of isomorphous replacement. This, and other techniques, is described in the following section.

6.16   Solving the phase problem

Initially, all crystal structures were solved by constructing a suitable model making use of any physical properties that reflect crystal symmetry and chemical intuition regarding formulae and bonding, computing F_{hkl} values and comparing them with observed F_{hkl} values. This 'trial and error' method was used to solve the structures of fairly simple crystals, such as the metals and minerals described in Chapter 1. (Many of the

structural relations described in the following chapter, Chapter 7, originated in attempts to arrive at good starting structures for subsequent X-ray analysis.) However, the method is very labour intensive, and to a large extent success depends upon the intuition (or luck) of the researcher.

One of the first mathematical tools to come to the aid of the crystallographer was the development of the **Patterson function**, described in 1934 and 1935. The Patterson function is:

$$P(u,v,w) = \frac{1}{V} \sum_h \sum_k \sum_l F(hkl)^2 \cos 2\pi (hu + kv + lw)$$

where the indices h, k and l run from $-\infty$ to $+\infty$. The Patterson function does not need phases, and the squares of the structure factors are available experimentally as the intensities of the hkl reflections. The Patterson function, when plotted on (u, v, w) axes, forms a map rather like the display of electron density. However, peaks in the Patterson map correspond to interatomic vectors and the peak heights are proportional to the product of the scattering factors of the two atoms at the vector head and tail. That is, suppose that there are atoms at A, B and C in a unit cell. The function $P(uvw)$ would show peaks at U, V, W, where $O - U$ is equal to $A - B$, $O - V$ is equal to $B - C$ and $O - W$ is equal to $A - C$, (Figure 6.25). The vectors can be related to the atom positions in the unit cell by a number of methods, including direct methods, described in the following paragraphs. The main drawback

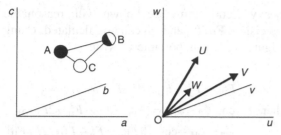

Figure 6.25 The Patterson function: (a) a molecule, containing atoms A, B, C, in a unit cell; (b) the corresponding Patterson function contains vectors OU, OV and OW which corresponds to the interatomic vectors A – B, B – C, A – C

of the Patterson function is that large numbers of similar atoms give overlapping peaks that cannot be resolved. It is of greatest use in locating a small number of heavy atoms in a unit cell. In this way, the location of the positions of Co atoms in vitamin B_{12}, lead to the successful resolution of the complete structure by 1957. This technique is also of importance in the multiple isomorphous replacement (MIR) method for the solution of protein structures described below.

Direct methods replaced trial and error and other methods of deducing model structures in the middle of the 20th century. In these techniques, statistical relationships between the amplitudes and phases of the strong reflections were established, and the mathematical methodology between these quantities was worked out, particularly by Hauptmann and Karle, in the 1940's and 1950's. (The Nobel Prize was awarded to these scientists in 1985 for these studies.) A number of algorithms, which exploited the growing power of electronic computers, used this mathematical framework to derive structures directly from the experimental data set of position, intensity and *hkl* index. The use of these programs allows the structures of molecular compounds with up to 100 or so atoms to be

derived routinely, but as more atoms are added the computations become increasingly lengthy and subject to error. The advent of increased computing power has lead to an extension of direct methods to molecules with up to about 500 atoms.

There are a number of computing approaches used in direct method techniques, one being the **'Bake and Shake'** method (see Bibliography). In this procedure, a model structure is derived, (the bake part). The phases are calculated and then perturburbed, that is, changed slightly, (the shake part), so as to arrive at a lower value, in accordance with the formulae implemented in the computer algorithms. When this is completed, a new set of atomic positions are calculated (a new bake) and the cycle is repeated for as long as necessary.

Although direct methods are being extended to larger and larger molecules, the structures of large proteins, with more than say 600 atoms, are still inaccessible to this technique. Such structures are mainly determined using the following two methods.

The first protein structures were derived using a technique called **isomorphous replacement** (IR), developed in the late 1950's. The materials used are heavy metal derivatives of protein crystals. To obtain a heavy metal derivative of a protein, the protein crystal is soaked in a solution of a heavy metal salt. The metals most used are Pt, Hg, U, lanthanides, Au, Pb, Ag and Ir. The heavy metal or a small molecule containing the heavy metal, depending upon the conditions used, diffuses into the crystal via channels created by the disordered solvent present. The aim is for the heavy metal to interact with some surface atoms on the protein, without altering the protein structure. This is never exactly achieved, but in suitable cases, the changes in structure are slight.

The experimental situation is now that two sets of data are available, F_P, a structure factor magnitude list for the native protein, and F_{HP}, a structure factor magnitude list for the isomorphous replacement protein. The vector relationships between the three structure factors, $\mathbf{F_P}$, $\mathbf{F_{PH}}$ and that of the heavy metal atoms alone, $\mathbf{F_H}$ is:

$$\mathbf{F_{PH}} = \mathbf{F_P} + \mathbf{F_H} \qquad (6.5)$$

The connection between these three pieces of data can be understood if the scattering factors are displayed in vector format, as described in Section 6.6. The result is shown in Figure 6.26, where $\mathbf{F_P}$ is the vector scattering factor of any *hkl* reflection from the protein, with magnitude F_P and phase angle ϕ_P, $\mathbf{F_H}$ is the vector scattering factor for the heavy atoms alone, with magnitude F_H and phase angle ϕ_H, and $\mathbf{F_{PH}}$ is the vector scattering factor of the heavy atom derivative of the protein, with magnitude F_{PH} and phase angle ϕ_{PH}.

The positions of the heavy metal atoms alone can be determined using techniques such as the Patterson function. Once the positions of the

heavy metal atoms are known with reasonable precision, $\mathbf{F_H}$, F_H and ϕ_H can be calculated. Using Figure 6.26, it is possible to write:

$$F_{PH}^2 = F_P^2 + F_H^2 + 2F_P F_H \cos(\phi_P - \phi_H)$$
$$\text{hence}: \ \cos(\phi_P - \phi_H) = (F_{PH}^2 - F_P^2 - F_H^2)/2F_P F_H$$
$$\phi_P = \phi_H + \cos^{-1}[(F_{PH}^2 - F_P^2 - F_H^2)/2F_P F_H]$$
$$(6.6)$$

Now this cannot be solved for ϕ_P unambiguously, as the $\cos^{-1}$ term has two roots (solutions), as displayed graphically in Figure 6.27.

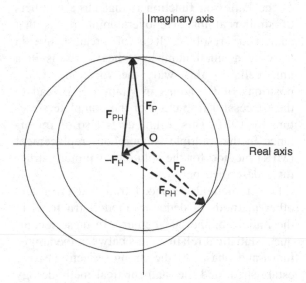

Figure 6.27 Representation in the Gaussian plane of the phase relationships derived by single isomorphous replacement, SIR, in a protein. The structure factor of the protein vector, $\mathbf{F_P}$, lies on a circle of radius F_P centred at O. The structure factor of the heavy metal derivative, $\mathbf{F_{PH}}$, lies on a circle of radius F_{PH}, with centre at the tip of the vector $-\mathbf{F_H}$. The intersection of the two circles represents the two solutions to equation (6.6). The resulting vector $\mathbf{F_P}$ can be drawn in two positions, corresponding to two different phase angles

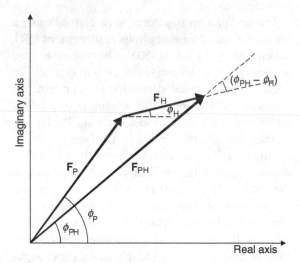

Figure 6.26 Representation in the Gaussian plane of the phase relationships between the structure factor of a pure protein, $\mathbf{F_P}$, a heavy atom, $\mathbf{F_H}$ and an isomorphic heavy atom derivative of the protein, $\mathbf{F_{PH}}$

A circle of radius F_P delimits the possible positions of the vector $\mathbf{F}_P$. A circle of radius F_{PH}, drawn from the head of the vector $-\mathbf{F}_H$, represents all possible positions of vector $\mathbf{F}_{PH}$ with respect to vector $\mathbf{F}_H$. The positions where these two circles intersect represents the solutions of equation (6.6), above. Although this does not give a unique value for the phase angle, it is sometimes possible to augment the data from elsewhere to give a preferred value. Once sufficient phase angles have been estimated, structure refinement can proceed. This technique, the **single isomorphous replacement (SIR)** method, has been successfully used to determine a number of protein structures.

The value of ϕ_P can be found unambiguously if another (different) heavy atom derivative is made, so that now $\mathbf{F}_{H1}$, $\mathbf{F}_{H2}$, $\mathbf{F}_{HP1}$, $\mathbf{F}_{HP2}$ and $\mathbf{F}_P$ are to be determined. This is called **multiple isomorphous replacement (MIR)** and is generally used rather than the single isomorphous replacement technique. The values for $\mathbf{F}_{H1}$ and $\mathbf{F}_{H2}$ can be determined using Patterson techniques. Two equations similar to equation 6.6 now exist:

$$\phi_P = \phi_{H1} + \cos^{-1}[(F_{PH1}^2 - F_P^2 - F_{H1}^2)/2F_PF_{H1}]$$
$$\phi_P = \phi_{H2} + \cos^{-1}[(F_{PH2}^2 - F_P^2 - F_{H2}^2)/2F_PF_{H2}]$$

This pair of simultaneous equations can be solved to give a unique solution. If a diagram similar to Figure 6.27 is drawn, incorporating the data for both derivatives, three circles can be plotted, which intersect at one point, corresponding to the unique value of ϕ_P, (Figure 6.28). Once again, when sufficient phase angles are known, refinement can proceed.

Protein structures are now more often determined by using **multiple anomalous dispersion (MAD)** techniques. In this method, scattering that does not obey the kinematic theory is used. The kinematic theory deals with single scatter-

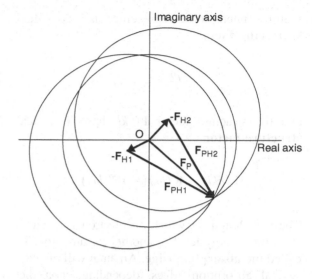

Figure 6.28 Representation in the Gaussian plane of the phase relationships derived by multiple isomorphous replacement, MIR, in a protein. The structure factor of the protein vector, $\mathbf{F}_P$, lies on a circle of radius F_P centred at O. The structure factor of the heavy metal derivatives, $\mathbf{F}_{PH1}$ and $\mathbf{F}_{PH2}$, lie on circles of radii F_{PH1} and F_{PH2} with centres at the tips of the vectors $-\mathbf{F}_{H1}$ and $-\mathbf{F}_{H2}$. The intersection of the three circles represents a unique solution to the position of $\mathbf{F}_P$, corresponding to a single phase angle

ing of X-ray photons which have low energy compared to the electron energy levels of the scattering atom. The scattered photon is supposed to suffer no loss of energy and no phase delay compared to the incident photon. This process is represented by the atomic scattering factor f_a, (Section 6.5).

However, if the energy of the X-ray photon is enough to excite an electron in the atom from one energy level to another, supplementary processes occur. In addition to the photons scattered 'normally', as described above, others will be absorbed and then re-emitted. To account for this extra detail, the scattering factor f_a needs to be represented as a **complex scattering factor**

$$f_a = f_a^0 + f_a' + if_a''$$

and the structure factor $\mathbf{F}(hkl)$ by a **complex structure factor**

$$\mathbf{F}(hkl) = \mathbf{F}'(hkl) + i\mathbf{F}''(hkl)$$

The wavelength at which an electron is excited from one energy level to another for any atom is called the **absorption edge**. An atom will display several absorption edges, depending upon the energy levels available to the excited electron, and **anomalous absorption** occurs when the wavelength of the X-ray beam is close to the absorption edge of an atom in the sample. Anomalous scattering causes small changes in intensities, and in particular, Friedel's law (see Section 6.9) breaks down and $F(hkl)$ is no longer equal to its Friedel opposite, $F(\bar{h}\,\bar{k}\,\bar{l})$ at the same wavelength. The difference, ΔF_b, is called the **Bijvoet difference**, and the reflection pairs are called **Bijvoet pairs**.

$$\Delta F_b = F(hkl) - F(\bar{h}\,\bar{k}\,\bar{l})$$

In addition, $F(hkl)$ is not constant but varies slightly with wavelength, to give an additional **dispersive difference**.

In general, the anomalous scattering from light atoms is negligible. However, anomalous scattering from heavy atoms, either native to the protein, or in isomorphous heavy atom derivative proteins, can be appreciable when the X-ray wavelength is chosen to be close to the absorption edge of the heavy atoms that have been used. The information provided by anomalous scattering can be used to solve the phase problem for proteins in the following way.

When anomalous scattering from Bijvoet pairs of a heavy metal atom derivative of a protein occurs, equation (6.5) now needs to be written:

$$\mathbf{F}_{PH}^+ = \mathbf{F}_P^+ + \mathbf{F}_H^+ = \mathbf{F}_P^+ + \mathbf{F}_H^{+\prime} + \mathbf{F}_H^{+\prime\prime}$$

$$\mathbf{F}_{PH}^- = \mathbf{F}_P^- + \mathbf{F}_H^- = \mathbf{F}_P^- + \mathbf{F}_H^{-\prime} + \mathbf{F}_H^{-\prime\prime}$$

where the superscripts $+$ and $-$ refer to the hkl and $\overline{hkl}$ Bijvoet pairs. If the protein contains only one type of anomalous scatterer, giving rise to **single anomalous dispersion**, the vectors $\mathbf{F}_H'$ and $\mathbf{F}_H''$ are perpendicular to each other, so if ϕ_H is the phase angle of $\mathbf{F}_H'$ the phase angle of $\mathbf{F}_H''$ is $\phi_H + \pi/2$. Figure 6.26 is now modified to Figure 6.29. Using Figure 6.29, it

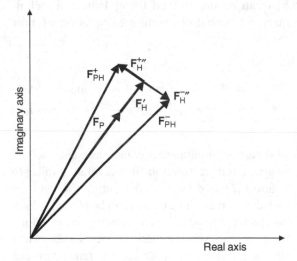

Figure 6.29   Representation in the Gaussian plane of the phase relationships between the structure factor of a protein, $\mathbf{F}_P$, and those from a single anomalous scattering heavy atom, $\mathbf{F}_H$ and an isomorphic heavy atom derivative of the protein, $\mathbf{F}_{PH}$, resulting in two scattering factors $\mathbf{F}_{PH}^+$ and $\mathbf{F}_{PH}^-$

is possible to write equations analogous to equation (6.6):

$$F_{PH}^+ - F_{PH}^- = 2F_H'' \sin(\phi_{PH} - \phi_H)$$

hence:

$$\phi_{PH} - \phi_H = \sin^{-1}[(F_{PH}^+ - F_{PH}^-)/2F_{PH}''] \quad (6.7)$$

Equation (6.7) like equation (6.6) gives two possible solutions for ϕ_{PH} but this can be used in place of a second heavy atom derivative, to

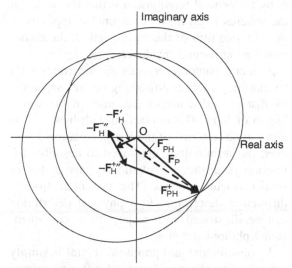

Figure 6.30 Representation in the Gaussian plane of the phase relationships derived by a single anomalous scattering atom in a heavy atom derivative protein. The structure factor of the protein vector, $\mathbf{F_P}$, lies on a circle of radius F_P centred at O. The structure factor of the heavy metal derivatives, $\mathbf{F_{PH}^+}$ and $\mathbf{F_{PH}^-}$, lie on circles of radii F_{PH}^+ and F_{PH}^- with centres at the tips of the vectors $-\mathbf{F_{PH}''}$ and $-\mathbf{F_{PH}^{+''}}$ as drawn. The intersection of the three circles represents a unique solution to position of $\mathbf{F_P}$, corresponding to a single phase angle

give a unique solution for the phase angle, (Figure 6.30). Naturally, data from two or more different anomalous scatterers, or data compiled from one anomalous scatterer at different wavelengths, (multiple anomalous dispersion), can also be used in a similar way.

Recently the bond valence method (see Section 7.8) has been applied successfully to the discrimination between various cation bonding sites in protein crystals (see Bibliography for the reference).

6.17 Photonic crystals

Photonic crystals are natural or artificial solids that are able to 'manipulate' light in a predetermined fashion, rather as X-rays are 'manipulated' by ordinary crystals. For this to be possible they must contain an array of scattering centres analogous to the atoms in ordinary crystals. Perhaps surprisingly, the diffraction phenomena are little different than that described for X-ray, electron and neutron diffraction, and the equations given above in this Chapter apply to photonic crystals as well as X-rays.

The two aspects can be compared via Bragg's law, (Section 6.1). The position of a strongly diffracted X-ray beam from a crystal is given by:

$$n\lambda = 2d_{hkl} \sin\theta \quad (6.1a)$$

where n is an integer, d_{hkl} is the separation of the $\{hkl\}$ planes of atoms which are responsible for the diffraction, λ is the X-ray wavelength and θ the angle between the X-ray beam and the atom planes. This equation holds good for any three-dimensional array no matter the size of the 'atoms'. Thus, an arrangement of particles, or even voids, which are spaced by distances, d, similar to the wavelength of light, will diffract light according to Bragg's Law. When white light is used, each wavelength will diffract at a slightly different (Bragg) angle and colours will be produced.

The best known natural photonic crystal is precious opal. The regions producing the colours are made up of an ordered packing of spheres of silica, (SiO_2), which are embedded in amorphous silica or a matrix of disordered spheres (Figure 6.31). These small volumes of ordered spheres resemble small crystallites. They interact with light because the spacing of the ordered spheres is similar to that of the wavelength of light.

Although the diffraction conditions are specified by the Bragg equation, because the diffraction takes place within a silica matrix, the real layer spacing, d nm, must be replaced by the optical thickness, $[d]$, equal to $n_s d$, where n_s is the refractive index of the silica in opal, about 1.45. The correct equation to use for opal is thus:

$$n\lambda = 2n_s d \sin\theta \sim 2.9 d \sin\theta$$

The maximum wavelength that will be diffracted is obtained when the opal 'crystal' is observed at normal incidence, found by substituting $\sin\theta = 1$ in gives:

$$\lambda_{max} = 2n_s d \approx 2.9 d$$

The relationship between the radius of the silica spheres, r, that make up the opal 'crystal', and the distance between the layers of spheres, d, will depend upon the geometry of the packing. If the spheres are arranged in hexagonal closest packing, the relationship between the sphere radius and the layer spacing is given in Section 7.2, as:

$$d = 1.633r$$

hence:

$$\lambda_{max} \approx 4.74r$$

A useful general relationship is that the radius of the spheres is given, to a reasonable approximation, by one fifth of the wavelength of the colour observed at normal incidence.

Artificial photonic crystals are structures built so that they contain diffracting centres separated by distances that are of the order of the wavelength of light. The interaction with light can be understood in terms of the Bragg equation. However, the terminology employed to describe diffraction in artificial photonic crystals is that of semiconductor physics. The transition from a diffraction description to a physical description can be illustrated with respect to a one-dimensional photonic crystal.

A one-dimensional photonic crystal is simply a stack of transparent layers of differing refractive indices. They are also called **Bragg stacks**, or, when built into an optical fibre, **fibre Bragg gratings**. The simplest model is that of a transparent material containing regularly spaced air voids, (Figure 6.32). When a beam of light is incident on such a grating, a wavelength λ will be diffracted when the Bragg law is obeyed:

$$n\lambda = 2[d]\sin\theta = 2n_m d \sin\theta$$

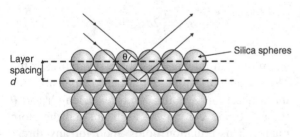

Figure 6.31 The structure of precious opal consists of ordered arrays of silica spheres with spacing of the order of the wavelength of light. These arrays diffract light and give rise to the colours seen in the gemstones

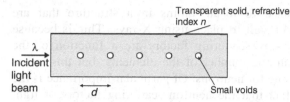

Figure 6.32 A simple one-dimensional photonic crystal can be imagined as an array of voids evenly spaced and separated by a distance approximately equal to the wavelength of light

where n is an integer, n_m is the refractive index of the material, d is the spacing of the voids, and $[d]$ is the optical thickness corresponding to the repeat spacing, $([d] = n_m d)$. For a beam normal to the voids, $\sin \theta = 1$, hence the diffraction condition is:

$$n\lambda = 2n_m d$$
$$\lambda_{max} = 2n_m d$$

Thus a beam of light with a wavelength given by λ_{max} will be diffracted back on itself and not be transmitted.

In terms of semiconductor physics, the array of voids has opened a **photonic band gap, (PBG)**, in the material. A photonic band gap blocks transmission of the light wave with an energy equal to the band gap. The energy, E, of the wave, wavelength λ is given by:

$$E = hc/\lambda$$

where h is the Planck constant and c is the speed of light in a vacuum. The energy gap of the array of voids, (Figure 6.32), is then:

$$E = hc/\lambda_{max} = hc/2n_m d$$

In real materials, the voids have thickness, and a range of wavelengths centred on λ_{max} is blocked. This range of wavelengths increases as the difference between the refractive index of the material and the voids increases. Similar effects will occur if the voids are replaced by spheres of a transparent material with a refractive index different from that of the matrix, or if alternating plane layers of transparent materials of differing refractive indices are used.

Two-dimensional photonic band gap crystals can be created by building a two-dimensional array of voids or 'atoms' in a transparent medium. Precious opal is an example of the three-dimensional photonic band gap crystal.

Answers to introductory questions

What crystallographic information does Bragg's law give?

A beam of radiation of a suitable wavelength will be diffracted when it impinges upon a crystal. Bragg's law defines the conditions under which diffraction occurs, and gives the position of a diffracted beam. Diffraction will occur from a set of (hkl) planes when:

$$n\lambda = 2d_{hkl} \sin \theta$$

where n is an integer, λ is the wavelength of the radiation, d_{hkl} is the interplanar spacing (the perpendicular separation) of the (hkl) planes and θ is the diffraction angle or Bragg angle. Note that the angle between the direction of the incident and diffracted beam is equal to 2θ. The geometry of Bragg diffraction is identical to that of reflection and diffracted beams are frequently called 'reflections' in X-ray literature.

What is an atomic scattering factor?

The atomic scattering factor is a measure of the scattering power of an atom for radiation such as X-rays, electrons or neutrons. Because each of these types of radiation interact differently with an atom, the scattering factor of an atom is different for X-rays, electrons and neutrons.

X-rays are mainly diffracted by the electrons on each atom. The scattering of the X-ray beam increases as the number of electrons, or equally, the atomic number (proton number), of the atom, Z, increases. Thus heavy metals scatter X-rays far more strongly than light atoms while the scattering from neighbouring atoms is similar in magnitude.

The X-ray scattering factor is strongly angle dependent, this being expressed as a function of $(\sin \theta)/\lambda$. The scattering factor curves for all atoms have a similar form, and at $(\sin \theta)/\lambda =$ zero the X-ray scattering factor is equal to the atomic number, Z, of the element in question.

What are the advantages of neutron diffraction over X-ray diffraction?

One of the greatest advantages of neutron diffraction over X-ray diffraction is that it can be used to locate atoms in a structure that are difficult to place using X-rays. This is because X-ray scattering factors are a function of the atomic number of the elements, but this is not true for neutrons. Of particular importance is the fact that the neutron scattering factors of light atoms, such as hydrogen, H, carbon, C, and nitrogen, N, are similar to those of heavy atoms, making it easier to determine their positions compared to X-ray diffraction. Neutron diffraction is the method of choice for the precision determination of the positions of light atoms in a structure, especially hydrogen atoms. Studies of hydrogen bonding often rely upon neutron diffraction data for accurate interatomic distances.

A second advantage is that in some instances neighbouring atoms have quite different neutron scattering capabilities, making them easily distinguished. This is not true for X-ray diffraction, in which neighbouring atoms always have very similar atomic scattering factors.

A third advantage is that neutrons have a spin and so interact with the magnetic structure of the solid, which arises from the alignment of unpaired electron spins in the structure. The magnetic ordering in a crystal is quite invisible to X-rays, but is revealed in neutron diffraction as extra diffracted beams and hence a change in unit cell.

Problems and exercises

Quick quiz

1 In crystallography, Bragg's law is written:
   (a) $2\lambda = d_{hkl} \sin \theta$
   (b) $\lambda = 2d_{hkl} \sin \theta$
   (c) $\lambda = nd_{hkl} \sin \theta$

2 The X-ray reflections from a mostly amorphous sample are:
   (a) Very sharp
   (b) Moderately well defined
   (c) Poorly defined or absent

3 The intensity of a beam of radiation diffracted by a set of (hkl) planes does NOT depend upon:

(a) The time of irradiation

(b) The nature of the radiation

(c) The chemical composition of the crystal

4 The atomic scattering factor for X-rays is greatest for:
(a) Metals

(b) Non-metals

(c) Heavy atoms, whether metal or not

5 When the diffracted beams scattered from two different atoms are exactly in step the phase difference is:

(a) 2π

(b) π

(c) $\pi/2$

6 The X-ray intensity scattered by a unit cell is given by the modulus of:
(a) The structure factor

(b) The square of the structure factor

(c) The square root of the structure factor

7 X-ray reflections with zero intensity because of the symmetry of the unit cell are called:
(a) Systematic absences

(b) Structural absences

(c) Symmetric absences

8 The compound SrF_2 crystallises with the same (fluorite) structure as CaF_2. The X-ray powder patterns will be:
(a) Identical

(b) Almost identical

(c) Similar but easily distinguished

9 In order to accurately locate light atoms in a crystal it is preferable to use:
(a) Electron diffraction

(b) Neutron diffraction

(c) X-ray diffraction

10 In protein crystallography, the technique of isomorphous replacement is used to:
(a) Determine the space group of the crystal

(b) Determine the structure factors of the reflections

(c) Determine the phases of the reflections

Calculations and Questions

6.1 The oxide $NiAl_2O_4$ adopts the spinel structure, with a cubic lattice parameter of 0.8048 nm. The structure is derived from a face-centred cubic lattice. Making use of Table 6.4, calculate the angles of diffraction of the first six lines expected on a powder diffraction pattern.

6.2 A small particle of CrN, which has the halite structure (see Chapter 1) with $a = 0.4140$ nm, gives a reflection from the (200) planes at an angle of 33.60°. The angular range over which the reflection occurs is approximately 1.55°. (a) Calculate the wavelength of the radiation used. (b) Estimate the size of the particle in a direction normal to the (200) planes.

6.3 Electron diffraction patterns recorded by tilting a crystal of niobium, Nb, which has the cubic A2 structure (see Chapter 1) with $a = 0.3300$ nm, are reproduced in the figure below. Using the partly indexed pattern, (a) complete the indexing; (b) determine λl for the microscope; (c), index the remaining patterns. [Note that not every lattice point will be present on these patterns, see Section 6.9. It may be helpful to determine first of all which reflections will be present using Table 6.8.]

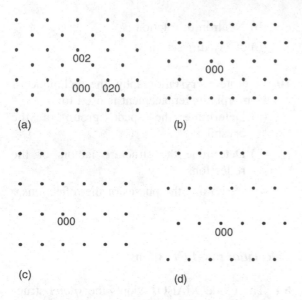

(a)

(b)

(c)

(d)

6.4 The Cromer-Mann coefficients for niobium, (λ in nm), are given in the table. Plot the scattering factor for niobium as a function of $\sin\theta / \lambda$.

Coefficient	Index			
	1	2	3	4
a	17.614	12.014	4.042	3.533
b	0.01189	0.11766	0.00205	0.69796
c	3.756			

These values adapted from http://www-structure.llnl.gov

6.5 Repeat the calculation in Q6.4 if the value of the atomic temperature factor B for niobium is (a), 0.1 nm^2 and (b) 0.2 nm^2. What are the mean displacements of the niobium atoms corresponding to these values of B?

6.6 Calculate the value of the scattering factor (magnitude and phase) for the 110 reflection for rutile, (TiO$_2$), using the data in Section 6.7. The scattering factors are f(Ti), 17.506, f(O), 6.407. Use this data to compute the scattering factor using vector diagrams.

6.7 Calculate the intensities of (a), the 200 and (b), the 110 reflections of rutile, (TiO$_2$) on an X-ray powder pattern, using the data in Sec-

tion 6.8 and Q6.6. Take the temperature factor, B, to be 0.1 nm^2 for each reflection and the wavelength of the X-rays as 0.15406 nm (Cu radiation).

6.8 Many inorganic crystals contain 1-dimensional chains of atoms that endow the material with interesting electronic properties. Using a computer, plot the electron density function:

$$\rho(x) = F(000) + \sum_{h=-\infty}^{+\infty} F(h) \cos 2\pi(hx - \phi_h)$$

for a chain of alternating Pt and Cl atoms, using first just F(000), 100 and $\bar{1}$00, then, using the first four reflections, F(000) up to 200 and −200, and so on, to see how the electron density pattern changes as the amount of information supplied increases. The repeat distance along the chain, a, is 1 nm, and the atomic positions are Cl, 0; Pt, ½. [Note that the 'lattice' fringe patterns observed in an electron microscope using these reflections will evolve in a similar manner.]

Structure factors and phase angles for a chain of Pt and Cl atoms

h	$|F(h)| = F(h)$	ϕ_h
8	60	0
7	46	π
6	68	0
5	52	π
4	78	0
3	57	π
2	89	0
1	60	π
0	95	0
$\bar{1}$	60	π
$\bar{2}$	89	0
$\bar{3}$	57	π
$\bar{4}$	78	0
$\bar{5}$	52	π
$\bar{6}$	68	0
$\bar{7}$	46	π
$\bar{8}$	60	0

7

The depiction of crystal structures

- *What is the size of an atom?*

- *How does the idea of bond valence help in structure determination?*

- *What is the secondary structure of a protein?*

Lists of atomic positions are not very helpful when a variety of structures have to be compared. This chapter describes attempts to explain and systematise the vast amount of structural data now available. The original aim of ways of comparing similar structures was to provide a set of empirical guidelines for use in the determination of a new crystal structure. With sophisticated computer methods now available, this approach is rarely employed, but one empirical rule, the bond-valence method, (see Section 7.8 below), is widely used in structural studies.

Most data is available for inorganic solids, as these have been studied longest and in greatest detail. The commonest ways of describing these structures is either as built up by packing together spheres, or else in terms of polyhedra linked by corners and edges. The description of structures

by nets or by tilings is also of value for some categories of structure. The study of large organic molecules, especially proteins, is of increasing importance. These are often described by stylised 'cartoons' in which sections of structure are drawn as coiled or folded ribbons.

7.1 The size of atoms

Quantum mechanics makes it clear that no atom has a fixed size. Electron orbitals extend from the nucleus to a greater or lesser extent, depending upon the chemical and physical environment in the locality of the atomic nucleus. Indeed, recent research on Bose-Einstein and Fermi condensation reveals that a collection of millions of atoms can enter an identical quantum state at temperatures just above 0 K and behave as a single atom, with a single wavefunction that spreads over the whole collection.

The concept of atomic size becomes more difficult to pin down in compounds. The interactions of nearest neighbours seriously perturb the electron charge clouds of the atoms, and this effect, chemical bonding, has a major influence upon the interatomic distances between bonded atoms.

It may be thought that the scattering of X-rays and electrons can give absolute values for the

Crystals and Crystal Structures. Richard J. D. Tilley
© 2006 John Wiley & Sons, Ltd

sizes of atoms in crystals, but this turns out to be incorrect. The scattering of both these types of radiation is due to interaction with the electrons surrounding each atomic nucleus. A diffraction experiment then gives details of a varying electron density throughout the unit cell volume. The electron density, $\rho(x, y, z)$, at a point (x, y, z) in the unit cell, is given by:

$$\rho(x, y, z) = \frac{1}{V} \sum_{h=-\infty}^{+\infty} \sum_{k=-\infty}^{+\infty} \sum_{l=-\infty}^{+\infty} \mathbf{F}(hkl)$$
$$\times \exp[-2\pi i(hx + ky + lz)]$$

where $\mathbf{F}(hkl)$ is the structure factor of each (hkl) reflection, and V is the unit cell volume, (see Chapter 6). In X-ray crystallography the electron density is computed mathematically while in an electron microscope, the operation is carried out by lenses. Naturally, the electron density is highest nearer to atomic cores, and least well away from atomic cores. Thus the diffraction experiment yields reasonable positions for the atomic cores rather than the atomic sizes, and strictly speaking, *interatomic distances* are obtained, *not* atomic radii, with these techniques.

Nevertheless, although it may be more correct to think of the contents of a unit cell in terms of a variation in electron density, the concept of an atom possessing a definite and fixed size is attractive, and is a useful starting point for the discussion of many chemical and physical properties, not least in the area of crystallography. A primary use of atomic radii is to derive structural detail, such as bond lengths, bond angles, coordination number and hence molecular geometry, in advance of any structural determination. Structural crystallography is able to help because it provides a set of interatomic distances in a crystal that can be used to derive values for the sizes of atoms. However, it is important to stress again that structure determination gives interatomic distances, and the division of these into parts 'belonging' to the individual

atoms is to some extent an arbitrary procedure. For example, the radii of similar atoms that are nearest neighbours, linked by strong chemical bonds, is different from the radii of the same atoms when they are next nearest neighbours, and are linked by so-called 'non-bonded interactions'. Such considerations lead to several different size scales.

Inorganic chemists have, by and large, used 'ionic radii', which are derived from the notion that anions are spherical and in contact in an inorganic crystal. Organic chemists attempting to model complex molecules use 'covalent radii' for neighbouring atoms, again presumed to be spherical, and 'van der Waals' radii for atoms at the periphery of molecules. Physicists using band theory to calculate the physical properties of metals find that the extension of the core electron orbitals provides a convenient idea of atomic size. Details of the derivation of these and other atomic radii, and critical discussions as to the rationale and self-consistency of the resulting values are given in the Bibliography.

7.2   Sphere packing

The structure of many crystals can conveniently be described in terms of an ordered packing of spheres, representing spherical atoms or ions, and the earliest sets of atomic radii were derived via this type of model. Although there are an infinite number of ways of packing spheres, only two main arrangements, called closest (or close) packing, are sufficient to describe many crystal structures. These structures can be thought of as built from layers of close-packed spheres. Each close-packed layer consists of a hexagonal arrangement of spheres just touching each other to fill the space as much as possible, (Figure 7.1).

These layers of spheres can be stacked in two principal ways to generate the structures. In the first of these, a second layer fits into the dimples in the first layer, and the third layer is stacked in dimples on top of the second layer to lie over the

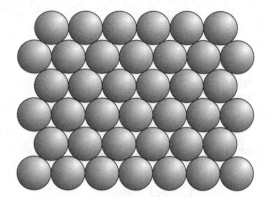

Figure 7.1 A single close-packed array of spheres

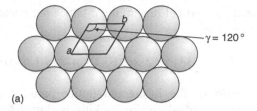

(a)

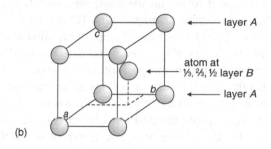

layer A

atom at
⅓, ⅔, ½ layer B

layer A

(b)

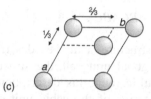

(c)

Figure 7.3 The hexagonal unit cell of hexagonal (*AB*) closest packing of spheres: (a) the (001) plane of the unit cell; (b) the unit cell in perspective; (c) projection of the unit cell along [001]

first layer, (Figure 7.2). This sequence is repeated indefinitely. If the position of the spheres in the first layer is labelled *A*, and the positions of the spheres in the second, *B*, the complete stacking is described by the sequence:

$$....ABABAB.....$$

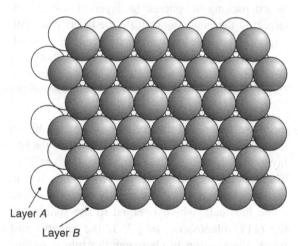

Layer *A*

Layer *B*

Figure 7.2 Hexagonal closest packing of spheres. All layers are identical to those in Figure 7.1. The position of the spheres in the lower layer is labelled *A* and the position of those in the second is labelled *B*. Subsequent layers follow the sequence ... *ABAB* ...

The structure has a **hexagonal** symmetry and unit cell. The **a**- and **b**-axes lie in the close-packed *A* sheet, and the hexagonal **c**-axis is perpendicular to the stacking and runs from one *A* sheet to the next above it, (Figure 7.3). There are two spheres (two atoms) in a unit cell, at positions 0, 0, 0 and ⅓, ⅔, ½. If the spheres just touch, the relationship between the sphere radius, *r*, and the lattice parameter *a*, is:

$$2r = a = b$$

The structural repeat normal to the stacking is two layers of spheres. The spacing of the close packed layers, *d*, is:

$$d = \sqrt{8}r/\sqrt{3} \approx 1.633\,r$$

so that the vertical repeat, the c-lattice parameter, is given by:

$$c \approx 2 \times 1.633\,r \approx 2 \times 1.633 \times a/2 \approx 1.633\,a$$

The ratio of the hexagonal lattice parameters, c/a, in this ideal sphere packing is thus 1.633. This model forms an idealised representation of the A3 structure of magnesium, (Chapter 1).

The second structure of importance also starts with two layers of spheres, A and B, as before. The difference lies in the position of the third layer. This fits onto the preceding B layer, but occupies a set of dimples that is not above the lower A level. This set is given the position label C, (Figure 7.4), and the three-layer stacking is repeated indefinitely, thus:

$$....ABCABC....$$

Although this structure can be described in terms of a hexagonal unit cell, the structure turns out to be **cubic**, and this description is always chosen. In terms of the cubic unit cell,

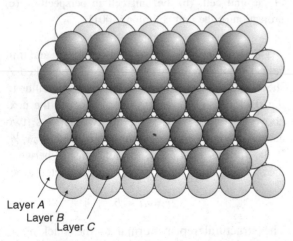

Layer A
Layer B
Layer C

Figure 7.4   Cubic closest packing of spheres. All layers are identical to those in Figure 7.1. The position of the spheres in the layers is labelled A in the lowest layer, B in the middle layer and C in the top layer. Subsequent layers follow the sequence ... $ABCABC$ ...

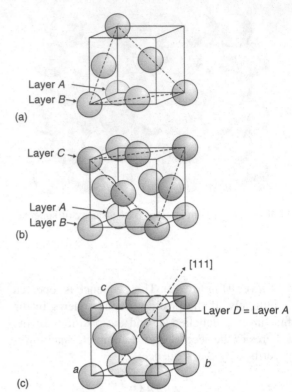

Layer A
Layer B
(a)

Layer C
Layer A
Layer B
(b)

[111]
Layer D = Layer A
c
a
b
(c)

Figure 7.5   The cubic unit cell of cubic (ABC) closest packing of spheres: (a) layers A and B with respect to the cubic unit cell; (b) the layers ABC with respect to the cubic unit cell; (c) the complete unit cell. The layers are packed perpendicular to the cubic [111] direction. The cubic unit cell has a sphere at each corner and one at the centre of each face

there are spheres at the corners of the cell and in the centre of each of the faces. The close packed layers lie along the [111] direction, (Figure 7.5). The spacing of the close packed planes for an ideal packing, d, is ⅓ of the body diagonal of the cubic unit cell, which is equal to the spacing of the (111) planes, i.e. $a/\sqrt{3}$. If the spheres just touch, the relationship between the sphere radius, r, and the cubic lattice parameter a, is:

$$r = a/\sqrt{8}$$

The relationship between the spacing of the close packed planes of spheres, d, the cubic unit cell parameter a, and r, is:

$$d = a/\sqrt{3} = (\sqrt{8}\,r)/\sqrt{3} \approx 1.633\,r$$

The cubic-close packed arrangement is almost identical to the A1 structure of copper, (Chapter 1).

Both the hexagonal closest packing of spheres and the cubic closest packing of spheres result in the (equally) densest packing of the spheres. The fraction of the total volume occupied by the spheres, when they touch, is 0.7405.

7.3 Metallic radii

Most pure metals adopt one of three crystal structures, A1, copper structure, (cubic close-packed), A2, tungsten structure, (body-centred cubic) or A3, magnesium structure, (hexagonal close-packed), (Chapter 1). If it is assumed that the structures of metals are made up of touching spherical atoms, (the model described in the previous section), it is quite easy, knowing the structure type and the size of the unit cell, to work out their radii, which are called **metallic radii**. The relationships between the lattice parameters, a, for cubic crystals, a, c, for hexagonal crystals, and the radius of the component atoms, r, for the three common metallic structures, are given below.

For the face-centred cubic [A1, copper structure], the atoms are in $ABCABC$ packing, and in contact along a cube face diagonal, so that

$$r = a/\sqrt{8}$$

The separation of the close packed atom planes, (along a cube body diagonal), is $a/\sqrt{3}$ i.e. d_{111}. Each atom has 12 nearest neighbours.

For the body-centred cubic [A2, tungsten structure] the atoms are in contact along a cube body diagonal. The body diagonal of the unit cell is thus equal to $4\,r$. This is also equal to $3d_{111}$, i.e. $3\,a/\sqrt{3}$, so that

$$r = \sqrt{3}\,a/4$$

Each atom has 8 nearest neighbours.

For the hexagonal close packed [A3, magnesium structure] the atoms are in $ABAB$ packing, and in contact along the **a**-axis direction, hence

$$r = a/2$$

The separation of the close packed atom planes is $c/2$, and the ratio of c/a in an ideal close packed structure is $\sqrt{8}/\sqrt{3} \approx 1.633$. The c/a ratio departs from the ideal value of 1.633 in most real structures. Each atom has 12 nearest neighbours.

Determination of these atomic radii, even in such an apparently clear cut situation, leads to contradictions. The radius determined experimentally is found to depend upon the **coordination number**, (the number of nearest neighbours), of the atom in question. That is, the radii derived for both the face-centred cubic and the hexagonal close packed structures, which have twelve nearest neighbours, (a coordination number of 12, CN12), are in agreement with each other, but differ from those derived for metals that adopt the body-centred cubic structure, in which each atom has eight nearest neighbours, (a coordination number of 8, CN8). The conversion between the two sets of radii can be made using the empirical formula

$$\text{radius[CN12]} = 1.032 \times (\text{radius[CN8]}) - 0.0006$$

where the radii are measured in nm. Metallic radii appropriate to CN12 are given in Figure 7.6.

There are a number of trends to note. In the well-behaved alkali metals and alkaline earth metals, the radius of an atom increases smoothly as the atomic number increases. The transition metals all have rather similar radii as one passes along the period, and these increase slightly with atomic number going down a group. The same is true for the lanthanides and actinides.

Li 0.1562	Be 0.1128											B	C
Na 0.1911	Mg 0.1602											Al 0.1432	Si
K 0.2376	Ca 0.1974	Sc 0.1641	Ti 0.1462	V 0.1346	Cr 0.1282	Mn 0.1264	Fe 0.1274	Co 0.1252	Ni 0.1246	Cu 0.1278	Zn 0.1349	Ga 0.1411	Ge
Rb 0.2546	Sr 0.2151	Y 0.1801	Zr 0.1602	Nb 0.1468	Mo 0.1400	Tc 0.1360	Ru 0.1339	Rh 0.1345	Pd 0.1376	Ag 0.1445	Cd 0.1568	In 0.1663	Sn 0.1545
Cs 0.2731	Ba 0.2243	La 0.1877	Hf 0.1580	Ta 0.1467	W 0.1408	Re 0.1375	Os 0.1353	Ir 0.1357	Pt 0.1387	Au 0.1442	Hg	Tl 0.1716	Pb 0.1750

Figure 7.6 Metallic radii for coordination number 12, [CN12]. Data from Teatum, Gschneidner and Waber, cited by W. Pearson, *The Crystal Chemistry and Physics of Metals and Alloys*, Wiley-Interscience, New York, 1971, p151

7.4   Ionic radii

As mentioned above, X-ray structures only give a precise knowledge of the distances between the atoms. While this is not so important for pure metals, as the interatomic distance is simply divided by two to obtain the metallic radius, this simple method will not work for ionic compounds. To begin, it is assumed that the individual ions are spherical and in contact. The strategy then used to derive ionic radii is to take the radius of one commonly occurring ion, such as the oxide ion, O^{2-}, as a standard. Other consistent radii can then be derived by subtracting the standard radius from measured inter-ionic distances.

The ionic radius quoted for any species depends upon the standard ion by which the radii were determined. This has lead to a number of different tables of ionic radii. Although these are all internally self-consistent, they have to be used with thought. Moreover, as with metallic radii, ionic radii are sensitive to the surrounding coordination geometry. The radius of a cation surrounded by six oxygen ions in octahedral coordination is different than the same cation surrounded by four oxygen ions in tetrahedral coordination. Similarly, the radius of a cation surrounded by six oxygen ions in octahedral coordination is different than the same cation surrounded by six sulphur ions in octahedral coordination. Ideally, tables of cationic radii should apply to a specific anion and coordination geometry. Representative ionic radii are given in Figure 7.7.

On this basis, several trends in ionic radius are apparent.

(i) Cations are usually regarded as smaller than anions, the main exceptions being the largest alkali metal and alkaline earth metal cations. The reason for this is that removal of electrons to form cations leads to a contraction of the electron orbital clouds due to the relative increase in nuclear charge. Similarly, addition of electrons to form anions leads to an expansion of the charge clouds due to a relative decrease in the nuclear charge.

+1	+2	+3	+4 / (+3)	+5 / (+4) / [3+]	+6 / (+4) / [3+]	+6 / (+4) / [3+] / {2+}	+4 / (+3) / [2+]	+4 / (+3) / [2+]	+4 / (+2)	+2 / (+1)	+2	+3	+4 / (2+)	+5 / (3+)	−2 / (6+)	−1
Li 0.088	Be 0.041(t)											B 0.026 (t)	C** 0.006	N*** 0.002	O 0.126	F 0.119
Na 0.116	Mg 0.086											Al 0.067	Si 0.040 (t) (0.056)	P 0.031 (t)	S 0.170	Cl 0.167
K 0.1521	Ca 0.114	Sc 0.0885	Ti 0.0745 (0.081)	V 0.068 (0.073) [0.078]	Cr [0.0755]	Mn (0.068) [0.079*] {0.097*}	Fe (0.079*) [0.092*]	Co (0.075*) [0.089*]	Ni (0.083)	Cu 0.087 (0.108)	Zn 0.089	Ga 0.076	Ge 0.068	As 0.064	Se (0.043 t)	Br 0.182
Rb 0.163	Sr 0.1217	Y 0.104	Zr 0.086	Nb 0.078	Mo 0.074	Tc (0.078)	Ru 0.076	Rh 0.0755	Pd (0.100)	Ag (0.129)	Cd 0.109	In 0.094	Sn 0.083 (0.105)	Sb 0.075	Te (0.068)	I 0.206
Cs 0.184	Ba 0.150	La 0.1185	Hf 0.085	Ta 0.078	W 0.074 (0.079)	Re 0.066	Os 0.077	Ir 0.077	Pt 0.077 (0.092)	Au	Hg 0.110	Tl 0.1025	Pb 0.0915 (0.132)	Bi 0.086 (0.116)	Po	At

** C in carbonate, CO_3^{2-}; *** N in nitrate, NO_3^-

Figure 7.7 Ionic radii. Values marked* are for high-spin states, which have maximum numbers of unpaired electrons. Data mainly from R. D. Shannon and C. T. Prewitt, *Acta Crystallogr.*, **B25**, 925–946 (1969); *ibid*, 1046–1048 (1970), with additional data from O. Muller and R. Roy, *The Major Ternary Structural Families*, Springer-Verlag, Berlin, (1974), p5

(ii) The radius of an ion increases with atomic number. This is simply a reflection that the electron cloud surrounding a heavier ion is of a greater extent than that of the lighter ion.

(iii) The radius decreases rapidly with increase of positive charge for a series of isoelectronic ions such as Na^+, Mg^{2+}, Al^{3+}, all of which have the electronic configuration [Ne], as the increase in the effective nuclear charge acts so as to draw the surrounding electron clouds closer to the nucleus.

(iv) Successive valence increases decrease the radius. For example, Fe^{2+} is larger than Fe^{3+}, for the same reason as in (iii).

(v) Increase in negative charge has a smaller effect than increase in positive charge. For example, F^- is similar in size to O^{2-} and Cl^- is similar in size to S^{2-}.

When making use of these statements, it is important to be aware that the potential experienced by an ion in a crystal is quite different from that of a free ion. This implies that all of the above statements need to be treated with caution. Cationic and anionic sizes in solids may be significantly different to those that free ion calculations suggest.

While the majority of the ions of elements can be considered to be spherical, a group of ions, the **lone pair ions**, found in at the lower part of groups 13, 14 and 15 of the Periodic Table, are definitely not so. These ions all take two ionic states. The high charge state, M^{n+} can be considered as spherical, but the lower valence state, $M^{(n-2)+}$, is definitely not so. For example, tin, Sn, has an outer electron configuration [Kr] $4d^{10}$ $5s^2$ $5p^2$. Loss of the two *p* electrons will leave the ion with a series of closed shells that is moderately stable. This is the Sn^{2+} state, with a configuration of [Kr] $4d^{10}5s^2$. The pair of *s* electrons – the *lone* *pair* – imposes important stereochemical constraints on the ion. However, loss of the lone pair *s* electrons will produce the stable configuration [Kr]$4d^{10}$ of Sn^{4+}, which can be considered spherical. The atoms that behave in this way are characterised by two valence states, separated by a charge difference of 2+. The atoms involved are indium, In (1+, 3+), thallium, Tl (1+, 3+), tin, Sn (2+, 4+), lead, Pb (2+, 4+), antimony, Sb (3+, 5+) and bismuth, Bi (3+, 5+). The lone pair ions, In^+, Tl^+, Sn^{2+}, Pb^{2+}, Sb^{3+} and Bi^{3+}, tend to be surrounded by an irregular coordination polyhedron of anions. This is often a distorted trigonal bipyramid, and it is hard to assign a unique radius to such ions.

7.5   Covalent radii

Covalent radii are mostly used in organic chemistry. The simplest way to obtain a set of covalent radii is to half the distance between atoms linked by a covalent bond in a homonuclear molecule, such as H_2. Covalent radii defined in this way frequently do not reproduce the interatomic distances in organic molecules very well, because these are influenced by double bonding and electronegativity differences between neighbouring atoms. In large molecules such as proteins, this has important structural consequences.

The estimation of covalent radii using data from large molecules poses another problem. The atomic positions given by different techniques are not identical. For example, neutron diffraction gives information on the positions of the atomic nuclei, whereas X-ray diffraction gives information on the electron density in the material. This is not serious for many atoms, but in the case of light atoms, especially hydrogen, the difference can be quite large, of the order of 0.1 nm. As hydrogen atoms play an important role in protein chemistry, such differences can be important. For this reason it is rather more difficult to arrive at a set of self-consistent radii than it is for either ionic or metallic systems. Nevertheless, the large numbers

of organic compounds studied have lead to a value for the covalent radius of single bonded carbon as 0.767 nm. Taking this as standard, other covalent radii can be calculated by subtracting the carbon radius from the bond length in organic molecules. Those best suited for this are the simple tetrahedral molecules with carbon as the central atom, such as carbon tetrachloride, CCl_4. In this way a consistent set of carbon based single bond **covalent radii** has been derived, (Figure 7.8). A comparison with molecules containing double or triple bonds then allows values for multiple bond radii to be obtained.

These radii are really applicable to isolated molecules. To a large extent, bond lengths between a pair of atoms $A - B$ in such molecules are fairly constant, because the energy of bond stretching and compression is usually high. Bond angles are also fairly constant, but less so than bond lengths, as the energy to distort the angle between three atoms is much less. Skeletal and 'ball and stick' models make use of this constancy to built

H 0.0299					
Be 0.106	B 0.083	C 0.0767 0.0661* 0.0591**	N 0.0702 0.0618* 0.0545**	O 0.0659 0.049*	F 0.0619
		Al 0.118	Si 0.109	P (III) 0.1088	S 0.1052	Cl 0.1023
		Ga 0.1411	Ge 0.122	As (III) 0.1196	Se (II) 0.1196	Br 0.1199
		In 0.141	Sn 0.139	Sb (III) 0.137	Te (II) 0.1391	I 0.1395

* Double bond radius   ** Triple bond radius

Figure 7.8   Covalent radii. Data mainly taken from N. W. Alcock, Structure and Bonding, Ellis Horwood, (1990), cited on http://www.iumsc.indiana.edu/radii.html

molecular structures with reliable shapes. However, bond distances and angles can become modified in crystals, where other interactions may dominate.

7.6 Van der Waals radii

The van der Waals bond is a weak bond caused by induced dipoles on otherwise neutral atoms. Atoms linked by this interaction have much larger interatomic distances, and hence radii, compared to those linked by strong chemical bonds. A measure of the van der Waals radius that is widely used, especially in organic chemistry, is the idea of a **non-bonded** radius. In this concept, the bond distance between an atom, X, and its *next nearest* neighbour, Z, in the configuration X-Y-Z, is measured. The van der Waals or non-bonded radius is then determined by assuming that the non-bonded atoms are hard spheres that just touch. Some values are given in Figure 7.9.

H 0.120							
			C 0.170	N 0.155	O 0.152	F 0.147	
			Si 0.210	P 0.180	S 0.180	Cl 0.175	
			Ge 0.195	As 0.185	Se 0.190	Br 0.185	
			Sn 0.210	Sb 0.206	Te 0.206	I 0.198	Xe 0.200
				Bi 0.215			

Figure 7.9 Van der Waals (non-bonded) radii. Data from N. W. Alcock, Structure and Bonding, Ellis Horwood, (1990), cited on http://www.iumsc.indiana.edu/radii.html

Organic chemists make considerable use of **space filling models**, also called **Corey-Pauling-Koltun** or **CPK** models, to represent organic molecules. The atoms in such models are given van der Waals radii, and are designed to give an idea of the crowding that may take place in a molecule. For example, a space filling model of an α-helix in a protein shows the central core of the helix to be fairly full, while a description of the helix as a coiled ribbon suggests that it is hollow (also see Section 7.15).

7.7 Ionic structures and structure building rules

Ionic bonding is non-directional. The main structural implication of this is that ions simply pack together to minimise the total lattice energy. [Note that the real charges on cations in solids are generally smaller than the formal ionic charges on isolated ions]. There have been many attempts to use this simple idea, coupled with chemical intuition, to derive the structures of inorganic solids. These resulted in a number of structure building 'rules', most famously Pauling's rules, (see Bibliography) which were used to derive model structures that were used as the starting point for the refinement of X-ray diffraction results in the early years of X-ray crystallography. These rules are no longer widely used for this purpose, as sophisticated computing techniques are able to derive crystal structures from raw diffraction data with minimal input of model structures. Nevertheless, these ideas are still useful in some circumstances. For example, although cation coordination is not determined quantitatively by the relative radii of anions and cations, by and large it is found that large cations tend to be surrounded by a cubic arrangement of anions, medium sized cations by an octahedral arrangement of anions and small cations by a tetrahedron of anions. The smallest cations are surrounded by a triangle of anions. In addition, the bond-valence model, described in the following section, which is an extension of Pauling's second rule, is widely used to elucidate cation location in inorganic structures.

7.8 The bond valence model

The intensities of X-rays scattered by atoms that are Periodic Table neighbours are almost identical so that it is often difficult to discriminate between such components of a structure. The same is true for the different ionic states of a single element. Thus, problems such as the distribution of Fe^{2+} and Fe^{3+} over the available sites in a crystal structure may be unresolved by conventional structure determination methods. The bond valence model[1], an empirical concept, is of help in these situations. The model correlates the notion of the relative strength of a chemical bond between two ions with the length of the bond. A short bond is *stronger* than a long bond. Because crystal structure determinations yield accurate interatomic distances, precise values of these relative bond strengths, called experimental **bond valence** values, can be derived.

Imagine a cation, i, surrounded by j anions. The formal valence of the central ion, V_i, is equal to the formal charge on the cation. Thus, the value of V_i for an Fe^{3+} ion is $+3$, for an Nb^{5+} ion is $+5$, and so on. This value, V_i, is also taken to be equal to the sum of all of the bond valences, v_{ij}, of the ions j of opposite charge within the first coordination shell of ion i thus:

$$\sum_j v_{ij} = V_i = \text{formal charge on the cation} \quad (7.1)$$

To make use of this simple idea, it is necessary to relate the experimentally observable bond length,

r_{ij}, between cation i and anion j with the bond valence, v_{ij}. Two empirical equations have been suggested, either of which can be employed.

$$v_{ij} = \left(\frac{r_{ij}}{r_0}\right)^{-N}$$
$$v_{ij} = e^{(r_0 - r_{ij})/B} \quad (7.2)$$

where r_0, N and B are empirical parameters[2]. These are derived from crystal structures that are considered to be particularly reliable. In general, equation (7.2) is most frequently used, and the value of B is often taken as 0.037 nm (0.37 Å) for all bonds, so that this equation takes the practical form:

$$v_{ij} = e^{(r_0 - r_{ij})/0.037} \quad (7.3)$$

Figure 7.10 gives graphs of this function for Zn^{2+}, Ti^{4+} and H^+ when bonded to O^{2-} ions. It is seen that the bond valence varies steeply with bond distance.

To illustrate the relationship between the valence of an ion and the bond length, imagine a Ti^{4+} ion surrounded by a regular octahedron of oxide ions, (Figure 7.10a). Ideally each bond will have a bond valence, v, of $4/6$, or 0.6667. Using a value of r_0 for Ti^{4+} of 0.1815 nm, each Ti – O bond is computed by equation (7.3) to have a length, r, of 0.1965 nm. If some of the bond lengths, r, are shorter than this, (Figure 7.11b), it can be seen from Figure 7.10, that the values of the bond valences, v, of these short bonds will be greater than 0.6667 each. When the six bond valences are added, they will come to more than 4.0. In this case the Ti^{4+} ion is said to be **over bonded**. In the reverse case, when some of the bond lengths, r, are longer than 0.1965 nm, (Figure 7.11c), the bond valences, v, will be less than 0.6667. When the six bond

---

[1] The bond valence model should not be confused with the quantum mechanical model of chemical bonding called the valence bond model, which describes covalent bonds in compounds.

[2] The units used here for r_{ij}, r_0, N and B are nm. Crystallographers often prefer the non-SI unit Å, where $1 \text{nm} = 10 \text{ Å}$.

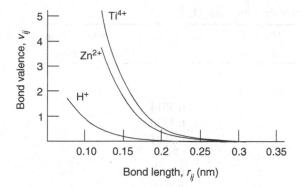

Figure 7.10 The relationship between bond valence, v, and ionic separation, (nm), for Ti^{4+}, Zn^{2+} and H^+, when bonded to oxide, O^{2-}

valences are added, the result will be less than 4.0, and the ion is said to be **under bonded**.

In order to apply the method to a crystal structure, take the following steps.

(i) Values of r_0 (and B, if taken as different from 0.037) are found from Tables, (see Biblography).

(ii) A list of bond valence (v_{ij}) values for the bonds around each cation are calculated, using the crystallographic bond lengths, r_{ij}.

(iii) The bond valence (v_{ij}) values are added for each cation to give a set of ionic valences.

If the structure is correct, the resulting ionic valence values should be close to the normal chemical valence.

The technique can be illustrated by describing the site allocation of cations in a crystal. The oxide $TiZn_2O_4$ crystallises with the AB_2O_4 *spinel* structure. In this structure, the two different cations, A and B, are distributed between two sites, one with octahedral coordination by oxygen, and one with tetrahedral coordination by oxygen. If all of the cations A are in the tetrahedral sites, (A), and all of the B cations are in octahedral sites, $[B]$, the formula is written $(A)[B_2]O_4$

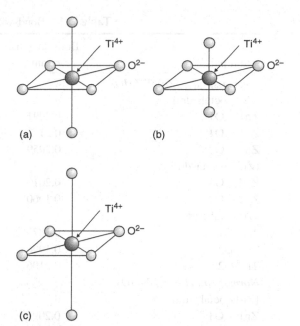

Figure 7.11 TiO_6 octahedra: (a) an ideal TiO_6 octahedron, with equal $Ti - O$ bond lengths of 0.1965 nm; (b) an over bonded TiO_6 octahedron, with two shorter bonds; (c) an under bonded TiO_6 octahedron with two longer bonds

and the compound is said to be a **normal** *spinel*. If all of the A cations are in the octahedral sites, $[A]$, and the B cations are distributed over the remaining octahedral and tetrahedral sites, (B) and $[B]$, the formula is written $(B)[AB]O_4$, and the compound is said to be an **inverse** *spinel*. The bond valence method can be used to differentiate between these two possibilities.

The inverse *spinel* structure $(Zn)[TiZn]O_4$ has Zn1 placed in octahedral sites, Zn2 placed in tetrahedral sites and Ti placed in octahedral sites. The normal *spinel* structure, $(Ti)[Zn_2]O_4$, exchanges the Ti in octahedral sites with Zn2 in tetrahedral sites. The bond valence sums for the two alternative structures are given in Table 7.1. The cation valences for the inverse structure, Zn1 = 2.13, Zn2 = 1.88, Ti = 3.85, are reasonably close to those expected, viz. Zn1 = 2.0,

Table 7.1   Bond-valence data for $TiZn_2O_4$

Atom pair	Bond length, r_{ij}/nm	Number of bonds	Bond valence, v_{ij}	Sum (cation valence)
Inverse spinel, (Zn)[TiZn]O₄				
[Zn1] octahedral				
Zn1 – O1	0.2091	2	0.3514	2.13
Zn1 – O1	0.2112	2	0.3320	
Zn1 – O2	0.2059	2	0.3831	
(Zn2) tetrahedral				
Zn2 – O1	0.2010	2	0.4373	1.88
Zn2 – O2	0.1960	2	0.5006	
[Ti] octahedral				
Ti – O1	0.1873	2	0.8558	3.85
Ti – O2	0.1996	2	0.6131	
Ti – O2	0.2106	2	0.4559	
Normal spinel, (Ti)[Zn₂]O₄				
[Zn1] octahedral				
Zn1 – O1	0.2091	2	0.3514	2.13
Zn1 – O1	0.2112	2	0.3320	
Zn1 – O2	0.2059	2	0.3831	
(Ti) tetrahedral				
Ti – O1	0.2010	2	0.5904	2.53
Ti – O2	0.1960	2	0.6758	
[Zn2] octahedral				
Zn2 – O1	0.1873	2	0.6340	2.85
Zn2 – O2	0.1996	2	0.4542	
Zn2 – O2	0.2106	2	0.3378	

Data taken from S. J. Marin, M. O'Keeffe and D. E. Partin, J. Solid State Chemistry, **113**, 413–419 (1994). The values of r_0 are Zn: 0.1704 nm; Ti, 0.1815 nm, from N. E. Breese and M. O'Keeffe, Acta Crystallogr., **B47**, 192–197 (1991).

Zn2 = 2.0, Ti = 4.0. In the case of the normal structure, the fit is nothing like so good, with Zn1 = 2.13, Zn2 = 2.85, Ti = 2.53. It is clear that the bond distances in the structure correspond more closely to the inverse cation distribution, (Zn)[TiZn]O₄. (Note that it may be possible to improve the fit further by assuming that the structure is not entirely inverse, but that a distribution holds so that most of the Ti ions occupy octahedral sites, but a small percentage occupy tetrahedral sites.)

7.9   Structures in terms of non-metal (anion) packing

The geometric problem of packing spheres is described in Section 7.2 above. In such structures, the spheres do not fill all the available volume. There are small holes between the spheres that occur in layers between the sheets of spheres. These holes, which are called **interstices**, **interstitial sites** or **interstitial positions**, are of two

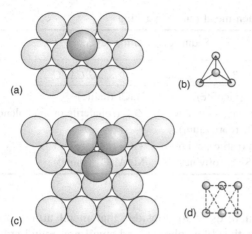

(a)

(b)

(c)

(d)

Figure 7.12 Tetrahedral and octahedral sites in closest-packed arrays of spheres: (a) a tetrahedral site between two layers and (b) the same site drawn as a tetrahedron; (c) an octahedral site between two layers and (d) the same site drawn as an octahedron

types (Figure 7.12). In one type of position, three spheres in the lower layer are surmounted by one sphere in the layer above, or vice versa. The geometry of this site is that of a **tetrahedron**. The other position is made up of a lower layer of three spheres and an upper layer of three spheres. The shape of the enclosed space is not so easy to see, but is found to have an **octahedral** geometry.

In the two closest packed sequences, ...$ABCABC$... and ...$ABAB$..., there are $2N$ tetrahedral interstices and N octahedral interstices for every N spheres. Ionic structures can be modelled on this arrangement by assuming that the anions pack together like spheres. The structure is made electrically neutral by placing cations into some of the interstices, making sure that the total positive charge on the cations is equal to the total negative charge on the anions. The formula of the structure can be found by counting up the numbers of each sort of ion present.

Consider the structures that arise from a cubic close packed array of X anions. If every octahedral position contains an M cation, there are equal numbers of cations and anions in the structure.

The formula of compounds with this structure is MX, and the structure corresponds to the *halite* (NaCl, B1) structure. If halide anions, X^-, form the anion array, to maintain charge balance, each cation must have a charge of $+1$. Compounds which adopt this arrangement include AgBr, AgCl, AgF, KBr, KCl, KF, NaBr, NaCl and NaF. Should oxide anions, O^{2-}, form the anion array, the cations must necessarily have a charge of $2+$, to ensure that the charges balance. This structure is adopted by a number of oxides, including MgO, CaO, SrO, BaO, MnO, FeO, CoO, NiO and CdO.

Should the anions adopt hexagonal close-packing and all of the octahedral sites contain a cation, the hexagonal analogue of the *halite* structure is produced. In this case, the formula of the crystal is again MX. The structure is the *nicolite*, (NiAs), structure, and is adopted by a number of alloys and metallic sulphides, including NiAs, CoS, VS, FeS and TiS.

If only a fraction of the octahedral positions in the hexagonal packed array of anions is filled a variety of structures results. The *corundum* structure is adopted by the oxides α-Al$_2$O$_3$, V$_2$O$_3$, Ti$_2$O$_3$ and Fe$_2$O$_3$. In this structure $\frac{2}{3}$ of the octahedral sites are filled in an ordered way. Of the structures that form when only half of the octahedral sites are occupied, those of rutile, (TiO$_2$) and α-PbO$_2$ are best known. The difference between the two structures lies in the way in which the cations are ordered. In the rutile form of TiO$_2$ the cations occupy straight rows of sites, while in α-PbO$_2$ the rows are staggered.

Structures containing cations in tetrahedral sites can be described in exactly the same way. In this case, there are twice as many tetrahedral sites as anions, and so if all sites are filled the formula of the solid will be M_2X. When half are filled this becomes MX, and so on.

A large number of structures can be generated by the various patterns of filling either the octahedral or tetrahedral interstices. The number can be extended if both types of position are

Table 7.2   Structures in terms of non-metal (anion) packing

Fraction of tetrahedral sites occupied	Fraction of octahedral sites occupied	Sequence of anion layers	
		...$ABAB$...	...$ABCABC$...
0	1	NiAs (nicolite)	NaCl (halite)
½	0	ZnO, ZnS (wurtzite)	ZnS (sphalerite or zinc blende)
0	⅔	Al_2O_3 (corundum)	—
0	½	TiO_2 (rutile), α-PbO_2	TiO_2 (anatase)
1/8	½	Mg_2SiO_4 (olivine)	$MgAl_2O_4$ (spinel)

occupied. One important structure of this type is the *spinel*, ($MgAl_2O_4$), structure, discussed above. The oxide lattice can be equated to a cubic close-packed array of oxygen ions, and the cubic unit cell contains 32 oxygen atoms. There are, therefore, 32 octahedral sites and 64 tetrahedral sites for cations. Only a fraction of these sites are filled, in an ordered way. The unit cell contains 16 octahedrally coordinated cations and 8 tetrahedrally coordinated cations, which are distributed in an ordered way over the available positions to give the formula $A_8B_{16}O_{32}$ or, as usually written, AB_2O_4. Provided that the charges on the cations A and B add to 8, any combination can be used. The commonest is $A = M^{2+}$, $B = M^{3+}$, as in spinel itself, $MgAl_2O_4$, but a large number of other possibilities exist including $A = M^{+6}$, $B = M^+$, as in Na_2WO_4, or $A = M^{+4}$, $B = M^{2+}$, as in $SnZn_2O_4$. In practice, the distribution of the cations over the octahedral and tetrahedral sites is found to depend upon the conditions of formation. The two extreme cation distributions, the *normal* and *inverse spinel* structures, are described above.

A small number of the many structures that can be described in this way is given in Table 7.2.

7.10   Structures in terms of metal (cation) packing

A consideration of metallic radii (Section 7.3), will suggest that an alternative way of modelling structures is to consider that metal atoms pack together like spheres, and small non-metal atoms fit into interstices so formed. In structural terms, this is the opposite viewpoint to that described in the previous section. In fact metallurgists have used this concept to describe the structures of a large group of materials, the **interstitial alloys**, for many years. The best known of these are the various steels, in which small carbon atoms occupy interstitial sites between large iron atoms in the A1 (cubic close-packed) structure adopted by iron between the temperatures of 912 °C and 1394 °C.

The simplest structures to visualise in these terms are those that can be derived directly from the structures of the pure metals, the A1 (cubic close-packed) or the A3, (hexagonal close-packed) structures. The *halite* structure of many solids of composition, *MX*, described above in terms of anion packing, can be validly regarded in terms of metal packing. For example, Ni, Ca and Sr adopt the A1 copper structure. The corresponding oxides, NiO, CaO and SrO can be built, at least conceptually, by the insertion of oxygen into every octahedral site in the metal atom array. Similarly, Zn adopts the A3 magnesium structure, and ZnO can be built by filling one half of the tetrahedral sites, in an ordered way, with oxygen atoms. The mechanism of oxidation of these metals often follows a route in which oxygen at first diffuses into the surface layers of the metal via octahedral sites, supporting a view of the structure in terms of large metal atoms and small non-metal atoms.

Many other inorganic compounds can be derived from metal packing in similar ways. For example, the structure of the mineral fluorite, CaF_2, can be envisaged as the A1 structure of the metal Ca, with every tetrahedral site filled by fluorine. Each Ca atom is surrounded by a distorted cube of fluorine atoms.

Metal packing models can account for more structures if alloys are chosen as the starting metal atom array. For example, the AB_2 metal atom positions in the normal spinel structure are identical to the metal atom positions in the cubic alloy $MgCu_2$. In spinel itself, $MgAl_2O_4$, the metal atom arrangement is identical to that of a cubic $MgCu_2$ alloy of formula $MgAl_2$. The spinel structure results if oxygen atoms are allocated to every tetrahedral site present in $MgAl_2$.

For many more examples see the Bibliography.

7.11 Cation-centred polyhedral representations of crystals

A considerable simplification in describing structural relationships can be obtained by drawing a structure as a set of linked polyhedra. The price paid for this simplification is that important structural details are often ignored, especially when the polyhedra are idealised. The polyhedra selected for such representations are generally metal – non-metal coordination polyhedra, composed of a central metal surrounded by a number of neighbouring non-metal atoms. To construct a metal-centred polyhedron, the non-metal atoms are reduced to points, which form the vertices of the anion polyhedron. These polyhedra are then linked together to build up the complete structure. As an example, the fluorite structure, described in Section 1.10, can be depicted as a three-dimensional packing of calcium centred CaF_8 cubes, (Figure 7.13). Each cube shares all its edges with neighbours, to

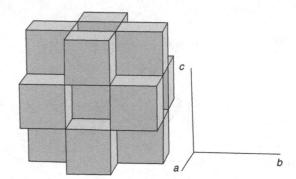

Figure 7.13 The fluorite (CaF_2) structure as an array of edge-shared CsF_8 cubes

create a 'three-dimensional chess board' array. A model can easily be constructed from sugar cubes.

Octahedral coordination is frequently adopted by the important 3d transition-metal ions. In this coordination polyhedron, each cation is surrounded by six anions, to form an octahedral $[MO_6]$ group, (Figure 7.14). The cubic structure of rhenium trioxide, ReO_3, $a = 0.3750$ nm, is

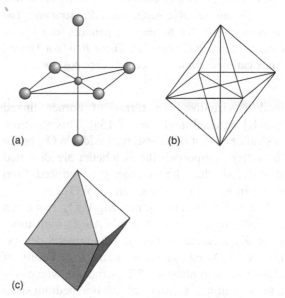

Figure 7.14 Representations of octahedra: (a) 'ball and stick', in which a small cation is surrounded by six anions; (b) a polyhedral framework; (c) a 'solid' polyhedron

ReO$_6$
octahedron →

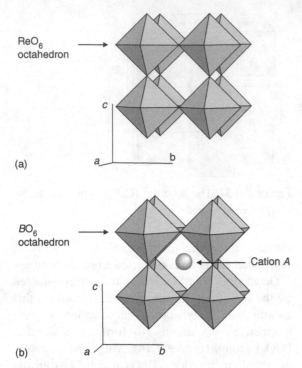

(a)

BO$_6$
octahedron →

← Cation A

(b)

Figure 7.15 (a) Perspective view of the cubic structure of ReO$_3$, drawn as corner-linked ReO$_6$ octahedra; (b) the idealised cubic ABO_3 *perovskite* structure. The framework, similar to that in part (a), consists of corner-shared BO_6 octahedra (note B is not boron). An A cation occupies the central cage position

readily visualised in terms of corner linked [ReO$_6$] octahedra, (Figure 7.15a). This structure is similar to that of tungsten trioxide, WO$_3$, but in the latter compound, the octahedra are distorted slightly, so that the symmetry is reduced from cubic in ReO$_3$, to monoclinic in WO$_3$.

Just as the mineral spinel, MgAl$_2$O$_4$, has given its name to a family of solids AB_2X_4 with topologically identical structures, the mineral perovskite, CaTiO$_3$ has given its name to a family of closely related phases ABX_3, structure, where A is a large cation, typically Sr^{2+}, B is a medium sized cation, typically Ti^{4+} and X is most often O^{2-}. The structures of these phases is often idealised to a cubic framework with a unit cell edge of

approximately 0.375 nm. It is seen that the skeleton of the idealised structure, composed of corner linked BO_6 octahedra, is identical to the ReO$_3$ structure, (Figure 7.15a). The large A cations sit in the cages that lie between the octahedral famework, (Figure 7.15b). Most real *perovskites*, including perovskite (CaTiO$_3$) itself, and barium titanate, BaTiO$_3$, are built of slightly distorted [TiO$_6$] octahedra, and as a consequence the crystal symmetry is reduced from cubic to tetragonal, orthorhombic or monoclinic.

The complex families of silicates are best compared if the structures are described in terms of linked tetrahedra. The tetrahedral shape used is the idealised coordination polyhedron of the [SiO$_4$] unit, (Figure 7.16). Each silicon atom is linked to four oxygen atoms by tetrahedrally

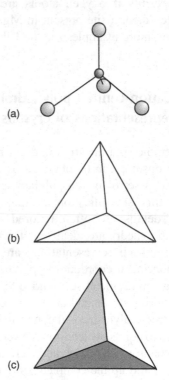

(a)

(b)

(c)

Figure 7.16 Representations of tetrahedra: (a) 'ball and stick', in which a small cation is surrounded by four anions; (b) a polyhedral framework; (c) a 'solid' polyhedron

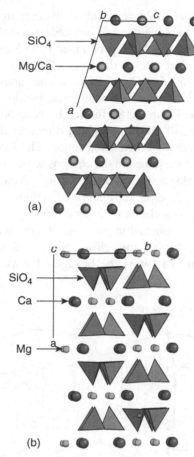

Figure 7.17 The structure of diopside: (a) projection down the monoclinic **b**-axis; (b) projection down the monoclinic **c**-axis

directed sp³-hybrid bonds. For example, Figure 7.17 shows the structure of diopside, $CaMgSi_2O_6$, drawn so as to emphasise the $[SiO_4]$ tetrahedra. These form chains linked by two corners, running in (100) layers, to provide the medium sized Mg^{2+} ions with octahedral (6-fold) coordination and the large Ca^{2+} ions with cubic (8-fold) coordination. For many purposes this depiction is clearer than that given in Chapter 5, (Figure 5.6). In particular, the $[SiO_4]$ units are very strong and persist during physical and chemical reactions, so that structural transformations of diopside

are more easily visualised in terms of the rearrangement of the $[SiO_4]$ tetrahedra.

Although the polyhedral representations described are generally used to depict structural relationships, they can also be used to depict diffusion paths in a compact way. The edges of a cation centred polyhedron represent the paths that a diffusing anion can take in a structure, provided that anion diffusion takes place from one normal anion site to another. Thus anion diffusion in crystals with the fluorite structure will be localised along the cube edges of Figure 7.13.

7.12 Anion-centred polyhedral representations of crystals

Just as viewing crystals in terms of metal atom packing instead of anion packing, (Section 7.10 *vs.* Section 7.9), gives a new perspective to structural relationships, so does viewing structures as made up of linked anion-centred polyhedra rather than cation-centred polyhedra. For example, the fluorite structure, described as corner-shared CaF_8 cubes above (Figure 7.15), can equally well be described in terms of linked FCa_4 tetrahedra.

The approach can be generalised to reveal structural relationships and cation diffusion paths in solids. As an example, the anion centred polyhedron formed by taking all of the possible cation sites in a face-centred cubic packing of anions as vertices is a rhombic dodecahedron, (Figure 7.18). The polyhedron can be related to the cubic axes by comparing Figure 7.18a with Figure 2.9d. However, it is often convenient to relate the representation to that of close packed anion layers, described in Section 7.9, and to do this it is best to make the direction normal to the layers, the [111] direction, vertical, (Figure 7.18b), by rotating the polyhedron anticlockwise.

The resulting anion-centred rhombic dodecahedron can be used to describe the possible structures derived from a face-centred cubic packing of anions. The octahedral sites in the structure are

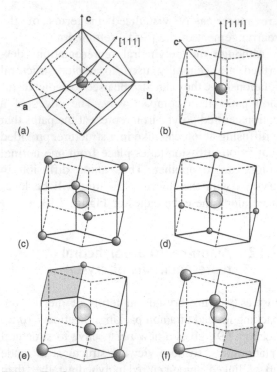

Figure 7.18 The anion-centred polyhedron (rhombic dodecahedron) found in the cubic closest-packed structure: (a) oriented with respect to cubic axes, the **c**-axis is vertical; (b) oriented with [111] vertical; (c) cation positions occupied in the *halite*, NaCl, structure; (d) cation positions occupied in the *zinc blende (sphalerite)*, ZnS, structure (e, f) the two anion-centred polyhedra needed to create the *spinel*, $MgAl_2O_4$, structure. Cations in tetrahedral sites are small and cations in octahedral sites are medium-sized. Adapted from E. W. Gorter, Int. Cong. for Pure and Applied Chemistry, Munich, 1959, Butterworths, London, 1960, p 303

represented by vertices at which four edges meet, while tetrahedral sites are represented by vertices at which three edges meet. The anion-centred polyhedral representation of the *halite* structure, in which all of the octahedral sites are filled, is drawn in Figure 7.18c. The corresponding anion-centred polyhedron for the cubic zinc blende (sphalerite) structure of ZnS, in which half of the tetrahedral sites are filled, is shown in Figure 7.18d.

The *spinel* structure, a cubic structure in which an ordered arrangement of octahedral and tetrahedral sites are filled, needs two identical blocks, one rotated with respect to the other, (Figure 7.18e, f), for construction. One block in the orientation Figure 7.18e is united with three blocks in the orientation Figure 7.18f, formed by rotation of 0°, 120° and 240° about the [111] direction through the tetrahedrally coordinated cation. The faces that are shaded are joined in the composite polyhedron.

The analogous anion-centred polyhedron for an hexagonal close-packed anion array, (Figure 7.19a), is similar to a rhombic dodecahedron, but has a mirror plane normal to the vertical axis, which is perpendicular to the close packed planes of anions and so forms the hexagonal **c**-axis. The

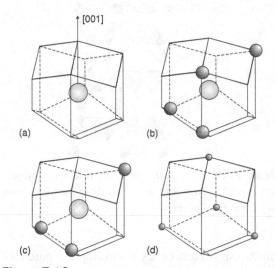

Figure 7.19 The anion-centred polyhedron found in the hexagonal closest-packed structure: (a) oriented with the hexagonal **c**-axis vertical; (b) cation positions occupied in the ideal *corundum*, Al_2O_3, structure; (c) cation positions filled in the idealised *rutile*, TiO_2, structure; (d) cation positions occupied in the *wurtzite*, ZnS, structure, the central anion is omitted for clarity. Cations in tetrahedral sites are small and cations in octahedral sites are medium-sized. Adapted from E. W. Gorter, Int. Cong. for Pure and Applied Chemistry, Munich, 1959, Butterworths, London, 1960, p 303

octahedral sites in the structure are again found at the vertices where four edges meet, and the tetrahedral sites at the vertices where three edges meet. The relationship between structures formed by an ordered filling these sites, such as corundum, Al_2O_3, (Figure 7.19b), in which ⅔ of the octahedral sites are filled in an ordered manner, and (idealised) rutile, TiO_2, (Figure 7.19c), in which half of the octahedral sites are filled in is readily portrayed. Filling of half the tetrahedral sites leads to the hexagonal wurtzite form of ZnS, (Figure 7.19d).

Just as the edges of cation-centred polyhedra represent anion diffusion paths in a crystal, the edges of anion centred polyhedra represent cation diffusion paths. The polyhedra shown in Figures 7.18 reveal that cation diffusion in cubic close-packed structures will take place via alternative octahedral and tetrahedral sites. Direct pathways, across the faces of the polyhedron, are unlikely, as these mean that a cation would have to squeeze directly between two anions. There is no pre-ferred direction of diffusion. For ions that avoid either octahedral or tetrahedral sites, for bonding or size reasons, diffusion will be slow compared to ions which are able to occupy either site. In solids in which only a fraction of the available metal atom sites are filled, such as the spinel structure, clear and obstructed diffusion pathways can easily be delineated.

The situation is different in the hexagonal close-packed structures. The octahedral and tetra-hedral sites are arranged in chains parallel to the c-axis, (Figure 7.19). A diffusing ion in one of these sites, either octahedral or tetrahedral, can jump to another without changing the nearest neighbour coordination geometry. However, dif-fusion perpendicular to the c-axis must be by way of alternating octahedral and tetrahedral sites. Thus a cation that has a preference for one site geometry, (for example, Zn^{2+} prefers tetrahedral coordination and Cr^{3+} prefers octahedral coordi-nation), will find diffusion parallel to the c-axis easier than diffusion parallel to the a- or b-axes. This effect is enhanced in structures in which

only a fraction of the cation sites are occupied. For example, cation diffusion parallel to the c-axis in the rutile form of TiO_2, (Figure 7.19c), is of the order of 100 times faster than perpendicular to the c-axis.

The use of anion-centred polyhedra can be parti-cularly useful in describing diffusion in fast ion conductors. These materials, which are solids that have an ionic conductivity approaching that of liquids, find use in batteries and sensors. An example is the high temperature form of silver iodide, α-AgI. In this material, the iodide anions form a body-centred cubic array, (Figure 7.20a).

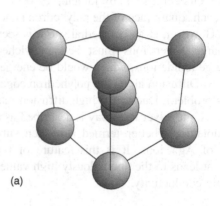

(a)

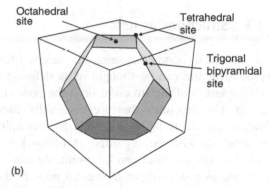

(b)

Figure 7.20 Perspective view of the structure of α-AgI: (a) the body-centred cubic arrangement of iodide ions in the unit cell; (b) the anion centred polyhedron, (a truncated octahedron), around the iodide ion at the unit cell centre. [Note that the ions in part (a) are depicted smaller than ionic radii suggest, for clarity]

The anion-centred polyhedron of this structure is a truncated octahedron, (Figure 7.20b). Each vertex of the polyhedron represents a tetrahedral cation site in the structure. Octahedral sites are found at the centres of each square face, and trigonal pyramidally coordinated cation sites occur at the midpoints of each edge, (Figure 7.20b). In α-AgI, two silver atoms occupy the twelve available tetrahedral sites on a statistical basis. These cations are continually diffusing from one tetrahedral site to another, along paths represented by the edges of the polyhedron in a sequence ... tetrahedral – trigonal bipyramidal – tetrahedral – trigonal bipyramidal . . . . Movement from one tetrahedral site to another via octahedral sites, by jumping across the centres of each square face of the polyhedron is not forbidden. However, at an octahedral site it is seen that the large silver ions must be sandwiched between large iodide ions, which would be energetically costly. Diffusion along the polyhedron edges avoids this problem. Due to the high diffusion rate of the cations, the silver ion array is described as a 'molten sublattice' (better termed a molten sub-structure), of Ag^+ ions. It is this feature of the structure that leads to the anomalously high values for the ionic conductivity.

7.13   Structures as nets

The chemical bonds between the atoms of a crystal structure can be thought of as defining a net. The atoms of the structure lie at the nodes of the net. (In structures that are essentially two-dimensional, many layer structures, for example, the nets can also be regarded as tilings.) For example, both graphite and boron nitride can be described as a stacking of hexagonal nets. In the graphite structure, (Figure 7.21 a, b), the nets are staggered in passing from one layer to the other, so as to minimise electron – electron repulsion between the layers. In the case of boron nitride, (Figure 7.21 c, d), the layers are stacked vertically over each other, with boron atoms lying over and

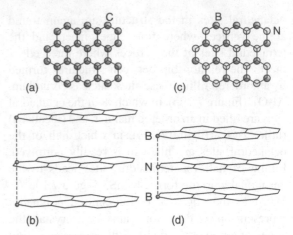

Figure 7.21 The structures of graphite and boron nitride: (a) a single layer of the graphite structure; (b) the stacking of layers, represented as nets, in graphite; (c) a single layer of the boron nitride structure; (d) the stacking of layers, represented as nets, in boron nitride

under nitrogen atoms in each layer, reflecting the different chemical bonding in this material. Naturally, these atomic differences may be lost when structures are represented as nets.

It is more difficult to represent truly three-dimensional structures by nets in plane figures, but the tetrahedral bond arrangement around carbon in diamond or around zinc in zinc sulphide and zinc oxide, are well suited to this representation. In the cubic diamond structure, (Figure 7.22a), each carbon atom is bonded to four others, at the vertices of a surrounding tetrahedron. The structure can also be described as a net with tetrahedral connections, (Figure 7.22b). If the carbon atoms are replaced by alternating sheets of zinc (Zn) or sulphur (S), the structure is that of cubic zinc blende (sphalerite), (Figure 7.22c, see also Figure 8.8). The net that represents this structure, Figure 7.22d, is exactly the same as the diamond net. Zinc sulphide also adopts a hexagonal symmetry in the wurtzite structure, identical to that of zincite, ZnO, (Figure 7.22e,

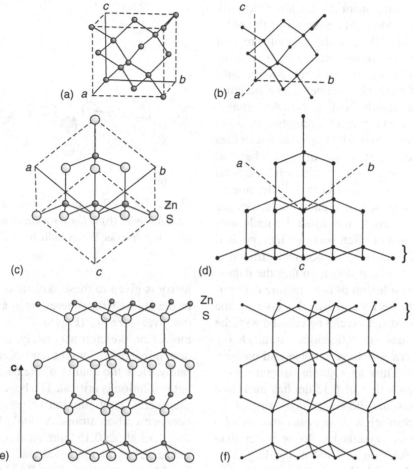

Figure 7.22 Three-dimensional nets: (a) the cubic diamond structure; (b) the net equivalent to (a); (c) the cubic zinc blende (sphalerite) structure; (d) the net equivalent to (c), which is identical to that in (b); (e) the hexagonal wurtzite structure; (f) the net equivalent to (e)

see also Figure 8.4). The net that represents this structure is the structure of 'hexagonal diamond' (Figure 7.22f).

If the puckered nets formed by the bases of the tetrahedra, bracketed (}) in Figure 7.22d, f, are flattened, they form hexagonal sheets identical to those in the boron nitride and graphite nets.

The representation of crystal structures as nets is often of help in demonstrating structural relationships that are not lucidly portrayed by the use of polyhedra or sphere packing. Research in this

area is active, and the portrayal of complex structures as (frequently, very beautiful) nets is important in the design and synthesis of new solids, as well as in the correlation of various structure types (see Bibliography).

7.14 The depiction of organic structures

The classical aspects of crystallography are centred upon inorganic or metallic crystals. This is because the component atoms are linked by strong chemical bonds with energies of the order of hundreds of kJ mol^{-1}, so that crystals are well formed and stable. Moreover, the early crystallographers could easily obtain large, perfect samples by way of mineral specimens. The situation with organic materials is different. An organic crystal structure, which can be loosely defined as the way in which individual organic molecules pack, is produced as a result of weak chemical bonds, such as dispersion forces, which have bond energies of only one or two kJ mol^{-1}. Because of this, crystals composed of organic molecules are often rather disordered, and when the molecules become large, solvent often becomes incorporated into the solid. This causes the spots on a diffraction pattern to blur and spread out, so that the data is degraded. The **resolution** of the structure determination is said to be reduced. In this context, the resolution of the structure can be equated with the lowest d_{hkl} values of reflections on an X-ray diffraction pattern that have reliable and measurable intensities. Thus an organic structure computed with a resolution of 0.15 nm has measured diffracted intensities down a d_{hkl} spacing of 0.15 nm. The resolution of polymer macromolecules and protein crystals has to be better than approximately 0.15 nm to avoid misleading interpretations. **High resolution** data in protein crystallography refers to data with a resolution of 0.12 nm or better.

The resolution needed for a particular structure determination will depend on the type of information sought. A determination of the overall crystal structure in terms of molecular packing is relatively undemanding. However, the object of many crystallographic studies of organic solids is to obtain information on the actual geometry of the molecules in the structure. Organic molecules are often represented by net-like fragments, in which the links between carbon atoms are represented by 'sticks' and the atoms by 'balls'. A more realistic impression of the molecular geo-

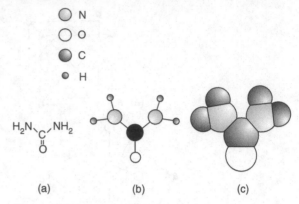

Figure 7.23   Depiction of urea: (a) chemical structural formula; (b) using covalent radii from figure 7.8; (c) using van der Waals radii from Figure 7.9

metry is given to these skeletal outlines by filling in the atoms with spheres sized according to van der Waals radii, (Figure 7.23). To depict the molecular skeleton accurately, it is necessary to determine bond lengths and bond angles accurately. Here the quality of the data and the resultant resolution is critical. The bonds that determine the overall configuration of these molecules are between carbon atoms. A single C-C bond has a length of about 0.154 nm, an aromatic bond, as in benzene, has a length of about 0.139 nm, a double bond has a length of about 0.133 nm, and a triple bond has a length of about 0.120 nm. That is, a total range of about 0.034 nm covers all of these bond lengths. The resolution of the diffraction data must be such as to reliably distinguish these bond types if useful structural and chemical information is to be produced.

The disposition of hydrogen atoms is another aspect of the crystallography of organic compounds that is vital. This is because these control the formation of hydrogen bonds in the crystal. Hydrogen bonds are bonds formed when a hydrogen atom lies close to, and more or less between, two strongly electronegative atoms. In organic crystals, hydrogen bonds usually involve oxygen or nitrogen atoms. The energy of a hydrogen bond is not all that great, of the order of $20\,kJ\,mol^{-1}$. Despite this, hydrogen bonding has a vital role to play in biological activity. For example, the base pairing in DNA and related molecules is mediated by hydrogen bonding. Similarly, the folding of protein chains into biologically active molecules is almost totally controlled by hydrogen bonding. As hydrogen is the lightest atom, great precision in diffraction data is needed to delineate probable regions where hydrogen bonding may occur. Indeed, neutron diffraction is often needed for this purpose, as neutrons are more sensitive to hydrogen than X-rays (see Section 6.14).

The ball and stick or space-filling approach to the depiction of organic structures is satisfactory when small molecules are concerned, but it becomes too cumbersome for the representation of large molecules. In the case of complex naturally occurring molecules, this level of detail may even obscure the way in which the molecule functions biologically. The problem of structural characterisation in such large molecules will be discussed with respect to proteins, in the following section.

7.15   The representation of protein structures

We have seen that in even very simple materials, the overall pattern of structure in the crystal is not always easy to appreciate just by plotting the contents of a single unit cell. This problem of visualisation is far greater when complex structures such as proteins are considered. One of the more important tasks for the scientists that determined the first protein structures was how to depict the data in such a way that biological activity could be inferred without a loss of chemical or structural information. The way in which this is achieved is to describe protein structures by way of a hierarchy of levels.

The sequence of amino acids present along the protein chain is known as the **primary structure** of the molecule, (see Section 6.15 and Scheme 6.2). The primary structure is written from left to right, starting with the amino ($-NH_2$) end of the molecular chain and ending with the carboxyl ($-COOH$) end.

Although the primary structure of a protein is of importance, the way in which the polypeptide chain folds controls the chemical activity of the molecule. The folding is generally brought about by hydrogen bonds between oxygen and nitrogen atoms in the chains, and mistakes in the folding lead to the malfunction of the protein and often to illness. The overall complexity of a protein molecule can be dissected by making use of the fact that, in all proteins, relatively short stretches of the polypeptide chains are organised into a small number of particular conformations. These local forms make up the **secondary structure** of the protein. They are often represented in a 'cartoon' form by coiled or twined ribbons. The most common secondary structures are helices and β-sheets.

The polypeptide chain can coil into a number of different helical arrangements, but the commonest is the α-**helix**, a right-handed coil. The backbone of the helix is composed of the repeating sequence: (i) carbon atom double bonded to oxygen, $C=O$; (ii) carbon atom bonded to a general organic group, $C-R$; (iii) nitrogen, N, (Figure 7.24a). The helical conformation is a result of the geometry of the bonds around the carbon atoms and hydrogen bonding between hydrogen atoms on N and oxygen atoms on $C-O$ groups. The repeat distance between any two corresponding atoms, the **pitch** of the helix, is 0.54 nm, and there are 3.6 amino acids, called **residues**, in each turn, (Figure 7.24b). The $-NH_2$ end of the helix has a positive charge relative to the $-COOH$ end, which is negative.

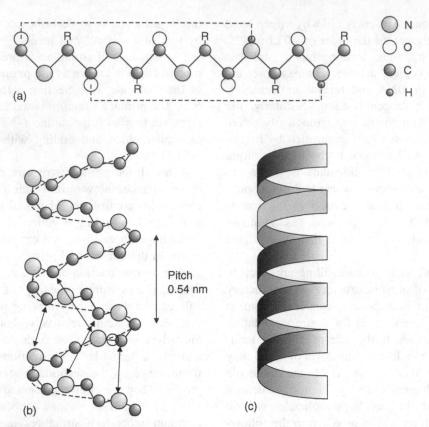

Pitch
0.54 nm

Figure 7.24 The α-helix: (a) the succession of carbon and nitrogen atoms along the backbone of an α-helix, hydrogen bonds form between O atoms on carbon and H atoms on nitrogen linked by dashed lines, R represents a general organic side-group and for clarity not all hydrogen atoms are shown; (b) schematic depiction of the coiled α-helix, with four hydrogen bonds indicated by double headed arrows; (c) 'cartoon' depiction of an α-helix as a coiled ribbon

The 'cartoon' depiction of the α-helix as a coiled ribbon is drawn in Figure 7.24c. The structure of the protein myoglobin, the first protein structure to be solved using X-ray diffraction, is built of a compact packing of eight α-helices, (Figure 7.25).

Sheets are formed by association of lengths of polypeptide chain called β-**strands**. The backbone of a β-strand is exactly the same as that of an α-helix, (Figure 7.26a). A single β-stand is not stable alone, but links to a parallel strand to form a pair, via hydrogen bonding. The direction of the strands runs from the −NH₂ termination to the −COOH termination. Pairs of strands can run in the same direction, to give **parallel chains**, or in opposite directions to give **antiparallel chains**, (Figure 7.26b, c). The antiparallel arrangement is more stable than the parallel arrangement, but both occur in protein structures. A β-strand is represented in 'cartoon' form by a broad ribbon arrow, with the head at the −COOH end, (Figure 7.26d). Several parallel or antiparallel β-strands in proximity form a β-**sheet** or β-**pleated sheet**, (Figure 7.26e). Stable sheets

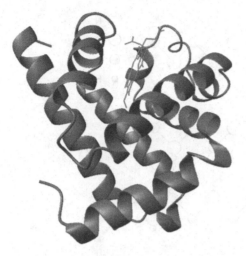

Figure 7.25 The structure of the protein myoglobin. This is composed of a compact packing of eight α-helices around the planar heme group. [This figure provided courtesy of D. Whitford, *Proteins, Structure and Function*, John Wiley & Sons, Ltd., Chichester, (2005), reproduced with permission]

can consist of just two or more antiparallel β-strands, or, due to the lower stability of the parallel arrangement, four or more parallel β-strands. A typical secondary structure fragment of a protein, containing helices and strands, is drawn in Figure 7.27a.

The **tertiary structure** of proteins, also called the **super-secondary structure**, denotes the way in which the secondary structures are assembled in the biologically active molecule, (Figure 7.27b). These are described in terms of **motifs**[3], that consist of a small number of secondary structure elements linked by **turns**, such as (α-helix – turn – α-helix). Motifs are further arranged into larger arrangements called **folds**, which can be, for example, a collection of β-sheets arranged in a

cylindrical fashion to form a β-barrel, or **domains**, which may be made up of helices, β-sheets or both.

This tertiary organisation governs molecular activity and is of the utmost importance in living systems. It is mediated by fairly weak chemical bonds between the components of the polymer chains, including hydrogen bonding, van der Waals bonds, links between sulphur atoms, called disulphide bridges, and so on. The bonds need to be relatively weak, because biological activity often requires the tertiary structure of the protein to change in response to biologically induced chemical signals. A mistake at a single amino acid position may prevent these weak interactions from operating correctly, leading to folding errors that can have disastrous implications for living organisms. The tertiary structure is easily disrupted, and the protein is said to become **denatured**. A typical example is given by the changes that occur when an egg is cooked. The white insoluble part of the cooked egg is denatured albumin. In denatured proteins, the primary structure of the polymer chains is maintained, but the coiling into a tertiary form is destroyed, and biological activity is lost.

Many proteins are built from more than one polypeptide chain or **subunit**. The bonding and arrangement of these is the **quaternary structure** of the protein, (Figure 7.27c). For example, the globular protein haemoglobin is composed of two distinct subunits, called α- and β-chains. These link to form αβ-dimers, and two of these, related by a two-fold axis, form the molecule, which thus consists of four subunits in all. When the molecule gains or looses oxygen, both the ternary and quaternary structures change significantly.

Protein studies make up one of the most important and dynamic areas of crystallography, and more information on how these immensely complex structures are described in this fast growing subject area is given in the Bibliography.

[3]Note that this use of the term motif is quite different than that used normally in crystallography - to indicate the unit of structure to be placed at each lattice point to recreate the whole structure.

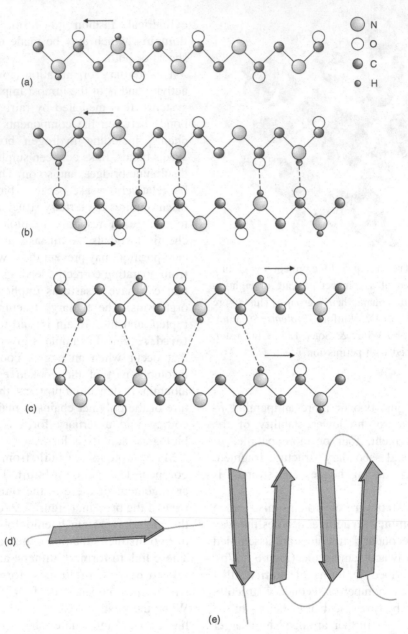

Figure 7.26   The β-sheet: (a) the succession of carbon and nitrogen atoms along the backbone of a β-strand, the arrow indicates the direction of the strand, from the −NH₂ termination to the −COOH termination, (for clarity not all hydrogen atoms and side groups are shown); (b) schematic depiction of a pair of antiparallel β-strands, hydrogen bonds form between O atoms on carbon and H atoms on nitrogen linked by dashed lines; (c) schematic depiction of a pair of parallel β-strands, hydrogen bonds form between O atoms on carbon and H atoms on nitrogen linked by dashed lines; (d) 'cartoon' depiction of a β-strand as a ribbon, in which the arrow head represents the −COOH termination; (e) a β-sheet composed of four antiparallel β-strands

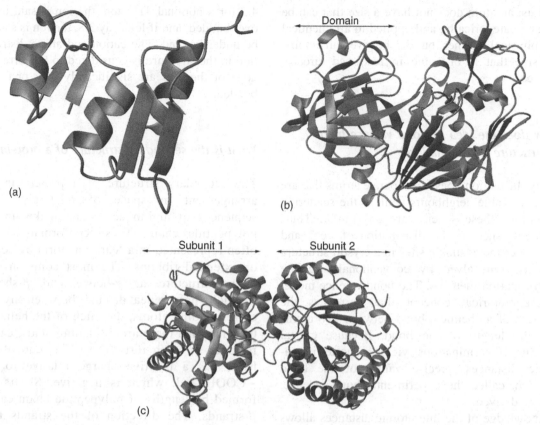

Figure 7.27 Organisational levels in protein crystallography: (a) secondary structure; (b) tertiary structure; (c) quaternary structure. [This figure provided courtesy of D. Whitford, *Proteins, Structure and Function*, John Wiley & Sons, Ltd., Chichester, (2005), reproduced with permission]

Answers to introductory questions

What is the size of an atom?

Quantum mechanics makes it clear that atoms do not have a fixed size. Electron orbitals extend from the nucleus to a greater or lesser extent, depending upon the chemical and physical environment in the locality of the atomic nucleus. The scattering of X-rays and electrons gives details of the varying electron density throughout the unit cell volume. As the electron density is highest nearer to atomic cores, and

least well away from atomic cores, these diffraction experiments yield the positions for the atomic cores, that is, interatomic distances are obtained, not atomic sizes. The tables of atomic radii found in textbooks are derived by dividing up the interatomic distances derived by diffraction methods to allocate radii to each atom involved. To some extent these radii are arbitrary, being chosen to reflect chemical or physical aspects of the crystal, which accounts for the differing values to be found for the radius of any atom.

Thus, an atom does not have a size that can be defined categorically, and applied to all chemical and physical regimes, but can be given an empirical size that is of value in restricted circumstances.

How does the idea of bond valence help in structure determination?

X-ray diffraction does not separate atoms that are Periodic Table neighbours well, as the scattering factors of these species are so similar. Thus, problems such as the distribution of Fe^{2+} and Fe^{3+} over the available sites in a crystal structure may be unresolved by conventional structure determination methods. The bond valence model is an empirical concept that correlates the strength of a chemical bond between two atoms and the length of the bond. Because crystal structure determinations yield accurate interatomic distances, precise values of the bond strength, called the experimental bond valence, can be derived.

Knowledge of the interatomic distances allows the experimental valence of an atom to be calculated using tables of bond-valence parameters. If the apparent valence of an atom is higher than that expected for an ionic bonding model, say

4.2 for a nominal Ti^{4+} ion, the ion is said to be over bonded, and if less, say 3.8, the ion is said to be under bonded. The cation and anion distribution in the structure is considered to be correct if none of the ions are significantly over or under bonded.

What is the secondary structure of a protein?

The secondary structure of a protein is the arrangement in space of relatively short sequences of amino acids that make up the polypeptide chain. These conformations are often represented in a 'cartoon' form by coiled or extended ribbons. The most common secondary structures are α-helices and β-sheets. The helix has a repeat distance between any two corresponding atoms, the pitch of the helix, of 0.54 nm, and there are 3.6 amino acids, called residues, in each turn. The $-NH_2$ end of the helix has a positive charge relative to the $-COOH$ end, which is negative. Sheets are formed by lengths of polypeptide chain called β-strands. The direction of the strands runs from the $-NH_2$ termination to the $-COOH$ termination. A single β-stand is not stable alone, but links to parallel strands to form sheets, via hydrogen bonding.

Problems and exercises

Quick quiz

1 X-ray diffraction is able quantify:
   (a) The volume of an atom
   (b) The radius of an atom
   (c) The distance between two atoms

2 The sphere packing that gives rise to a cubic structure is described as:
   (a) ... ABABAB ...
   (b) ... ABCABCABC ...
   (c) ... ABACABAC ...

3 The radius of a cation M^{4+} is:
   (a) Greater than the radius of M^{2+}

(b) Less than the radius of M^{2+}

(c) The same as the radius of M^{2+}

4 Space filling models of organic molecules use:
(a) Van der Waals radii

(b) Covalent radii

(c) Ionic radii

5 The largest cations tend to be surrounded by anions at the vertices of:
(a) A tetrahedron

(b) An octahedron

(c) A cube

6 In a cubic close-packed array of N spheres there are:
(a) N octahedral interstices

(b) $2N$ octahedral interstices

(c) $4N$ octahedral interstices

7 In a solid represented by packing of anion-centred polyhedra, the polyhedron edges represent possible:
(a) Cation diffusion paths

(b) Anion diffusion paths

(c) Both cation and anion diffusion paths

8 The resolution quoted for a protein structure corresponds to:
(a) The wavelength of the radiation used in the diffraction experiment

(b) The smallest d_{hkl} value for which reliable intensity data is available

(c) The lowest diffracted intensity measured

9 Protein structures are described in a hierarchy of:
(a) Two levels

(b) Three levels

(c) Four levels

10 In a protein structure, a β-pleated sheet is regarded as part of the:
(a) Primary structure

(b) Secondary structure

(c) Ternary structure

Calculations and Questions

7.1 Estimate the ideal lattice parameters of the room temperature and high temperature forms of the following metals, ignoring thermal expansion: (a) Fe [r(CN12) = 0.1274 nm] adopts the A2 (body-centred cubic) structure at room temperature and the A1 (face-centred cubic) structure at high temperature; (b) Ti [r(CN12) = 0.1462 nm] adopts the A3 (hexagonal close packed) structure at room temperature and the A2 (body-centred cubic) structure at high temperature; (c) Ca [r(CN12) = 0.1974 nm] adopts the A1 (face-centred cubic) structure at room temperature and the A2 (body-centred cubic) structure at high temperature. The radii given, r(CN12), are metallic radii appropriate to 12 coordination.

7.2 The following compounds all have the *halite* (NaCl) structure (see Chapter 1). LiCl, a = 0.512954 nm; NaCl, a = 0.563978 nm; KCl, a = 0.629294 nm; RbCl, a = 0.65810 nm; CsCl, a = 0.7040 nm. Make a table of the ionic radii of the ions assuming that the anions are in contact in the compound with the smallest cations.

7.3 Because the X-ray scattering factors of Mg and Al are similar, it is not easy to assign the cations in the mineral spinel, $MgAl_2O_4$ to either octahedral or tetrahedral sites (see Section 7.8 for more information). The bond lengths around the tetrahedral and octahedral positions are given in the table. Use the bond valence method to determine whether the spinel is normal or inverse. The values of r_0 are: r_0 (Mg^{2+}) = 0.1693 m, r_0 (Al^{3+}) = 0.1651 nm, B = 0.037 nm, from

N. E. Breese and M. O'Keeffe, Acta Crystallogr., **B47**, 192–197 (1991).

	Number of bonds	r_{ij}/nm
Tetrahedral site		
$M - O$	4	0.19441
Octahedral sites		
$M - O$	6	0.19124

Data from N. G. Zorina and S. S. Kvitka, Kristallografiya, **13**, 703–705 (1968), provided by the EPSRC's Chemical Database Service at Daresbury, (see Bibliography).

7.4 The tetragonal tungsten bronze structure is adopted by a number of compounds including $Ba_3(ZrNb)_5O_{15}$. There are two different octahedral positions in the structure, which are occupied by the ions Zr^{4+} and Nb^{5+}. The compound metal – oxygen bond lengths, r_{ij}, for the two octahedral sites are given in the table. Use the bond-valence method to determine which ion occupies which site. The values of r_0 are: r_0 (Nb^{5+}) $= 0.1911$ m, r_0 (Zr^{4+}) $= 0.1937$ nm, $B = 0.037$ nm, from N. E. Breese and M. O'Keeffe, Acta Crystallogr., **B47**, 192–197 (1991).

	Number of bonds	r_{ij}/nm*
Site 1		
$M - O1$	1	0.1858
$M - O2$	1	0.2040
$M - O3$	4	0.2028
Site 2		
$M - O1$	1	0.1982
$M - O2$	1	0.1960
$M - O3$	1	0.2125
$M - O4$	1	0.2118
$M - O5$	1	0.2123
$M - O6$	1	0.2098

*These bond lengths are averages derived from a number of different crystallographic studies.

7.5 The following solids are described in terms of close-packed anion arrays. What is the formula of each?

(a) Lithium oxide: anion array, cubic close-packed; Li^+ in all tetrahedral sites.

(b) Iron titanium oxide, ilmenite: anion array, hexagonal close-packed; Fe^{3+} and Ti^{4+} share ⅔ of the octahedral sites.

(c) Gallium sulphide: anion array, hexagonal close-packed; Ga^{3+} in ⅓ of the tetrahedral sites.

(d) Cadmium chloride: anion array, cubic close-packed; Cd^{2+} in every octahedral site in alternate layers.

(e) Iron silicate (fayalite): anion array, hexagonal close-packed; Fe^{2+} in ½ of the octahedral sites, Si^{4+} in ⅛ of the tetrahedral sites.

(f) Chromium oxide: anion array, cubic close-packed; Cr^{3+} in ⅙ of the octahedral sites, Cr^{6+} in ⅛ of the tetrahedral sites.

7.6 Draw the urea molecule CH_4N_2O (see Chapter 1), using the metallic, ionic, covalent and van der Waals radii, bond angles and bond lengths given in the Table below. The crystallographic data is from: V. Zavodnik, A. Stash, V. Tsirelson, R. de Vries and D. Feil, Acta Crystallogr., **B55**, (1999), p45. The ionic radii are from O. Muller and R. Roy, *The Major Ternary Structural Families*, Springer-Verlag, Berlin (1974), p5. The metallic radii are from Teatum, Gschneidner and Waber, cited in *The Crystal Chemistry and Physics of Metals and Alloys*, Wiley – Interscience, New York, (1972), p 151. Other data are from Figure 7.7 and Figure 7.8. Comment on the validity of using metallic or ionic radii in such depictions.

Data for urea

Bond lengths: N – H, 0.100 nm; C – N, 0.134 nm; C – O, 0.126 nm.

Bond angles: O – C – N, 121.6°; N – C – N, 116.8°; H – N – H, 174.0°.

Atom	Metallic radius/nm	Ionic radius/nm	Covalent radius/nm	Van der Waals radius/nm
H	0.078	0.004 (H^+)	0.0299	0.120
C	0.092	0.006 (C^{4+})	0.0767	0.170
N	0.088	0.002 (N^{5+})	0.0702	0.155
O	0.089	0.121 (O^{2-})	0.0659	0.152

8

Defects, modulated structures and quasicrystals

- *What are modular structures?*

- *What are incommensurately modulated structures?*

- *What are quasicrystals?*

In this chapter, the concepts associated with classical crystallography are gradually weakened. Initially the effects of introducing small defects into a crystal are examined. These require almost no modification of the ideas already presented. However, structures with enormous unit cells pose more severe problems, and incommensurate structures are known in which a diffraction pattern is best quantified by recourse to higher dimensional space. Finally, classical crystallographic ideas break down when quasicrystals are examined. These structures, related to the Penrose tilings described in Chapter 2, can no longer be described in terms of the Bravais lattices described earlier.

8.1 Defects and occupancy factors

In previous discussions of crystal structures, each atom was considered to occupy a crystallographic position completely. For example, in the crystal structure of Cs_3P_7, described in Chapter 5, the Cs1 atoms occupied completely all of the positions with a Wyckoff symbol $4a$. There are four equivalent Cs1 atoms in the unit cell. In such (normal) cases, the **occupancy** of the positions is said to be 1.0.

Not all materials are so well behaved. For example, many metal alloys have considerable composition ranges and a correct calculation of the intensities of diffracted beams needs inclusion of a site **occupancy factor**. For example, the disordered gold-copper alloy Au_xCu_{1-x} is able to take compositions with x varying from 1, pure gold, to 0, pure copper. The structure of the alloy is the copper (A1) structure, (see Chapter 1), but in the alloy the sites occupied by the metal atoms contain a mixture of Cu and Au, (Figure 8.1). This situation can be described by giving a site occupancy factor to each type of atom. For example,

Crystals and Crystal Structures.   Richard J. D. Tilley
© 2006 John Wiley & Sons, Ltd

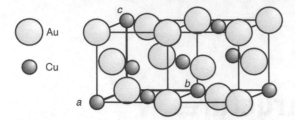

Figure 8.1 The structure of the disordered alloy CuAu. The Cu and Au atoms are distributed at random over the available metal sites

the alloy $Au_{0.5}Cu_{0.5}$ would have occupancy factors of 0.5 for each atom. The occupancy factor for an alloy of x Au and $(1-x)$ Cu would simply be x and $(1-x)$.

The same is true for compounds that form a **solid solution**. A solid solution, as the name implies, is a crystal in which some or all of the atoms are distributed at random over the various sites available, just as molecules are distributed throughout the bulk of a solvent. Many alloys are simply metallic solid solutions. Another example is provided by the oxides that form when mixtures of Al_2O_3 and Cr_2O_3 are heated together at high temperatures. The crystal structure of both of these oxides is the same, *corundum* (Al_2O_3) structure, and the final product is an oxide in which the Al and Cr atoms are distributed over the available metal atom positions (with Wyckoff symbol 12c), while the oxygen sites remain unchanged, give a material of formula $Cr_xAl_{2-x}O_3$. The metal site occupancy factor for a material containing x Cr and $(2-x)$ Al would be $x/2$ and $(1-x/2)$, as the total site occupancy is 1 not 2.

In many compounds that are not obviously solid solutions or alloys some sites will be occupied normally, with occupancy of 1, while other sites may accommodate a mixture of atoms. The *spinel* structure, described earlier, (see Chapter 7, Section 7.8), can be used to illustrate this. Normal *spinels* are written $(A)[B_2]O_4$, and inverse *spinels* are written $(B)[AB]O_4$, where ( ) represents metal

atoms in tetrahedral sites and [ ] represents metal atoms in octahedral sites. Now the forces that produce this ordering are not strong, and in many *spinels* the cation distribution is not clear cut so that the tetrahedral sites are not filled by one atom type alone, but by a mixture. In such cases, satisfactory structure determination will require that appropriate site occupancy factors are used. The oxygen atoms, however, are not subject to mixing, and the occupancy factor of these sites will always be 1.

The atomic scattering factor applicable to such alloys or solid solutions is an average value, referred to as the **site scattering factor**, f_{site}. In general, if two atoms, A and B, with atomic scattering factors f_A and f_B fully occupy a single site in a structure, the site scattering factor is

$$f_{site} = xf_A + (1-x)f_B$$

where x is the occupancy of A and $(1-x)$ the occupancy of B.

In some structures, all positions require site occupancy factors different from 1 if the diffracted intensities are to be correctly reproduced. This is so when structures contain defects at all atomic sites in the structure, called **point defects**. Cubic calcia-stabilised zirconia, which crystallises with the *fluorite* (CaF_2), structure, provides an example. The parent structure is that of zirconia, ZrO_2. The stabilised phase has Ca^{2+} cations in some of the positions that are normally filled by Zr^{4+} cations, that is, cation substitution has occurred. As the Ca^{2+} ions have a lower charge than the Zr^{4+} ions, the crystal will show an overall negative charge if we write the formula as $Ca_x^{2+}Zr_{1-x}^{4+}O_2$. The crystal compensates for the extra negative charge by leaving some of the anion sites unoccupied. The number of vacancies in the anion sub-structure needs to be identical to the number of calcium ions present for exact neutrality, and the correct formula of the crystal is $Ca_x^{2+}Zr_{1-x}^{4+}O_{2-x}$. In order to successfully model the intensities of the diffracted beams, it is necessary to include a site occupancy factor of less than one for the oxygen sites, as well as fractional site occupancies for the Ca and Zr ions.

Many other examples of the use of site occupancy factors could be cited, especially in mineralogy, where most natural crystals have a mixed population of atoms occupying the various crystallographic cites.

8.2 Defects and unit cell parameters

The positions of the lines or spots on the diffraction pattern of a single phase are directly related to the unit cell dimensions of the material. The unit cell of a solid with a fixed composition varies with temperature and pressure, but under normal circumstances is regarded as constant. If the solid has a composition range, as in a solid solution or an alloy, the unit cell parameters will vary. **Vegard's law**, first propounded in 1921, states that the lattice parameter of a solid solution of two phases with similar structures will be a *linear function* of the lattice parameters of the two end members of the composition range (Figure 8.2a).

$$x = (a_{ss} - a_1)/(a_2 - a_1)$$
$$\text{i.e.} \quad a_{ss} = a_1 + x(a_2 - a_1)$$

where a_1 and a_2 are the lattice parameters of the parent phases, a_{ss} is the lattice parameter of the solid solution, and x is the mole fraction of the parent phase with lattice parameter a_2. (The relationship holds for the a-, b- and c-lattice parameters, and any interaxial angles.) This 'law' is simply an expression of the idea that the cell parameters are a direct consequence of the sizes of the component atoms in the solid solution. Vegard's law, in its ideal form, is almost never obeyed exactly. A plot of cell parameters that lies below the ideal line, (Figure 8.2b), is said to show a *negative* deviation from Vegard's law, and a plot

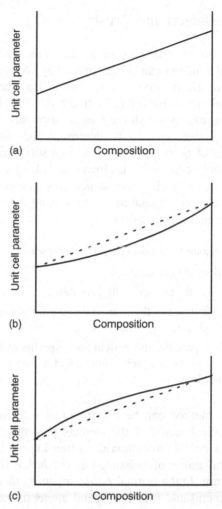

Figure 8.2 Vegard's law relating unit cell parameters to composition for solid solutions and alloys: (a) ideal Vegard's law behaviour; (b) negative deviation from Vegard's law; (c) positive deviation from Vegard's law

that lies above the ideal line, (Figure 8.2c), is said to show a *positive* deviation from Vegard's law. In these cases, atomic interactions, which modify the size effects, are responsible for the deviations. In all cases, a plot of composition versus cell parameters can be used to determine the formula of intermediate compositions in a solid solution.

8.3   Defects and density

The theoretical density of a solid with a known crystal structure can be determined by dividing the mass of all the atoms in the unit cell by the unit cell volume, (Chapter 1, Section 1.12). This information, together with the measured density of the sample, can be used to determine the notional species of point defect present in a solid that has a variable composition. However, as both techniques are averaging techniques they say nothing about the real organisation of the point defects.

The general procedure is:

(i) Measure the composition of the solid.

(ii) Measure the density.

(iii) Measure the unit cell parameters.

(iv) Calculate the theoretical density for alternative point defect populations.

(v) Compare the theoretical and experimental densities to see which point defect model best fits the data.

The method can be illustrated by reference to a classical study of the defects present in iron monoxide[1]. Iron monoxide, often known by its mineral name of wüstite, has the *halite* (NaCl) structure. In the normal *halite* structure, there are four metal and four non-metal atoms in the unit cell, and compounds with this structure have an ideal composition $MX_{1.0}$, (see Chapter 1, Section 1.8). Wüstite has an oxygen-rich composition compared to the ideal formula of $FeO_{1.0}$. Data for an actual sample found an oxygen: iron ratio of 1.059, a density of $5728\,kg\,m^{-3}$, and a cubic lattice parameter, a, of 0.4301 nm. Because there is more oxygen present than iron, the real composition can be obtained by assuming either that there are extra oxygen atoms in the unit cell, as interstitials, or that there are iron vacancies present.

[1]The data is from the paper: E. R. Jette and F. Foote, J. Chem. Phys., **1**, 29 (1933).

Model A: Assume that the iron atoms in the crystal are in a perfect array, identical to the metal atoms in the *halite* structure and an excess of oxygen is due to interstitial oxygen atoms over and above those on the normal anion positions. The ideal unit cell of the structure contains 4 Fe and 4 O, and so, in this model, the unit cell must contain 4 atoms of Fe and $4(1+x)$ atoms of oxygen. The unit cell contents are Fe_4O_{4+4x} and the composition is $FeO_{1.059}$.

The mass of 1 unit cell is m_A:

$$m_A = [(4 \times 55.85) + (4 \times 16 \times (1+x))]/N_A$$
$$[(4 \times 55.85) + (4 \times 16 \times 1.059)]/N_A \text{ grams}$$
$$= 4.835 \times 10^{-25}\,kg$$

The volume, V, of the cubic unit cell is given by a^3, thus:

$$V = (0.4301 \times 10^{-9})^3 m^3 = 7.9562 \times 10^{-29}\,m^3.$$

The density, ρ, is given by the mass m_A divided by the unit cell volume, V:

$$\rho = 4.835 \times 10^{-25}/7.9562 \times 10^{-29}$$
$$= 6077\,kg\,m^{-3}$$

Model B: Assume that the oxygen array is perfect and identical to the non-metal atom array in the *halite* structure and that the unit cell contains some vacancies on the iron positions. In this case, one unit cell will contain 4 atoms of oxygen and $(4-4x)$ atoms of iron. The unit cell contents are $Fe_{4-4x}O_4$ and the composition is $Fe_{1/1.058}O_{1.0}$ or $Fe_{0.944}O$.

The mass of one unit cell is m_B:

$$m_B = [(4 \times (1-x) \times 55.85) + (4 \times 16)]/N_A$$
$$= [(4 \times 0.944 \times 55.85) + (4 \times 16)]/N_A \text{ grams}$$
$$= 4.568 \times 10^{-25}\,kg$$

The density, ρ, is given by m_B divided by the volume, V, to yield

$$\rho = 4.568 \times 10^{-25}/7.9562 \times 10^{-29}$$
$$= 5741 \, \text{kg} \, \text{m}^{-3}$$

The difference in the two values is surprisingly large. The experimental value of the density, $5728 \, \text{kg} \, \text{m}^{-3}$, is in good accord with Model B, which assumes vacancies on the iron positions. This indicates that the formula should be written $Fe_{0.945}O$, and that the point defects present are iron vacancies.

8.4 Modular structures

In this and the following three sections, structures that can be considered to be built from **slabs** of one or more parent structures are described. These materials pose particular crystallographic difficulties. For example, some possess enormous unit cells, of some 100's of nm in length, whilst maintaining perfect crystallographic translational order. They are frequently found in mineral specimens, and the piece-meal way in which early examples were discovered has lead to a number of more or less synonymic terms for their description, including 'intergrowth phases', 'composite structures', 'polysynthetic twinned phases', 'polysomatic phases' and 'tropochemical cell-twinned phases'. In general they are all considered to be **modular structures**.

Modular structures can be built from slabs of the same or different compositions, and the slab widths can be disordered, or ordered in a variety of ways. The simplest situation corresponds to a material built from slabs of only a single parent phase and in which the slab thicknesses vary widely. In this case, the slab boundaries will not fall on a regular lattice, and they then form **planar defects**, (Figure 8.3), which are two-dimensional analogues of the point defects described above.

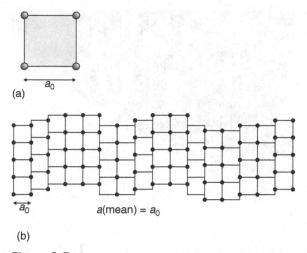

Figure 8.3 A crystal containing random planar faults: (a) a unit cell of the parent structure; (b) the same crystal containing a random distribution of planar faults. The average unit cell of this material has the same dimensions as that of the parent structure

The unit cell of the disordered material is the same as the unit cell of the parent phase, although the diffraction pattern will reveal evidence of this disorder in the form of diffuse or streaked diffraction spots, (see also Section 8.6).

An example of such a disordered crystal is shown in Figure 8.4a. The material is $SrTiO_3$, which adopts the *perovskite* structure. The faulted crystal can be best described with respect to an idealised cubic *perovskite* structure, with a lattice parameter of 0.375 nm, (Figure 8.4b). The skeleton of the material is composed of corner-linked TiO_6 octahedra, with large Sr cations in the cages that lie within the octahedral framework. The planar defects, which arise where the perovskite structure is incorrectly stacked, lie upon [110] referred to the idealised cubic structure (Figure 8.4c).

If the width of the component slabs in a modular structure is constant, *no* defects are present in the structure, because the unit cell includes the planar boundaries between the slabs as part of the structure. Take a case where two different slab widths

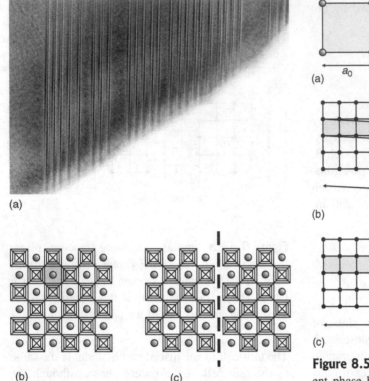

Figure 8.4 (a) Electron micrograph showing a flake of $SrTiO_3$ containing disordered planar defects lying upon {110} planes with respect to the idealised cubic unit cell. Lattice fringes resolved near the crystal edge are (110) planes with a spacing of 0.265 nm. (b) The idealised structure of $SrTiO_3$ projected down the cubic **c**-axis. Each square represents a TiO_6 octahedron and the circles represent Sr atoms. The unit cell is shaded. (c) Idealised representation of a (110) fault in $SrTiO_3$

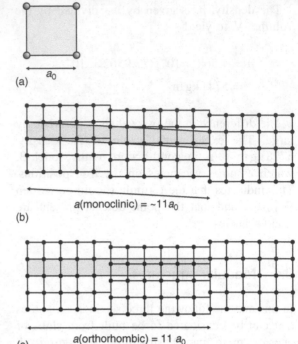

Figure 8.5 Structures derived from that of the parent phase by stacking alternating slabs of material 5 and 6 unit cells in thickness: (a) a unit cell of the parent structure; (b) the displacement of the slabs is the same in every case, often resulting in a monoclinic unit cell; (c) the displacement of the slabs alternates in direction, often resulting in an orthorhombic unit cell

of the parent structure are present, (Figure 8.5). In the case where the displacement between adjacent slabs is always the same, (and not equal to ½ of the cell parameter in the direction of the displacement), the unit cell will (often) be monoclinic, and such structures are given the prefix 'clino' in mineralogical literature, (Figure 8.5b). The case in which slab displacement alternates in direction

produces an orthorhombic unit cell, the 'ortho' form, (Figure 8.5c).

An example of this behaviour is provided by $SrTiO_3$ described above. In cases where the boundaries are ordered, new structures form with well-defined unit cells and precise compositions. Figure 8.6a shows the structure of the phase constructed by stacking slabs four octahedra in width, and Figure 8.6b a structure built of slabs five octahedra in width. A vast number of structures can be envisaged to form between these two. Figure 8.6c shows the simplest, consisting of alternating slabs of widths four and five octahedra

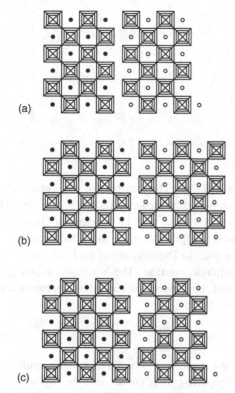

(a)

(b)

(c)

Figure 8.6 Schematic representation of the structures of the $Sr_n(Nb,Ti)_nO_{3n+2}$ phases: (a) $Sr_4Nb_4O_{14}$; (b), $Sr_5(Ti,Nb)_5O_{17}$; (c) $Sr_9(Ti,Nb)_9O_{29}$. The squares represent $(Nb,Ti)O_6$ octahedra and the circles represent Sr atoms. In (a) the perovskite slabs are four octahedra wide, and in (b) they are five octaheda wide. In (c) slabs of four and five octahedra alternate

in width, written (4,5), but regular repeating sequences such as (4,4,4,5), (4,4,5), (4,5,5) and so on have been characterised. The multiplicity of structures is increased enormously when other slab widths are also considered.

An examination of these structures shows that the overall composition is no longer $SrTiO_3$. The faulting arises because the crystal contains some Nb_2O_5. This gives each crystal a composition $Sr(Nbi_xTi_{1-x})O_{3+x}$. Each Nb^{5+} ion that replaces a Ti^{4+} ion requires that extra oxygen

is incorporated into the crystal. The planar defects open the structure and allow this extra oxygen to be accommodated without introducing oxygen interstitials. The formula of each ordered structure is given by a series formula $Sr_n(Nb,Ti)_nO_{3n+2}$, where n takes integer values from 4 upwards.

A change in chemical composition is not a mandatory prerequisite for the formation of modular structures. A typical example of a structure built from displaced slabs but without change of composition is provided by the ordered alloy phase CuAu II, (Figure 8.7a). This can be regarded as made up of slabs of another ordered alloy, CuAu I, each of which is five unit cells in thickness, (Figure 8.7b).

As these two examples show, besides altered unit cell dimensions, a series of modular structures may or may not show a regular variation of composition. When only one slab type is present this will depend upon the nature of the planar boundaries involved. However, a change of composition must occur if the series of

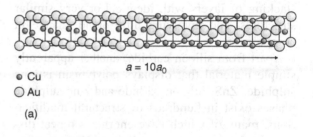

$a = 10a_0$

⊙ Cu
◯ Au

(a)

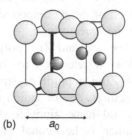

(b) a_0

Figure 8.7 The structure of ordered copper – gold alloys: (a) CuAu II; (b) CuAu I

structures is composed of two or more slabs, each with a different composition. In the following three sections these three aspects are described in more detail, and deal with structures built of slabs of a single parent type showing no overall composition variation, polytypes, (Section 8.5), structures built of slabs of a single parent and showing regular composition variation, the crystallographic shear phases, (Section 8.6), and structures built of slabs of two different structures, intergrowths or polysomes, (Section 8.7).

8.5   Polytypes

The first long period structures to be characterised were polymorphs of the superficially simple material silicon carbide, or carborundum, SiC. These structures were called **polytypes**, to distinguish them from more normal and commonly occurring polymorphs that are typical of many minerals, such as the aragonite and calcite forms of calcium carbonate, $CaCO_3$. Polytypes are now considered to be long period structures built by stacking of layers with identical or very similar structures, and involving little or no change in composition.

Apart from silicon carbide, another apparently simple material that displays polytypism is zinc sulphide, ZnS. Silicon carbide and zinc sulphide phases exist in hundreds of structural modifications, many of which have enormous repeat distances along one unit cell axis. Despite this complexity, the composition of these compounds never strays from that of the parent phase.

The structure of both the SiC and ZnS polytypes can be illustrated with reference to the crystalline forms of ZnS. Zinc sulphide crystallises in either of two structures, one of which is cubic and given the mineral name zinc blende (sphalerite) while the other is hexagonal and given the mineral name wurtzite. The relationship

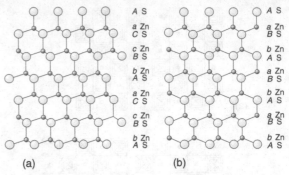

Figure 8.8   (a) The structure of the cubic form of zinc sulphide, zinc blende (sphalerite). The cubic [111] direction is vertical and the structure is viewed down the cubic [110] direction, i.e. projected onto the cubic (110) plane. (b) The structure of the hexagonal form of zinc sulphide, wurtzite. The hexagonal **c**-axis is vertical, and the structure is viewed as if projected onto $(110) = (11\bar{2}0)$

between these two structures can be understood by comparing the cubic and hexagonal close packing of spheres, (see Chapter 7, Section 7.9). When the zinc blende structure is viewed along the cube face-diagonal [110] direction, and oriented with the cubic [111] direction vertical, to make the zinc and sulphur layers horizontal, the layers of both zinc and sulphur are packed in the cubic closest packing arrangement ... *aAbBc-CaAbBcC* ..., where the lowercase letters stand for the Zn layers and the uppercase letters for the S layers, (Figure 8.8a). The wurtzite structure, viewed with the hexagonal **c**-axis vertical and projected onto the hexagonal unit cell diagonal, i.e. the hexagonal $(11\bar{2}0)$ plane, has a layer packing ... *aAbBaAbB* ... where the lowercase letters stand for Zn layers and the uppercase letters for S layers, (Figure 8.8b). The zinc sulphide polytypes are complex arrangements of layers of both wurtzite (hexagonal) and zinc blende (cubic) types. Some zinc sulphide polytypes are listed in Table 8.1.

Table 8.1   Some polytypes of SiC and ZnS

Ramsdell notation	Zhdanov notation*	Stacking sequence*
2H	(11)	h
3C (3R)	(1)	c
4H	(22)	hc
6H	(33)	hcc
8H	(44)	hccc
10H	(82)	$(hc_7)(hc)$
	(55)	hcccc
14H	(5423)	$(hc_4)(hc_3)(hc)(hc_2)$
	(77)	$(hc_6)_2$
16H	(88)	$(hc_7)_2$
	(14, 2)	$(hc_{13})(hc)$
	(5335)	$(hc_4)_2(hc_2)_2$
	$(3223)_2$	$(hcc)_4(hc)_2$
24H	(15, 9)	$(hc_{14})(hc)_8$
24R	$(53)_2$	$[(hc_4)(hc_2)]_2$
36R	$(6222)_3$	$[(hc_5)(hc)_3]_3$
48R	$(7423)_3$	$[(hc_6)(hccc)(hc)(hcc)]_3$
72R	$(6, 11, 5, 2)_3$	$[(hc_5)(hc_{10})(hc_4)(hc)]_3$

*Subscript numbers have the same significance as in ordinary chemical nomenclature. Thus $(53)_2 = (5353)$, $hc_7 = hccccccc$ and $(hcc)_2 = hcchcc$.

Silicon carbide, carborundum, also crystallises in two forms, of which β-SiC has the cubic *zinc blende (sphalerite)* structure (Figure 8.8a). When viewed along the cube face-diagonal [110] direction, the layers of both silicon and carbon are packed in the cubic closest packing arrangement ... *aAbBcCaAbBcC* ..., where the uppercase and lowercase letters stand for layers of Si and C. The other form of silicon carbide, α-SiC, is a collective name for the various silicon carbide polytypes, which consist of complex arrangements of zinc blende and wurtzite slabs. Some of these are known by names such as carborundum I, carborundum II, carborundum III, and so on. One of the simplest structures is that of carbo-

rundum I, which has a packing ... *aAbBaAc-CaAbBaAcC* ..., where the upper and lower case letters refer to the two atom types respectively. Some silicon carbide polytypes are listed in Table 8.1.

The complexity of the structures has lead to a number of proposed compact forms of nomenclature. The most widely used of these is the **Ramsdell** notation, which gives the number of layers of the stacked slabs in the crystallographic repeat, together with the symmetry of the unit cell, specified by *C* for cubic, *H* for hexagonal and *R* for rhombohedral. The repeating slab in the silicon carbide polytypes as taken as one sheet of Si atoms plus one sheet of C atoms, (Si + C) and in the zinc sulphide polytypes as one sheet of Zn atoms plus one sheet of S atoms, (Zn + S). The packing of the (Si + C) layers in cubic β-SiC, with the *zinc blende* structure, is ...*ABC*..., and the Ramsdell notation is 3*C*. The same symbol would apply to cubic ZnS. The Ramsdell notation for the wurtzite form of ZnS, in which the (Zn + S) layers are in hexagonal ...*AB*... stacking, would be 2*H*. The same term would apply to a pure hexagonal form of SiC. The Ramsdell notation for the form of silicon carbide called carborundum I, the packing of which is described above, is 4*H*.

While the Ramsdell notation it is compact, it does not give the layer stacking sequence of the polytype, and a number of other terminologies have been invented to overcome this shortcoming. The **Zhdanov** notation, one way of specifying the stacking sequence, is derived in the following way. The translation between an *A* layer and a *B* layer, or a *B* layer and a *C* layer can be represented by a rotation of +60° or a translation of $+\frac{1}{3}$. Reverse transformations, *C* to *B* or *B* to *A*, are then represented by −60° or $-\frac{1}{3}$. The stacking sequence of any polytype can then be written as a series of + and − signs. The Zhdanov symbol of the polytype gives the number of signs of the same type in the sequence.

Thus the cubic *zinc blende* structure of ZnS or SiC has a stacking sequence ...+++++ .... The Zhdanov symbol is written as (1), rather than (∞). The hexagonal *wurtzite* structure has a stacking sequence ...+ − + − ... so the Zhdanov symbol is (11). Similarly, carborundum I has a stacking sequence ... ++−− ..., which is represented by the Zhdanov symbol (22).

The Zhdanov notation also provides a pictorial representation of a structure. Draw structures in the same projection as used in Figure 8.8, and represent each layer pair in the stacking, say (Zn + S), by a single "atom". The result is a zig-zag pattern. The Zhdanov notation specifies the zig-zag sequence. Thus hexagonal ZnS, (11) structure is a simple zig-zag, (Figure 8.9a), carborundum I, (22), has a double repeat, Figure 8.9b, and the structure with a Zhdanov symbol (33) has a triple repeat, (Figure 8.9c).

A third widely used terminology relies upon specifying the relative position of layers. Once again it is convenient to consider two layers, (Si + C) or (Zn + S) as a unit. A middle bilayer sandwiched between sheets of the same type, say *BAB* or *CAC*, is labelled *h*. Similarly, a layer sandwiched between sheets of different type, say *ABC* or *BCA*, is labelled *c*. The *4H* polytype of silicon carbide is then labelled as (*hc*), (Table 8.1).

The complexity of polytypes is enormous, as can be judged from the few examples given in Table 8.1, and it must not be forgotten that polytypes exist in many other chemical systems, from chemically simple CdI_2 to the complex mica silicates. The fact that the stacking sequences are able to repeat so accurately over such long distances is still puzzling, despite the many attempts at theoretical explanations.

8.6 Crystallographic shear phases

Like polytypes, crystallographic shear (*CS*) phases are built from slabs of a single parent

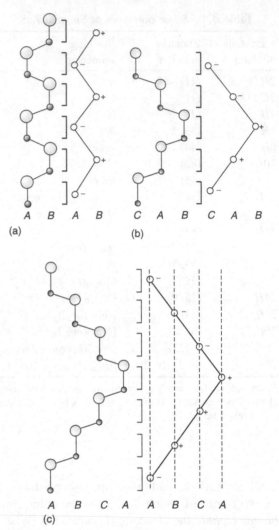

Figure 8.9 Simplified representations of polytypes: (a) wurtzite, 2*H*, (11); (b) carborundum I, 4*H*, (22); (c) 6*H*, (33). The zig-zag designs to the right of the structure representations show the sequence of translations in a direction, + or −, summarised in the Zhdanov symbol. In the zig-zags, each circle represents the position of a composite (Zn + S) or (Si + C) layer

structure, but now joined in such a way as to produce a change in the overall composition of the solid. This effect can be illustrated with respect to a set of structurally simple examples, the *CS* phases that form when tungsten trioxide, WO_3, is reduced.

Tungsten trioxide has a unit cell that is monoclinic at room temperature, with lattice parameters, $a = 0.7297$ nm, $b = 0.7539$ nm, $c = 0.7688$ nm, $\beta = 90.91°$, which is most easily visualised as a three-dimensional array of slightly distorted corner-shared WO_6 octahedra. For the present discussion, the structure can be idealised as cubic, (Figure 8.10a), with a cell parameter of $a = 0.750$ nm, and viewed in projection down

one of the cubic axes, as an array of corner-linked squares, (Figure 8.10b). Very slight reduction, to a composition of $WO_{2.9998}$, for example, results in a crystal containing a low concentration of faults which lie on {120} planes. Structurally these faults consist of lamellae made up of blocks of four edge-shared octahedra in a normal WO_3-like matrix of corner sharing octahedra (Figure 8.10c). A slightly reduced crystal contains *CS* planes lying upon many {120} planes, Figure 8.11.

Continued reduction causes an increase in the density of the *CS* planes and as the composition approaches $WO_{2.97}$, they tend to become ordered. In this case the structure will possess a (generally monoclinic) unit cell with a long axis, which will be more or less perpendicular to the *CS* planes, (Figure 8.12). The length of the long axis, which can be taken as the **c**-axis, will be equal to an integral number, *m*, of the d_{102} spacing. The **a**-axis is roughly perpendicular to the **c**-axis, and the monoclinic **b**-axis will be perpendicular to the plane of the figure and approximately equal to the **c**-axis of WO_3.

The *CS* planes reduce the amount of oxygen in the structure. The composition of a crystal

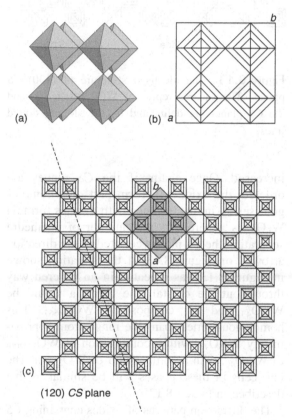

(120) *CS* plane

Figure 8.10 Crystallographic shear in WO_3: (a) perspective view of idealised cubic WO_3; (b) projection of the idealised structure down the cubic **c**-axis, the cubic unit cell is outlined; (c) an idealised (120) *CS* plane, the WO_3 unit cell is shaded. The squares represent WO_6 octahedra

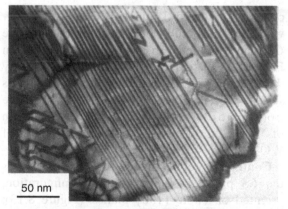

50 nm

Figure 8.11 Electron micrograph of slightly reduced WO_3 showing disordered {120} *CS* planes, imaged as dark lines

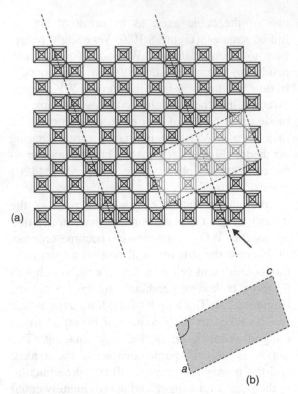

Figure 8.12 (a) the idealised structure of $W_{11}O_{32}$; (b) the monoclinic unit cell of $W_{11}O_{32}$. The squares represent WO_6 octahedra, and the *CS* planes are delineated by the sloping dotted lines

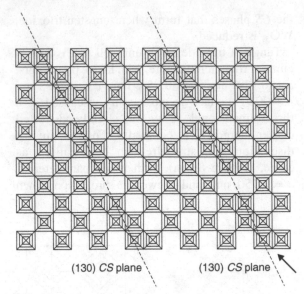

(130) *CS* plane   (130) *CS* plane

Figure 8.13 The idealised structure of (130) *CS* planes. The squares represent WO_6 octahedra, and the *CS* planes are delineated by the sloping dotted lines

containing ordered *CS* planes is given by W_nO_{3n-1} where n is the number of octahedra separating the *CS* planes (counted in the direction arrowed on Figure 8.12). The family of oxides represented by the formula W_nO_{3n-1} is known as a **homologous series**. This homologous series spans the range from approximately $W_{30}O_{89}$, with a composition of $WO_{2.9666}$, to $W_{18}O_{53}$, with a composition of $WO_{2.9444}$.

Continued reduction of the oxide causes the *CS* planes to adopt another configuration, oriented on cubic {130} planes, (Figure 8.13). The CS plane is now made of blocks of six edge-shared octahedra, and the degree of reduction per unit length of *CS* plane is consequently

increased. Once again, if the *CS* planes are ordered, the oxide will form part of a homologous series, in this instance with a series formula W_nO_{3n-2}, where n is the number of octahedra separating the *CS* planes counted in the direction arrowed on Figure 8.13. If the situation shown in Figure 8.13 was repeated in an ordered way throughout the crystal, the formula would be $W_{15}O_{43}$ and the composition $WO_{2.8667}$. This homologous series spans the range from approximately $W_{25}O_{73}$, with a composition of $WO_{2.9200}$, to $W_{16}O_{46}$, with a composition of $WO_{2.8750}$. The unit cells of these phases will be similar to that described in Figure 8.12.

The diffraction patterns of oxides containing *CS* planes evolve in a characteristic way from that of the parent structure. This can be explained using WO_3 as an example. As the parent structure forms the greatest component of a *CS* structure, the diffraction pattern of these materials will closely resemble that of WO_3 itself, (Figure 8.14a).

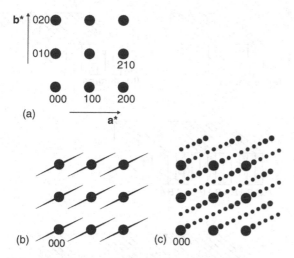

(a)

(b) 000   (c) 000

Figure 8.14 The evolution of diffraction patterns of materials containing *CS* planes: (a) idealised cubic WO$_3$; (b) idealised WO$_3$ containing disordered (210) *CS* planes, giving rise to streaking on the diffraction pattern; (c) idealised WO$_3$ containing ordered (210) *CS* planes. Note that the real structures are less symmetrical and not all reflections shown in these diagrams may be present in experimental patterns

However, the *CS* planes impose a new set of diffraction conditions. Suppose that a crystal contains parallel but disordered (210) CS planes. The resulting diffraction pattern will be similar to that of the parent structure, but streaks will now appear, parallel to the line joining the origin to the 210 reflection, and passing through each reciprocal lattice point, (Figure 8.14b). This, in fact, is an expression of the form factor for the crystal (see Sections 6.3 and 6.4). The planar faults, in effect, break the large crystal into a number of small slabs and the streaks run normal to disordered slab boundaries. As the *CS* planes become ordered, the streaks begin to show maxima and minima, and fully ordered *CS* planes give sharp spots (Figure 8.14c). The number of extra reflections will be equal to n in the homologous series formula, or else a multiple of it, depending upon

the true symmetry of the unit cell. These extra reflections are called **superlattice** or **superstructure reflections**, and because the spacing between them will be $1/md_{210}$, where m is an integer, they fit exactly into the WO$_3$ reciprocal lattice, and are said to be **commensurate** with it. Examples of these types of diffraction patterns are given in Figure 8.15.

The discussion of these diffraction patterns is quite general and not just restricted to *CS* planes. This means that the polytypes and other phases described above, as well as the long period

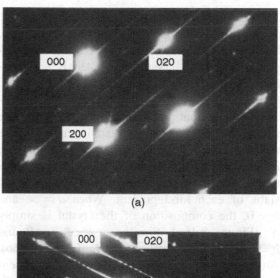

(a)

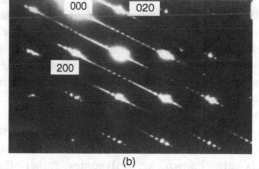

(b)

Figure 8.15 Electron diffraction patterns from: (a) a crystal containing disordered ($\bar{1}$20) *CS* planes, showing continuous streaks; (b) a crystal containing more ordered (120) *CS* planes, showing the streaks breaking up into rows of superlattice reflections

structures described in the following section, will yield similar diffraction patterns. In them, the diffraction pattern of the parent structure will be well emphasised, together with streaks or commensurate rows of superlattice reflections on the diffraction pattern, running perpendicular to the fault planes that break up the structure. To obtain the intensities of these reflections the same structure factor calculations described earlier can be carried out.

8.7   Planar intergrowths and polysomes

A large number of solids encompass a range of composition variation by way of **intergrowth**. Unlike the two previous examples, in these compounds, slabs of crystalline materials with **different** compositions, say A and B, interleave to give a compound of formula A_aB_b. This implicitly entails that they each have at least one structurally compatible crystallographic plane that is shared between the two slabs and so forms the interface between them. The composition of any phase will be given by the numbers of slabs of each kind present. When $a = \infty$ and $b = 0$, the composition of the crystal is simply A, (Figure 8.16a). Similarly, when $a = 0$ and $b = \infty$, the composition is B, (Figure 8.16b). An enormous number of ordered stacking sequences can be imagined, including $ABAB$, composition AB, (Figure 8.16c), $ABBABB$, composition AB_2, (Figure 8.16d) and $ABAAB$, composition A_3B_2, (Figure 8.16e). The overall composition can vary virtually continuously from that of one of the parents to the other, depending upon the proportions of each phase present. These materials are known as **polysomes** rather than polytypes.

Examples abound, especially in minerals. A classic polysomatic series is given by the minerals that are formed by an ordered intergrowth

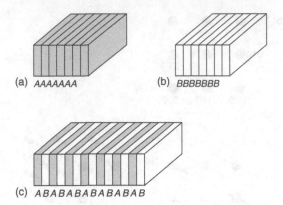

(a) *AAAAAA*   (b) *BBBBBBB*

(c) *ABABABABABAB*

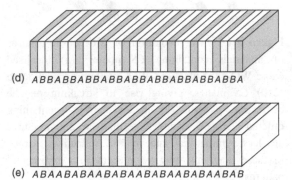

(d) *ABBABBABBABBABBABBABBA*

(e) *ABAABABAABABAABABAABABAABAB*

Figure 8.16   The stacking of slabs with two different compositions, A and B: (a) ... AAA ..., composition A; (b) ... BBB ..., composition B; (c) ... $ABAB$ ..., composition AB; (d) ... $ABBABB$ ..., composition AB_2; (e) ... $ABAAB$ ..., composition A_3B_2

of slabs of the *mica*[2] and *pyroxene* structures. Both are built from silicate layers that are only weakly bonded, so that they cleave easily to form thin plates. The micas have formulae typified by phlogopite, $KMg_3(OH)_2Si_3AlO_{10}$,

---

[2]Mica also forms polytypes. These occur when the component layers of a mica stack in alternative ways. In these polytypes the composition does not change from one member of the series to another and remains that of the parent phase.

where the K^+ ions lie between the silicate layers, the Mg^{2+} ions are in octahedral sites and the Si^{4+} and Al^{3+} ions occupy tetrahedral sites within the silicate layers. The pyroxenes can be represented by enstatite, $MgSiO_3$, where the Mg^{2+} ions occupy octahedral sites and the Si^{4+} ions occupy tetrahedral sites within the silicate layers. Both these materials have layer structures which fit together parallel to the layer planes. If an idealised *mica* slab is represented by M, and an idealised *pyroxene* slab by P, the polysomatic series can span the range from ... MMM ..., representing pure mica, to ... PPP ..., representing pure pyroxene. Many intermediates are known, for example, the sequence $MMPMMP$ is found in the mineral jimthompsonite, $(Mg,Fe)_{10}(OH)_4Si_{12}O_{32}$, and the sequence ... $MPMMPMPMMP$ ... corresponds to the mineral chesterite, $(Mg,Fe)_{17}(OH)_6Si_{20}O_{54}$.

There are many intergrowth phases that are composed of slabs of the ABO_3 perovskite structure. One family, $Sr_n(Ti,Nb)_nO_{3n+2}$, related to $SrTiO_3$, was described in Section 8.4. Another structurally simple series built from slabs of $SrTiO_3$ is represented by the oxides Sr_2TiO_4, $Sr_3Ti_2O_7$ and $Sr_4Ti_3O_{10}$, with general formula $Sr_{n+1}Ti_nO_{3n+1}$, known as **Ruddleston – Popper** phases. In these materials, slabs of $SrTiO_3$, (Figure 8.4b), are cut parallel to the idealised cubic *perovskite* {100} planes, (rather than {110} as described above), and stacked together, each slab being slightly displaced in the process, (Figure 8.17a–d). The structure of the junction regions is identical to that of lamellae of the *halite* (NaCl) structure, and so the series can be thought of as being composed of intergrowths of varying thicknesses of *perovskite* $SrTiO_3$, linked by identical slabs of *halite* SrO.

Many other examples of Ruddleston-Popper phases have been synthesised. The generalised homologous series formula for these compounds is $A_{n+1}B_nO_{3n+1}$ $[=(AO)(ABO_3)_n$ or $(H)(P)_n]$ where A is a large cation, typically an alkali

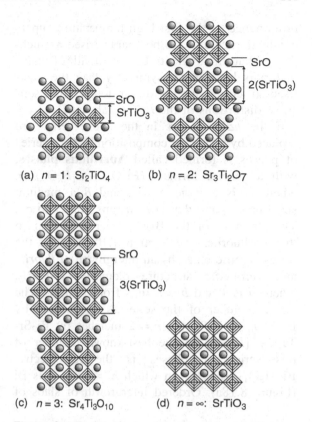

(a) $n = 1$: Sr_2TiO_4 (b) $n = 2$: $Sr_3Ti_2O_7$

(c) $n = 3$: $Sr_4Ti_3O_{10}$ (d) $n = \infty$: $SrTiO_3$

Figure 8.17 Idealised structures of the Ruddleston-Popper phases $Sr_{n+1}Ti_nO_{3n+1}$: (a) Sr_2TiO_4, $n = 1$; (b) $Sr_3Ti_2O_7$, $n = 2$; (c) $Sr_4Ti_3O_{10}$, $n = 3$; (d) $SrTiO_3$, $n = \infty$. The structures can be regarded as built from slabs of $SrTiO_3$ separated by slabs of SrO

metal, alkaline earth or rare earth and B is a medium sized cation, typically a 3d transition metal cation, and H and P stand for *halite* and *perovskite*. The $Sr_{n+1}TiO_{3n+1}$ series is then composed of $(SrO)_a(SrTiO_3)_b$, or $(H)_a(P)_b$ where a is 1 and b runs from 1 to 3. The structure of first member of the series, corresponding to $n = 1$, is adopted by many compounds, and is often referred to as the K_2NiF_4 structure. One of these, the phase La_2CuO_4, when doped with Ba^{2+} to form the oxide $(La_xBa_{1-x})_2CuO_4$, achieved prominence as the first high temperature super-conducting ceramic to be characterised as such. [Note that the real structures of all of these materials have lower symmetry than the idealised structures, mainly due to temperature sensitive distortions of the BO_6 octahedra.]

If the *halite* layers in the above series are replaced by a layer of composition Bi_2O_2 a series of phases is formed called **Aurivillius phases**, with a general formula $(Bi_2O_2)(A_{n-1}B_nO_{3n+1})$, where A is a large cation, and B a medium sized cation, and the index n runs from 1 to ∞. The structure of the Bi_2O_2 layer is similar to that of fluorite (see Section 1.10), so that the series is represented by an intergrowth of *fluorite* and *perovskite* structure elements, $(F)_a(P)_b$, where a is 1 and $b = n$ runs from 1 to ∞. The $n = 1$ member of the series is represented by Bi_2WO_6 or $[FP]$, the $n = 2$ member by $Bi_2Sr_2Ta_2O_9$, $[FPP]$, but the best-known member of this series of phases is the ferroelectric $Bi_4Ti_3O_{12}$, $[FPPP]$, in which $n = 3$ and A is Bi (Figure 8.18a). Ordered intergrowth of slabs of more than one thickness, especially n and $(n+1)$, are often found in this series of phases, for example, $n = 1$ and $n = 2$, $[FPFPP]$, typified by $Bi_5TiNbWO_{15}$, (Figure 8.18b). [As with the other structures described, the symmetry of these structures is lower than that implied by the idealised figures, mainly due to distortions of the metal – oxygen octahedra.]

The high temperature superconductors are similar to both of these latter examples in many ways, as they are composed of slabs of *perovskite* structure intergrown with slabs that can be thought of as structurally related to *halite* or *fluorite*. To illustrate these materials, the idealised structures of the phases $Bi_2Sr_2CuO_6$ ($= Tl_2Ba_2CuO_6$), $Bi_2CaSr_2Cu_2O_8$ ($= Tl_2CaBa_2Cu_2O_8$) and $Bi_2Ca_2Sr_2Cu_3O_{10}$ ($= Tl_2Ca_2Ba_2Cu_3O_{10}$) are shown in Figure 8.19. The single perovskite sheets in the idealised structure of $Bi_2Sr_2CuO_6$ are complete, (Figure 8.19a). These are separated by Bi_2O_2 (or Tl_2O_2) layers similar, but not identical, to those in the Aurivillius phases. In the other compounds, the oxygen structure needed to form the perovskite framework is incomplete. The nominal double layer of CuO_6 octahedra needed to form a sheet of the idealised perovskite structure is replaced by square pyramids in $Bi_2CaSr_2Cu_2O_8$, (Figure 8.19b). To make the relationship clearer, the octahedra are completed in faint outline in Figure 8.19c. In $Ba_2Ca_2Sr_2Cu_3O_{10}$ the three CuO_6 octahedral perovskite layers have been replaced by two sheets of square pyramids and middle layer by a sheet of CuO_4 squares, (Figure 8.19d). The octahedra are completed in faint outline in Figure 8.19e.

8.8 Incommensurately modulated structures

In the examples just presented the reciprocal lattices and the resulting diffraction patterns

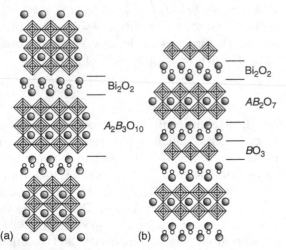

Figure 8.18 The idealised structure of the Aurivillius phases viewed along [110]: (a) $n = 3$, typified by $Bi_4Ti_3O_{12}$; (b) ordered $n = 1$, $n = 2$ intergrowth, typified by $Bi_5TiNbWO_{15}$

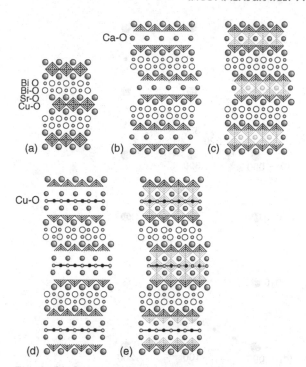

Ca-O

Bi O
Bi-O
Sr-O
Cu-O

(a)   (b)   (c)

Cu-O

(d)   (e)

Figure 8.19 The idealised structures of some high temperature superconductors: (a) $Bi_2Sr_2CuO_6 (= Tl_2Ba_2CuO_6)$; (b) $Bi_2CaSr_2Cu_2O_8 (= Tl_2CaBa_2Cu_2O_8)$; (c) as (b), but with the nominal CuO_6 octahedra completed in faint outline; (d) $Bi_2Ca_2Sr_2Cu_3O_{10} (= Tl_2Ca_2Ba_2Cu_3O_{10})$; (e) as (d), but with the nominal CuO_6 octahedra completed in faint outline.

resemble that of a parent structure, say *perovskite*, together with arrays of 'extra' superstructure reflections, running between all of the main reflections, (Figure 8.20a). These additional reflections can be indexed via a unit cell with one or two long cell axes, so that all reflections have conventional *hkl* values, (Figure 8.20b).

In recent years, a rapidly increasing number of crystalline solids have been characterised in which the spots on the diffraction pattern cannot be indexed in this way. As before, the main reflections, those of the parent structure, have rows of superlattice reflections associated with them. However, it is found that the spacing

between the superlattice spots is sometimes 'anomalous', and the superlattice reflections do not quite fit in with the parent reflections, either in spacing, (Figure 8.20c), or in both the spacing and the orientation of the rows, (Figure 8.20d). [Note that in Figures 8.20 c, d, for clarity, only two rows of superlattice spots are shown, that associated with 000 as filled circles and that through 210 as open circles. Similar rows pass through all of the reflections from the parent structure.]

The diffraction patterns, and the structures giving rise to this feature, are both said to be **incommensurate**. Crystallographic techniques have now been developed that are able to resolve such ambiguities. In general, these materials have a structure that can be divided into two parts. To a reasonable approximation, one component is a conventional structure that behaves like a normal crystal, but an additional part exists that is **modulated**[3] in one, two or three dimensions. The fixed part of the structure might be, for example, the metal atoms, while the anions might be modulated in some fashion, as described below. [In some examples of these structures, the modulated component of the structure can also force the fixed part of the structure to become modulated in turn, leading to considerable crystallographic complexity.]

The diffraction pattern from a normal crystal is characterised by an array of spots separated by a distance $1/a = a^*$ that arise from the parent structure, together with a set of commensurate superlattice reflections that arise as a consequence of the additional ordering. In this case the spot spacing is $1/na = a^*/n$, where n is an integer, (Figure 8.21a, b). In modulated structures, the modulation might be in the position of the atoms, called a **displacive modulation**, (Figure 8.21c). Displacive modulations

---

[3]Take care to note that a *modulated* structure is *not* the same as a *modular* structure.

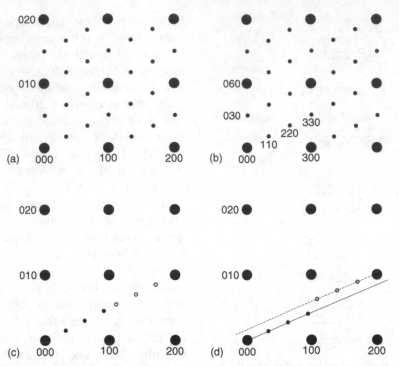

Figure 8.20 Superlattice reflections on diffraction patterns: (a) strong reflections are indexed in terms of the parent structure; (b) normal superlattice reflections can be indexed in terms of a larger 'supercell'; (c) co-linear; incommensurate superlattice reflections, (associated with the 000, filled circles, and 210 reflections, open circles), (d) non-co-linear incommensurate superlattice reflections, (associated with the 000, filled circles, and 210 reflections, open circles)

sometimes occur when a crystal structure is transforming from one stable structure to another as a result of a change in temperature. Alternatively, the modulation might be in the occupancy of a site, called **compositional modulation**, for example, the gradual replacement of O by F in a compound $M(O,F)_2$, (Figure 8.21d). In such a case the site occupancy factor would vary in a regular way throughout the crystal. Compositional modulation is often associated with solids that have a composition range.

As with normal superlattices, the diffraction patterns of modulated structures can also be divided into two parts. The diffraction pattern of the unmodulated component stays virtually unchanged in the modulated structure and gives

rise to a set of strong 'parent structure' reflections of spacing $1/a = a^*$. The modulated part gives rise to a set of superlattice reflections 'attached' to each parent structure reflection. In cases where the dimensions of the modulation are incommensurate, (that is do not fit), with the underlying structure, the phase is an **incommensurately modulated phase**. The spot spacing is equal to $1/\lambda$, where λ is the wavelength of the modulation, (Figure 8.21e). The positions of these reflections change smoothly as the modulation varies. In such cases, the diffraction pattern will show rows of superlattice spots that cannot be indexed in terms of the diffraction pattern of the parent phase. In cases where the modulation wave runs at an angle to the fixed part of the structure,

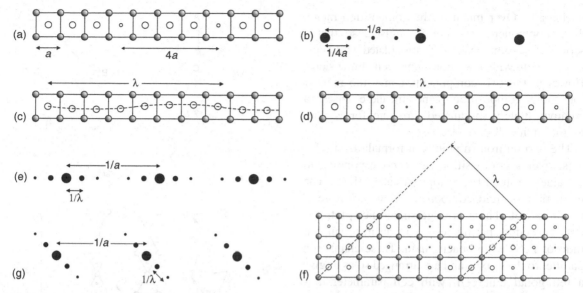

Figure 8.21 Schematic representations of normal and modulated crystal structures and diffraction patterns: (a) a normal superlattice, formed by the repetition of an anion substitution; (b) part of the diffraction pattern of (a); (c) a crystal showing a displacive modulation of the anion positions; (d) a crystal showing a compositional modulation of the anion conditions, (the change in the average chemical nature of the anion is represented by differing circle diameters); (e) part of the diffraction pattern from (c) or (d); (f) a modulation wave at an angle to the unmodulated component; (g) part of the diffraction pattern from (f). Metal atoms are represented by shaded circles and non-metal atoms by open circles

the extra spots will fall on lines at an angle to the main reflections, to give orientation anomalies, (Figure 8.21f, g).

When the modulation wavelength exactly fits a number of parent unit cells, (that is, is commensurate with it), it will be possible to index the reflections in terms of a normal super-lattice. Recently a number of structures that were once described in this way have been restudied and found to be better described as modulated structures with commensurate modulation waves – **commensurately modulated structures**.

There are a number of ways in which a structure can be modulated and by and large these were characterised little by little, a procedure that gave rise a number of different naming schemes. Thus, among these sub-sets, classes of compounds described as **vernier structures**, **chimney-ladder structures** and **layer misfit structures** can be found. In addition to these principally chemical modulations, magnetic spins, electric dipoles and other physical attributes of atoms can be modulated in a commensurate or incommensurate fashion.

The crystal structures of modulated materials can be illustrated by reference to the barium iron sulphides of formula $Ba_xFe_2S_4$, and the similar strontium titanium sulphides, Sr_xTiS_3. These phases fall into the sub-set of modulated structures called vernier structures, which have two interpenetrating chemical sub-structures. One of these expands or contracts as a smooth function of composition, while the other remains relatively unchanged. Over much of the composition range, these components are not in register, so that a series of incommensurately modulated structures form, often with enormous unit cell dimensions. However, the two components come into register for some compositions, and in these cases, a commensurately modulated structure forms with normal unit cell dimensions.

The barium iron sulphides of formula $Ba_xFe_2S_4$, exist over a composition range corresponding to x values from 1.0 to approximately 1.25. The more or less rigid component of the structure is composed of chains of edge-shared FeS_4 tetrahedra. The edge-sharing results in an iron: sulphur ratio of FeS_2 within each chain. These chains run parallel to the **c**-axis, and are arranged so as to give a tetragonal unit cell, with cell parameter a_{FeS} $(= b_{FeS})$, (Figure 8.22 a–c). The Ba atoms form the second component. These atoms lie between the chains of tetrahedra and also form rows parallel to the **c**-axis. The Ba array can also be assigned to a tetragonal unit cell, with cell parameter a_{Ba} $(= b_{Ba})$. The cell edges a_{FeS} and a_{Ba} are identical for the two subsystems, (Figure 8.22d).

The c-parameters of the FeS_2 and Ba components are different. The c-parameter of the FeS_2 chains, c_{FeS}, is constant, but the c-parameter of the Ba component, c_{Ba}, changes smoothly as the Ba content varies. This modulation falls in step with the FeS_2 component at regular intervals governed by the modulation wavelength. Broadly speaking, the spacing of the Ba atoms is equal to the spacing of two FeS_2 units, so that the overall formula of the phases is close to $Ba(Fe_2S_4)$. As an example, the c-axis projection of the phase $Ba_{1.1250}Fe_2S_4$, is shown in Figure 8.22f. The coincidence of the repeat is 9 Ba atoms to 16 FeS_2 units, so that the formula can be written $Ba_9(Fe_2S_4)_8$ and represents a commensurately modulated structure.

The general compositions of these structures can be written as $Ba_p(Fe_2S_4)_q$. The c-parameter of the complete structure, c_S, is given by the period at which the two subcells fit, that is, by

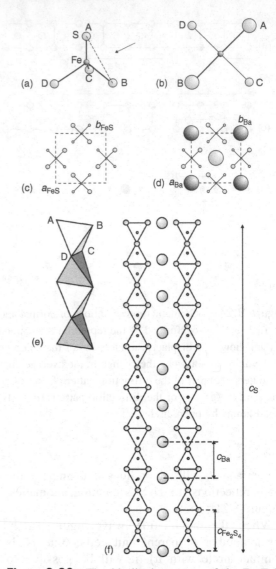

Figure 8.22   The idealised structure of the Ba_xFeS_2 incommensurate phases: (a) an idealised FeS_4 tetrahedron; (b) the same tetrahedron as (a), viewed in the direction of the arrow, perpendicular to an edge; (c) the tetragonal unit cell of the edge-shared FeS_4 tetrahedral chains viewed along [001]; (d) the tetragonal unit cell including Ba atoms; (e) chains of edge-shared FeS_4 tetrahedra; (f) a projection of the idealised structure of $Ba_9(Fe_2S_4)_8$, along [110]; (the tetrahedra are viewed in a direction C – D with respect to (a). Only two of the four FeS_4 tetrahedral chains are shown and one set of Ba atoms

the least common multiple of the subcell dimensions, so that:

$$c_S = qc_{FeS} = pc_{Ba}$$

where c_{FeS} is the average dimension of a pair of (Fe_2S_4) tetrahedra and c_{Ba} is the average separation of the Ba atoms. For the phase $Ba_{1.1250}Fe_2S_4$, which can be written as $Ba_9(Fe_2S_4)_8$, $c_S = 8c_{FeS} = 9c_{Ba}$. It can be appreciated that if the Ba content is changed infinitesimally, the two subcells will only come into register after an enormous distance. This is true, of course, for most compositions, and the commensurately modulated compositions are the exceptions.

The strontium titanium sulphides, Sr_xTiS_3, which exist between the x values of 1.05 and 1.22, are structurally similar to the barium iron sulphides, but in this system, both subsystems are modulated. In *idealised* Sr_xTiS_3 columns of TiS_6 ocahedra, which share faces, to give a repeat composition of $TiS_{6/2}$ or TiS_3, replace the edge-shared tetrahedra in $Ba_xFe_2S_4$, (Figure 8.23 a, b). These TiS_3 columns are arranged to give a hexagonal unit cell, (Figure 8.23c), and chains of Sr atoms lie between them to complete the idealised structure, (Figure 8.23d). The a-parameters of the TiS_3 and Sr arrays are equal. The Sr chains are flexible, and expand or contract along the **c**-axis, as a smooth function of the composition x in Sr_xTiS_3. The real structures of these phases are much more complex. The coordination of the Ti atoms is always six, but the coordination polyhedron of sulphur atoms around the metal atoms is in turn modulated by the modulations of the Sr chains. The result of this is that some of the TiS_6 polyhedra vary between ocahedra and a form some way between an octahedron and a trigonal prism, (Figure 8.23e). One of the simpler commensurately modulated structures reported is $Sr_8(TiS_3)_7$, the idealised structure of which is drawn in Figure 8.23f. The vast majority of compositions give incommensurately modulated structures with enormous unit cells.

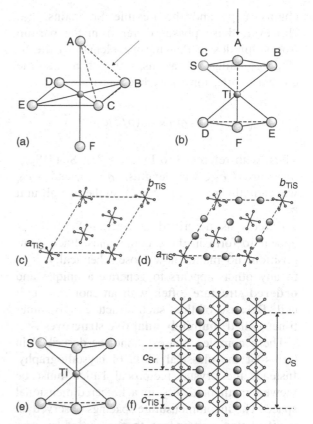

Figure 8.23 The idealised structure of the Sr_xTiS_3 incommensurate phases; (a) an idealised TiS_6 octahedron; (b) the same octahedron as in (a), viewed in the direction of the arrow, approximately normal to a triangular face. The chains of face-shared octahedra in Sr_xTiS_3 are linked by faces arrowed in (b); (c) the idealised hexagonal unit cell, formed by columns of edge sharing TiS_6 octahedra; (d) the idealised unit cell of Sr_xTiS_3 showing the location of the Sr atoms between the TiS_3 chains; (e) a trigonal prism; (f) the idealised structure of Sr_xTiS_3 projected onto (110)

As in the case of the barium iron sulphides, the compositions of the (real or idealised) materials can be written as $Sr_p(TiS_3)_q$. The c-lattice parameter of the complete structure, c_S, is given by the least common multiple of the c-lattice parameters of the (more or less fixed) TiS_3

chains, c_{TiS}, and the flexible Sr chains, c_{Sr}. However, these phases differ from the barium iron sulphides in that the wavelength of the Sr chain contains 2 Sr atoms, rather than 1, so the correct expression for c_S becomes:

$$c_S = qc_{TiS} = (p/2)c_{Sr}$$

Thus, with reference to Figure 8.23f, $Sr_8(TiS_3)_7$, there are 7 c_{TiS} wavelengths, $(q = 7)$, and 4 c_{Sr} wavelengths, $(p = 8, p/2 = 4)$, in the overall unit cell.

Defects as described in Section 8.1 do not appear to form in these compounds. Each composition, no matter how close, chemically, it is to any other, appears to generate a unique and **ordered** structure, often with an enormous unit cell. Because of this, such structures are sometimes called '**infinitely adaptive structures**'.

These complex phases cannot be described in terms of the classical ideas of crystallography. Instead, the crystal reciprocal lattice must be viewed mathematically in a higher dimensional space. The incommensurate spacings observed on a diffraction pattern are then regarded as projections from this higher dimension. The mathematical aspects of the crystallography of these interesting materials are beyond the scope of this book, but further information is given in the Bibliography.

8.9   Quasicrystals

Experimentally, a perfect crystal gives a diffraction pattern consisting of sharp reflections or spots. This is the direct result of the translational order that characterises the crystalline state. The translational order allowed in a crystal (in classical crystallography) has been set out earlier in this book. To recapitulate, a crystal can only be built from a unit cell consistent with the seven crystal systems and the 14 Bravais lattices. The unit cell can be translated to build up the crystal, but no other transformation, such as reflection or rotation, is allowed. This totally rules out a unit cell with overall five-fold, or greater than six-fold, rotation symmetry, just as a floor cannot be tiled completely with regular pentagons, (see Chapters 3 and 4).

As discussed earlier, the symmetry of the structure plays an important part in modifying the intensity of diffracted beams, one consequence of which is that the intensities of a pair of reflections hkl and $\bar{h}\bar{k}\bar{l}$ are equal in magnitude. This will cause the diffraction pattern from a crystal to appear centrosymmetric even for crystals that lack a centre of symmetry and the point symmetry of any sharp diffraction pattern will belong to one of the 11 Laue classes, (see Section 4.7 and Chapter 6, especially Section 6.9).

In 1984, this changed, with (at the time, the rather shocking) discovery of a metallic alloy of composition approximately $Al_{88}Mn_{12}$ which gave a sharp electron diffraction pattern that clearly showed the presence of an apparently ten-fold symmetry axis. The crystallographic problem is that the *sharp* diffraction pattern indicated long-range translational order. However, this was *quite incompatible* with the 10-fold rotation axis, which indicated that the unit cell possessed a forbidden rotation axis. Moreover, the overall diffraction pattern did *not* belong to one of the 11 Laue classes, and worse still, seemed to suggest that the unit cell showed *icosahedral* symmetry, (Figure 4.5d). Initially effort was put into trying to explain the rotational symmetry as an artefact due to the presence of defects in 'normal' crystals. However, it was soon proven that the material really did have ten-fold rotational symmetry, and belonged to the icosahedral point group $m\bar{3}\bar{5}$. This totally violated the laws of classical crystallography.

Since then, many other alloys that give rise to sharp diffraction patterns and which show five-fold, eight-fold, ten-fold and twelve-fold rotation symmetry have been discovered. The

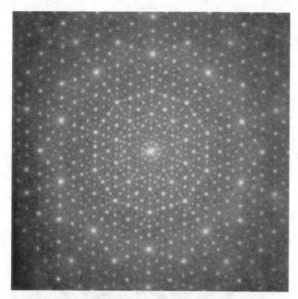

Figure 8.24 Electron diffraction pattern from the quasicrystalline alloy $Al_{72}Ni_{20}Co_8$, showing ten-fold rotation symmetry. (Photograph courtesy of Dr. Koh Saitoh, see K. Saitoh, M. Tanaka, A. P. Tsai and C. J. Rossouw, ALCHEMI studies on quasicrystals, JEOL News, **39**, No. 1, p20 (2004), reproduced with permission

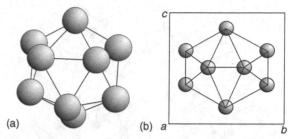

Figure 8.25 Icosahedra: (a) small numbers of spheres or metal atoms pack preferentially with an icosahedral geometry; (b) part of the As_3Co skutterudite structure, projected down [100] showing an icosahedron of As atoms. [Note that only a fraction of the As atoms and none of the Co atoms in the As_3Co unit cell are shown. The atoms in both (a) and (b) touch. They are drawn smaller to show the icosahedral geometry]

resulting materials rapidly became known as **quasiperiodic crystals**, or more compactly, as **quasicrystals**. Figure 8.24 shows the electron diffraction pattern from a quasicrystalline alloy of composition $Al_{72}Ni_{20}Co_8$, clearly showing ten-fold rotation symmetry.

There are a number of ways that the contradiction between classical crystallography and the structures of quasicrystals can be resolved, all of which require a slight relaxation in the strict rules of conventional crystallography. The simplest is to consider a quasicrystal as an imperfectly crystallised liquid. The liquid state of many metal alloys contain metal icosahedra that form in a transitory fashion, because this polyhedral arrangement represents a very efficient way of packing a small number of spheres together, (Figure 8.25a). Thus a liquid metal can often be

regarded as being composed of icosahedral clusters of metal atoms that are continually forming and breaking apart. On cooling a liquid metal the icosahedra freeze. If the cooling rate is slow enough, the icosahedra order onto a lattice, say cubic or tetragonal, which then destroys the *overall* icosahedral symmetry of the structure. A large number of metal alloy structures are known which contain ordered icosahedra, including $CoAs_3$, with the cubic skutterudite structure, in which As atoms take on this geometry, (Figure 8.25b). On the other hand, if the cooling rate is very fast the icosahedra freeze at random, to give an **icosahedral glass**. Over a range of intermediate temperatures, the icosahedra all freeze in the same orientation, but they do not have the energy or time to order on a lattice. A quasicrystal can then be regarded as made up of icosahedral clusters of metal atoms, all oriented in the same way, and separated by variable amounts of disordered material, (Figure 8.26). That is, the materials show orientational order but not translational order.

Quasicrystals can also be considered as three-dimensional analogues of Penrose tilings.

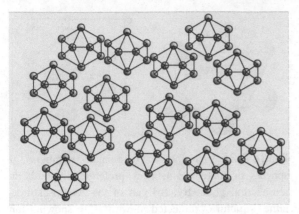

Figure 8.26 Icosahedra, arranged with the same orientation, but not on a lattice, can be taken as one model of a quasicrystal alloy

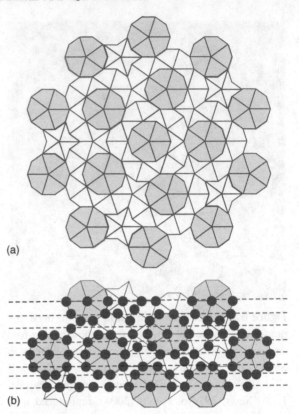

(a)

(b)

Figure 8.27 A Penrose tiling: (a) the decagons are all oriented in the same way, although they do not lie upon a lattice: (b) 'atoms' placed on the nodes of (a) fall onto a fairly well defined set of planes, capable of giving sharp diffracted reflections

Penrose tilings are aperiodic: they cannot be described as having unit cells and do not show translational order (see Chapter 3). However, a Penrose tiling has a sort of translational order, in the sense that parts of the pattern, such as all of the decagons, are oriented identically, but they are not spaced in such a way as to generate a unit cell, (Figure 8.27a). Moreover, if the nodes in a Penrose tiling are replaced by 'atoms', there are well defined regularly spaced atom planes running through the tilings that would suffice to give sharp diffraction spots, following Bragg's law, (Figure 8.27b). In fact, the diffraction patterns computed from an array of atoms placed at the nodes of Penrose tilings show sharp spots and five- and ten-fold rotation symmetries.

The Penrose model of quasicrystals consists of a three-dimensional Penrose tiling using effectively two 'unit cells', corresponding to the dart and kite units of the plane patterns. These can be joined to give an aperiodic structure which has the same sort of order as the two-dimensional tiling. That is, all of the icosahedra are in the same orientation, but not arranged on a lattice, and once again the model structures show orientational order but not translational order. As with the planar tilings, these three-dimensional

analogues give diffraction patterns that show both sharp spots and forbidden rotation symmetries.

In reality, structures that combine features of both of these models fit the observed properties best.

Answers to introductory questions

What are modular structures?

Modular structures are those that can be considered to be built from slabs of one or more parent structures. Slabs can be sections from just one parent phase, as in many *perovskite*-related structures, *CS* phases and polytypes, or they can come from two or more parent structures, as in the mica – pyroxene intergrowths. Some of these crystals possess enormous unit cells, of some 100's of nm in length. In many materials the slab thicknesses may vary widely, in which case the slab boundaries will not fall on a regular lattice, and so form planar defects.

What are incommensurately modulated structures?

In general, incommensurately modulated structures have two fairly distinct parts. One part of the crystal structure is conventional, and behaves like a normal crystal. An additional, more or less independent part exists that is modulated in one, two or three dimensions. For example, the fixed part of the structure might be the metal atom array, while the modulated part might be the anion array. The modulation might be in the position of the atoms, called a displacive modulation or the modulation might be in the occupancy of a site, for example, the gradual replacement of O by F in a compound $M(O,F)_2$, called compositional modulation. In some more complex crystals modulation in one part of the structure induces a corresponding modulation in the "fixed" part.

In cases where the wavelength of the modulation fits exactly with the dimensions of the underlying structure, a commensurately modulated crystallographic phase forms. In cases where the dimensions of the modulation are incommensurate (that is, do not fit) with the underlying structure, the phase is an incommensurately modulated phase. The modulation produces sets of extra reflections on the diffraction pattern that may (commensurate) or may not (incommensurate) fit with the reflections from the parent non-modulated component.

What are quasicrystals?

Quasicrystals or quasiperiodic crystals are metallic alloys which yield sharp diffraction patterns that display 5-, 8-, 10- or 12-fold symmetry rotational axes - forbidden by the rules of classical crystallography. The first quasicrystals discovered, and most of those that have been investigated, have icosahedral symmetry. Two main models of quasicrystals have been suggested. In the first, a quasicrystal can be regarded as made up of icosahedral clusters of metal atoms, all oriented in the same way, and separated by variable amounts of disordered material. Alternatively, quasicrystals can be considered to be three-dimensional analogues of Penrose tilings. In either case, the material does not possess a crystallographic unit cell in the conventional sense.

Problems and exercises

Quick Quiz

1 The oxides Al_2O_3 and Cr_2O_3 form a complete solid solution $Al_xCr_{2-x}O_3$. The site occupancy factor for the cations in the crystal $AlCrO_3$ is:
(a) 1.0

(b) 0.5

(c) 0.3

2 The *fluorite* structure non-stoichiometric oxide $Ca^{2+}_{0.1} Zr^{4+}_{0.9} O_{1.9}$ has a site occupancy factor for O equal to:
(a) 1.9

(b) 0.95

(c) 0.1

3 A solid containing interstitial atom point defects will have a theoretical density:
(a) Higher than the parent crystal
(b) Lower that the parent crystal
(c) The same as the parent crystal.

4 A solid is prepared by heating a 1 : 1 mixture of $SrTiO_3$ and $Sr_4Nb_4O_{14}$ to give a member of the $Sr_n(Nb,Ti)_nO_{3n+2}$, series of phases. The value of n of the new material is:
(a) 2.5
(b) 5
(c) 10

5 Carborundum is a form of:
(a) Al_2O_3
(b) ZnS
(c) SiC

6 The Ramsdell symbol $2H$ applies to polytypes with the same structure as:
(a) Wurzite
(b) Zinc blende
(c) Carborundum I

7 A family of phases with compositions represented by a general formula such as W_nO_{3n-2} is called:
(a) A homogeneous series
(b) A heterogeneous series
(c) A homologous series

8 Ruddleston-Popper phases, Aurivillius phases and high temperature superconductors *all* contain slabs of structure similar to that of:
(a) Rutile
(b) Perovskite
(c) Wurtzite

9 If the positions of a set of atoms in a structure follows a wave-like pattern, the modulation is described as:
(a) Compositional
(b) Displacive
(c) Incommensurate

10 Quasicrystals are alloys that are characterised by 5-fold or higher symmetry and:
(a) Orientational but not translational order
(b) Glass – like disorder
(c) Modulated order

Problems and Exercises

8.1 Both TiO_2 and SnO_2 adopt the rutile structure (see Chapter 1) and crystals with intermediate compositions $Ti_xSn_{1-x}O_2$ are often found in nature. A crystal of composition $Ti_{0.7}Sn_{0.3}O_2$ has lattice parameters $a = 0.4637$ nm, $c = 0.3027$ nm. The scattering factors appropriate to the (200) reflection from this unit cell are $f_{Ti} = 15.573$, $f_{Sn} = 39.405$ and $f_O = 5.359$. (a). Calculate the metal atom site scattering factor, the structure factor for the 200 reflection and the intensity of the reflection, (see Chapter 6). Repeat the calculations for (b), pure TiO_2 and (c), SnO_2 samples, assuming that the lattice parameter of these materials is identical to that of the solid solution.

8.2 The compound $Y_3Ga_5O_{12}$ crystallises with the garnet structure[4]. A solid solution,

---

[4]The garnet structure is cubic, with a lattice parameter of approximately 1.2 nm. The atoms are situated in three different coordination sites, and the formula is conveniently represented as $\{A^{3+}\}_3[B^{3+}]_2(C^{3+})_3O_{12}$, where $\{A\}$ represents cations in cubic sites, $[B]$ represents cations in octahedral sites and (C) represents cations in tetrahedral sites. In the garnets $Y_3Ga_5O_{12}$ and $Tm_3Ga_5O_{12}$, the cations Y or Tm are in cubic sites and the Ga ions occupy both the octahedral and tetrahedral sites.

Data for the solid solution $(Y_xTm_{1-x})_3$ Ga_5O_{12}[*]

Composition, x	Lattice parameter/nm
0	—
0.20	1.22316
0.35	1.22405
0.50	1.22501
0.65	1.22563
0.80	1.22638
1.00	1.22734

[*]Adapted from data given by F. S. Liu *et al.*, J. Solid State Chemistry, **178**, 1064–1070 (2005).

$(Y_xTm_{1-x})_3Ga_5O_{12}$, forms when Y is replaced by Tm. The lattice parameters of several members of the solid solution series are given in the Table. Plot these against composition to determine how well Vegard's law is followed. (a). Estimate the lattice parameter of $Tm_3Ga_5O_{12}$. (b). What is the composition of the phase with a lattice parameter of 1.22400 nm?

8.3 The compound $LaMnO_{3.165}$ adopts a distorted perovskite structure, in which the unit cell has rhombohedral symmetry, with hexagonal lattice parameters a = 0.55068 nm, c = 1.33326 nm, Z = 6 $LaMnO_{3.165}$. The oxygen excess can arise either as oxygen interstitials or as metal atom vacancies. (a). Write the formula of each alternative; (assume that metal vacancies occur in equal numbers on both the La and Mn positions). (b). Determine the theoretical density of each. (Data adopted from A. Barnabé *et al.*, Materials Research Bulletin, **39**, 725–735 (2004).

8.4 (a). The packing symbol of a 15*R* ZnS polytype is (*cchch*). Sketch the zig-zag structure (projected onto the hexagonal (11$\bar{2}$0) plane) and determine the Zhdanov symbol for the phase. (b). The Zhdanov symbol for a 24*H* ZnS polytype is (7557). Sketch the zig-zag

structure (projected onto the hexagonal (11$\bar{2}$0) plane) and determine the packing sequence. (c). The zig-zag structure (projected onto the hexagonal (11$\bar{2}$0) plane) of a ZnS polytype is shown in the Figure below. Determine the Zhdanov symbol and the stacking sequence.

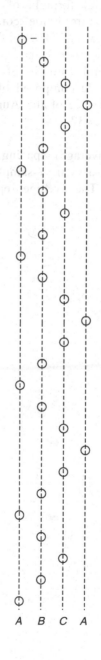

A   B   C   A

8.5 (a). *CS* planes that lie between {102} and {103} are known in the reduced tungsten oxides. What plane would the *CS* plane composed of alternating blocks of 4 and 6 edge-sharing octahedra lie along, and what is the series formula of materials containing ordered arrays of these *CS* planes? (b). What is the general formula of a Ruddleston-Popper structure oxide composed of the slab sequence ... 344 ... of the Ruddleston-Popper phases $n = 3$ and $n = 4$? (c). What is the general formula of an Aurivillius structure oxide composed of the slab sequence ... 23 ... of the Aurivillius phases $n = 2$ and $n = 3$?

8.6 (a). The (average) spacing of Ba atoms along the *c*-axis in a sample of $Ba_xFe_2S_4$ is 0.5 nm. The average repeat dimension of a pair of FeS_4 tetrahedra in the same direction is 0.56 nm. What is the formula of the phase and its *c*- lattice parameter? [Data adapted from I. E. Grey, *Acta Crystallogr.* **B31**, 45 (1975)]. (b). Assuming the same dimensions for barium atom separation and iron sulphur tetrahedra hold, express the formula of the phase $Ba_{1.077}(Fe_2S_4)$ in terms of integers p and q, and estimate the *c*-lattice parameter. (c). The composition of a strontium titanium sulphide is $Sr_{1.145}TiS_3$. The repeat distance of the TiS_3 chains, c_{TiS}, was found to be 0.2991 nm and that of the Sr chains, c_{Sr}, was found to be 0.5226 nm. Express the formula in terms of (approximate) indices $Sr_p(TiS_3)_q$, and determine the approximate *c*-parameter of the phase. [Data adopted from M. Onoda *et al.*, *Acta Crystallogr.*, **B49**, 929–936 (1993).]

Appendix 1   Vector addition and subtraction

Vectors are used to specify quantities that have a direction and a magnitude. Unit cell edges are specified by vectors **a**, **b** and **c**, which have a magnitude, which is a scalar (an ordinary number), a, b and c, and a direction.

A vector is often represented by an arrow, (Figure A1.1a).

A vector **a** multiplied by a scalar $+a$ is a vector with the same direction and a times as long, (Figure A1.1b).

A vector **a** multiplied by a scalar $-a$ is a vector pointing in the opposite direction to **a** and a times as long, (Figure A1.1c).

Two vectors, **a** and **b**, can be **added** to give **a** + **b** by linking them so that the tail of the second vector, **b**, is placed in contact with the head of the first one, **a**. The vector joining the tail of **a** to the head of **b** is the **vector sum**, or **resultant**, **r**, which is also a vector, (Figure A1.1d). The resultant, **r**, of a large number of vector additions is found by successive application of this process, (Figure A1.1e).

Two vectors, **a** and **b**, can be **subtracted** to give **a** − **b** by 'adding' the negative vector, −**b** to **a**, (Figure A1.1f). The resultant, **r**, of a large number of vector subtractions is found by successive application of this process.

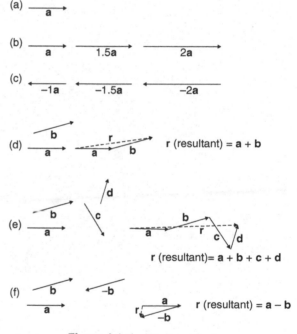

Figure A1.1   Vector notation

Crystals and Crystal Structures.   Richard J. D. Tilley
© 2006 John Wiley & Sons, Ltd

Appendix 2   Data for some inorganic crystal structures

Copper (A1)

Structure: cubic; $a = 0.3610$ nm; $Z = 4$; Space group, $Fm\bar{3}m$, (No. 225);

Atom positions: Cu: $4a$ 0, 0, 0; ½, ½, 0; 0, ½, ½; ½, 0, ½;

Tungsten (A2)

Structure: cubic; $a = 0.3160$ nm; $Z = 2$; Space group, $Im\bar{3}m$, (No. 229);

Atom positions: W: $2a$ 0, 0, 0; ½, ½, ½;

Magnesium (A3)

Structure: hexagonal; $a = 0.3210$ nm, $c = 0.5210$ nm; $Z = 2$; Space group, $P6_3/mmc$, (No. 194);

Atom positions: Mg: $2d$ ⅔, ⅓, ¼; ⅓, ⅔, ¾;

Diamond (A4)

Structure: cubic; $a = 0.3560$ nm; $Z = 8$; Space group, $Fd\bar{3}m$, (No. 227);

Atom positions: C: $8a$ 0, 0, 0; ½, ½, 0; 0, ½, ½; ½, 0, ½; ¼, ¼, ¼; ¾, ¾, ¼; ¾, ¼, ¾; ¼, ¾, ¾;

Graphite

Structure: hexagonal, $a = 0.2460$ nm, $c = 0.6701$ nm; $Z = 4$; Space group, $P6_3/mmc$, (No. 186);

Atom positions: C1 $2a$ 0, 0, 0; 0, 0, ½; C2 $2b$ ⅓, ⅔, 0; ⅔, ⅓, ½;

Halite, NaCl (B1)

Structure: cubic; $a = 0.5640$ nm; $Z = 4$; Space group, $Fm\bar{3}m$, (No. 225);

Atom positions: Na: $4a$ 0, 0, 0; ½, ½, 0; ½, 0, ½; 0, ½, ½; Cl: $4b$ ½, 0, 0; 0, 0, ½; 0, ½, 0; ½, ½, ½;

(or *vice versa*)

Caesium chloride, CsCl (B2)

Structure: cubic $a = 0.4110$ nm; $Z = 2$; Space group, $Pm\bar{3}m$, (No. 221);

Crystals and Crystal Structures.   Richard J. D. Tilley
© 2006 John Wiley & Sons, Ltd

Atom positions: Cs: 1a 000;
                   Cl: 1b ½, ½, ½;

(or *vice versa*)

Zinc blende, (sphalerite), ZnS (B3)

Structure: cubic; $a = 0.5420$ nm, $Z = 4$;
                    Space group, $F\bar{4}3m$, (No. 216);
Atom positions: Zn: 4a 0, 0, 0; ½, ½, 0;
                    ½, 0, ½; 0, ½, ½;
                    S: 4c ¼, ¼, ¼; ¾, ¾, ¼;
                    ¾, ¼, ¾; ¼, ¾, ¾;

Wurtzite, ZnS (B4)

Structure: hexagonal; $a = 0.3810$ nm, $c = 0.6230$ nm; $Z = 2$;   Space
                    group, $P6_3mc$, (No. 186);
Atom positions: Zn: 2b ⅓, ⅔, ½; ⅔, ⅓, 0;
                    S: 2b ⅓, ⅔, ⅜; ⅔, ⅓, ⅞;

Nickel arsenide, NiAs

Structure: hexagonal; $a = 0.3610$ nm,
                    $c = 0.5030$ nm; $Z = 2$; Space
                    group, $P6_3/mmc$, (No. 194);
Atom positions: Ni: 2b 0, 0, ¼; 0, 0, ¾;
                    As: 2c ⅓, ⅔, ¼; ⅔, ⅓, ¾;

Boron nitride

Structure: hexagonal; $a = 0.2500$ nm,
                    $c = 0.6660$ nm; $Z = 2$; Space
                    group, $P6_3/mmc$, (No. 194);
Atom positions: B: 2c ⅓, ⅔, ¼; ⅔, ⅓, ¾;
                    N: 2d ⅓, ⅔, ¾; ⅔, ⅓, ¼;

Corundum, Al₂O₃

Structure: trigonal, hexagonal axes; $a = 0.4763$ nm, $c = 1.3009$ nm;

                    $Z = 6$; Space group, $R\bar{3}c$,
                    (No. 167);
Atom positions: each of (0, 0, 0); (⅔, ⅓, ⅓);
                    (⅓, ⅔, ⅔); plus
                    Al: 12c 0, 0, z; 0, 0, $\bar{z}$ + ½;
                    0, 0, $\bar{z}$; 0, 0, z + ½;
                    O: 18e x, 0, ¼; 0, x, ¼; $\bar{x}$, $\bar{x}$,
                    ¼; $\bar{x}$, 0, ¾; 0, $\bar{x}$, ¾; x, x, ¾;

The x coordinate (O) and the z coordinate (Al) can be approximated to ⅓. In general these can be written $x = ⅓ + u$ and $z = ⅓ + w$, where u and w are both small. Taking typical values of $x = 0.306$, $z = 0.352$, the positions are:

Al: 12c 0, 0, 0.352; 0, 0, 0.148; 0, 0, 0.648;
0, 0, 852;
O: 18e 0.306, 0, ¼; 0, 0.306, ¼; 0.694,
0.694, ¼; 0.694, 0, ¾; 0, 0.694, ¾;
0.306, 0.306, ¾;

Fluorite, CaF₂ (C1)

Structure: cubic; $a = 0.5463$ nm, $Z = 4$;
                    Space group, $Fm\bar{3}m$, (No. 225);
Atom positions: Ca: 4a 0, 0, 0; ½, ½, 0; ½,
                    0, ½; 0, ½, ½;
                    F: 8c ¼, ¼, ¼; ¼, ¾, ¾;
                    ¾, ¼, ¾; ¾, ¾, ¼; ¼, ¼, ¾;
                    ¼, ¾, ¼; ¾, ¼, ¼; ¾, ¾, ¾.

Pyrites, FeS₂ (C2)

Structure: cubic; $a = 0.5440$ nm, $Z = 4$;
                    Space group, $Pa\bar{3}$, (No. 205);
Atom positions: Fe: 4a 0, 0, 0; ½, ½, 0;
                    ½, 0, ½; 0, ½, ½;
                    S: 8c x, x, x; $\bar{x}$ + ½, $\bar{x}$, x + ½; $\bar{x}$,
                    x + ½, $\bar{x}$ + ½; x + ½, $\bar{x}$ + ½, $\bar{x}$;
                    $\bar{x}$, $\bar{x}$, $\bar{x}$; x + ½, x, $\bar{x}$ + ½; x,
                    $\bar{x}$ + ½, x + ½; $\bar{x}$ + ½, x + ½,
                    x;

The x coordinate for O is close to $\frac{1}{3}$. Taking a typical value of x = 0.375, the positions are:

S: 8c 0.378, 0.378, 0.378; 0.122, 0.622, 0.878;
0.622, 0.878, 0.122;   0.578, 0.122, 0.622;
0.622, 0.622, 0.622;   0.878, 0.378, 0.122;
0.378, 0.122, 0.878; 0.122, 0.878, 0.378;

Rutile, TiO_2

Structure:	tetragonal; $a = 0.4594$ nm, $c = 0.2959$ nm, Z = 2; Space group, $P4_2/mnm$, (No. 136);
Atom positions:	Ti: 2a 0, 0, 0; $\frac{1}{2}$, $\frac{1}{2}$, $\frac{1}{2}$;
	O: 4f x, x, 0; $\bar{x}$, $\bar{x}$, 0; $\bar{x} + \frac{1}{2}$, $x + \frac{1}{2}$, $\frac{1}{2}$; $x + \frac{1}{2}$, $\bar{x} + \frac{1}{2}$, $\frac{1}{2}$;

The x coordinate for O is close to $\frac{1}{3}$. Taking a typical value of $x = 0.305$, the positions are:

O: 4f 0.305, 0.305, 0;   0.695, 0.695, 0;
0.195, 0.805, $\frac{1}{2}$; 0.805, 0.195, $\frac{1}{2}$;

Rhenium trioxide, ReO_3

Structure:	cubic; $a = 0.3750$ nm, Z = 1; Space group, $Pm\bar{3}m$, (No. 221);
Atom positions:	Re: 1a 0, 0, 0;
	O: 3d $\frac{1}{2}$, 0, 0; 0, $\frac{1}{2}$, 0; 0, 0, $\frac{1}{2}$;

Strontium titanate (ideal perovskite), $SrTiO_3$

Structure:	cubic; $a = 0.3905$ nm, Z = 1; Space group, $Pm\bar{3}m$, (No. 221);
Atom positions:	Ti: 1a 0, 0, 0;
	Sr: 1b $\frac{1}{2}$, $\frac{1}{2}$, $\frac{1}{2}$;

O: 3c 0, $\frac{1}{2}$, $\frac{1}{2}$; $\frac{1}{2}$, 0, $\frac{1}{2}$;
$\frac{1}{2}$, $\frac{1}{2}$, 0;

Spinel, $MgAl_2O_4$

Structure:	cubic; $a = 0.8090$ nm, Z = 8; Space group, $Fd\bar{3}m$, (No. 227);
Atom positions:	each of (0, 0, 0); (0, $\frac{1}{2}$, $\frac{1}{2}$); ($\frac{1}{2}$, 0, $\frac{1}{2}$); ($\frac{1}{2}$, $\frac{1}{2}$, 0); plus
	Mg: 8a 0,0,0; $\frac{3}{4}$, $\frac{1}{4}$, $\frac{3}{4}$;
	Al: 16d $\frac{5}{8}$, $\frac{5}{8}$ $\frac{5}{8}$; $\frac{3}{8}$, $\frac{7}{8}$, $\frac{1}{8}$; $\frac{7}{8}$, $\frac{1}{8}$, $\frac{3}{8}$; $\frac{1}{8}$, $\frac{3}{8}$, $\frac{7}{8}$;
	O: 32e x, x, x; $\bar{x}$, $\bar{x} + \frac{1}{2}$, $x + \frac{1}{2}$; $\bar{x} + \frac{1}{2}$, $x + \frac{1}{2}$, $\bar{x}$; $x + \frac{1}{2}$, $\bar{x}$, $\bar{x} + \frac{1}{2}$; $x + \frac{3}{4}$, $x + \frac{1}{4}$, $\bar{x} + \frac{3}{4}$; $\bar{x} + \frac{1}{4}$, $\bar{x} + \frac{1}{4}$, $\bar{x} + \frac{1}{4}$; $x + \frac{1}{4}$, $\bar{x} + \frac{3}{4}$, $x + \frac{3}{4}$; $\bar{x} + \frac{3}{4}$, $x + \frac{3}{4}$, $x + \frac{1}{4}$;

The x coordinate for O is approximately equal to $\frac{3}{8}$, and is often given in the form $\frac{3}{8} + u$, where u is of the order of 0.01. Taking a typical value of $x = 0.388$, in which $u = 0.013$, the positions are:

O: 32e 0.388, 0.388, 0.388;   0.612, 0.112, 0.888; 0.112, 0.888, 0.612; 0.888, 0.612, 0.112; 0.138, 0.638, 0.862; 0.862, 0.862, 0.862; 0.368, 0.362, 0.138; 0.362, 0.138, 0.638;

The structural details given above are representative and taken from several sources within the EPSRC's Chemical Database Service at Daresbury, (see Bibliography). The Strukturbericht symbols are given for some compounds.

Appendix 3   Schoenflies symbols

Schoenflies symbols are widely used to describe molecular symmetry, the symmetry of atomic orbitals, and in chemical group theory. The terminology of the important symmetry operators and symmetry elements used in this notation are given in Table A3.1.

Table A3.1   Symmetry operations and symmetry elements

Symmetry element	Symmetry operation	Symbol
Whole object	Identity	E
n-fold axis of rotation	Rotation by $2\pi/n$	C_n
Mirror plane	Reflection	σ
Centre of symmetry	Inversion	i
n-fold improper axis	Rotation by $2\pi/n$ plus *reflection*	S_n

The symbol E represents the identity operation, that is, the combination of symmetry elements that transforms the object (molecule say) into a copy identical in every way to the original. There is one important feature to note. The improper rotation axes defined here are **not** the same as the improper rotation axes defined via Hermann-Mauguin symbols, but are rotoreflection axes (see Section 4.3 for details).

An object such as a molecule can be assigned a collection of symmetry elements that characterise the **point group** of the shape. The main part of the Schoenflies symbol describing the point group is a letter symbol describing the principle rotation symmetry, as set out in Table A3.2.

Table A3.2   Schoenflies symbols for point groups

Rotation group	Symbol
Cyclic	C
Dihedral	D
Tetrahedral	T
Octahedral	O

The symbol C represents a (proper) rotation axis. The symbol D represents a (primary) rotation axis together with another (supplementary) rotation axis normal to it. The symbol T represents tetrahedral symmetry, essentially the presence of four three-fold axes and three two-fold axes. The symbol O represents octahedral symmetry, essentially four three-fold axes and three four-fold axes.

These main symbols are followed by one or two subscripts giving further information on the order of the rotation and position. For example, the symbol C_n represents a (proper) rotation axis with a rotation of $2\pi/n$ around the axis. The symbol D_n represents a rotation axis with a rotation of $2\pi/n$ around the axis together with another rotation axis normal to it. Subscripts are also added to other symbols in the same way. Thus, a mirror plane

Crystals and Crystal Structures.   Richard J. D. Tilley
© 2006 John Wiley & Sons, Ltd

perpendicular to the main rotation axis, which is regarded as vertical, is written as σ_h. A mirror plane that includes the vertical axis can be of two types. If all are identical, they are labelled σ_v, and if both types are present they are labelled σ_v and σ_d, where the d stands for dihedral. In general σ_v planes include the horizontal two-fold axes, while σ_d planes lie between the horizontal 2-fold axes. The nomenclature in brief is set out in Table A3.3.

Table A3.3   Combinations of symmetry elements and group symbols

Symmetry elements	Group designation
E only	C_1
σ only	C_s
i only	C_i
C_n only	C_n
S_n only, n even	S_n $(S_2 \equiv C_i)$
S_n only, n odd	C_{nh} (C_n axis plus horizontal mirror)
C_n + perpendicular 2-fold axes	D_n
$C_n + \sigma_h$	C_{nh} (C_n is taken as vertical)
$C_n + \sigma_v$	C_{nv}
C_n + perpendicular σ_h	D_{nh}
C_n + perpendicular 2-fold axes + σ_d	D_{nd}
linear molecule with no symmetry plane perpendicular to molecule axis	$C_{\infty v}$
linear molecule with symmetry plane perpendicular to molecule axis	$D_{\infty h}$
3 mutually perpendicular 2-fold axes	T
3 mutually perpendicular 4-fold axes	O
C_5 axes + i	I_h

The correspondence between the Schoenflies and Hermann-Mauguin notation for the 32 crystallographic point groups is given in Table A3.4.

Table A3.4   Crystallographic point group symbols

Hermann-Mauguin full symbol	Hermann-Mauguin short symbol[*]	Schoenflies symbol
1		C_1
$\bar{1}$		C_i
2		C_2
m		C_s
$2/m$		C_{2h}
222		D_2
$mm2$		C_{2v}
$2/m\ 2/m\ 2/m$	mmm	D_{2h}
4		C_4
$\bar{4}$		S_4
$4/m$		C_{4h}
422		D_4
$4mm$		C_{4v}
$\bar{4}2m$ or $\bar{4}m2$		D_{2d}
$4/m\ 2/m\ 2/m$	$4/mmm$	D_{4h}
3		C_3
$\bar{3}$		C_{3i}
32 or 321 or 312		D_3
$3m$ or $3m1$ or $31m$		C_{3v}
$\bar{3}\ 2/m$ or $\bar{3}\ 2/m\ 1$ or $\bar{3}\ 1\ 2/m$	$\bar{3}m$ or $\bar{3}m1$ or $\bar{3}1m$	D_{3d}
6		C_6
$\bar{6}$		C_{3h}
$6/m$		C_{6h}
622		D_6
$6mm$		C_{6v}
$\bar{6}m2$ or $\bar{6}2m$		D_{3h}
$6/m\ 2/m2/m$	$6/mmm$	D_{6h}
23		T
$2/m\ \bar{3}$	$m\bar{3}$	T_h
432		O
$\bar{4}3m$		T_d
$4/m\ \bar{3}\ 2/m$	$m\bar{3}m$	O_h

[*] Only given if different from the full symbol

A full description of the Schoenflies notation is contained in the references given in the Bibliography.

Appendix 4   The 230 space groups

Point group[*]	Space group[**]			
	Triclinic			
$1, C_1$	$1\ P1$ C_1^1			
$\bar{1}, C_i$	$2\ P\bar{1}$ C_i^1			
	Monoclinic			
$2, C_2$	$3\ P\ 1\ 2\ 1$ $P2, C_2^1$	$4\ P\ 1\ 2_1\ 1$ $P2_1, C_2^2$	$5\ C\ 1\ 2\ 1$ $C2, C_2^3$	
m, C_s	$6\ P\ 1\ m\ 1$ Pm, C_s^1	$7\ P\ 1\ c\ 1$ Pc, C_s^2	$8\ C\ 1\ m\ 1$ Cm, C_s^3	$9\ C\ 1\ c\ 1$ Cc, C_s^4
$2/m, C_{2h}$	$10\ P\ 1\ 2/m\ 1$ $P2/m, C_{2h}^1$	$11\ P\ 1\ 2_1/m\ 1$ $P2_1/m, C_{2h}^2$	$12\ C\ 1\ 2/m\ 1$ $C2/m, C_{2h}^3$	$13\ P\ 1\ 2/c\ 1$ $P2/c, C_{2h}^4$	$14\ P\ 1\ 2_1/c\ 1$ $P2_1/c, C_{2h}^5$
	$15\ C\ 1\ 2/c\ 1$ $C2/c, C_{2h}^6$			
	Orthorhombic			
$222, D_2$	$16\ P\ 2\ 2\ 2$ D_2^1	$17\ P\ 2\ 2\ 2_1$ D_2^2	$18\ P\ 2_1\ 2_1\ 2$ D_2^3	$19\ P\ 2_1 2_1 2_1$ D_2^4	$20\ C\ 2\ 2\ 2_1$ D_2^5
	$21\ C\ 2\ 2\ 2$ D_2^6	$22\ F\ 2\ 2\ 2$ D_2^7	$23\ I\ 2\ 2\ 2$ D_2^8	$24\ I\ 2_1\ 2_1 2_1$ D_2^9	
$mm2,$ C_{2v}	$25\ P\ m\ m\ 2$ C_{2v}^1	$26\ P\ m\ c\ 2_1$ C_{2v}^2	$27\ P\ c\ c\ 2$ C_{2v}^3	$28\ P\ m\ a\ 2$ C_{2v}^4	$29\ P\ c\ a\ 2_1$ C_{2v}^6
	$30\ P\ n\ c\ 2$ C_{2v}^6	$31\ P\ m\ n\ 2_1$ C_{2v}^7	$32\ P\ b\ a\ 2$ C_{2v}^8	$33\ P\ n\ a\ 2_1$ C_{2v}^9	$34\ P\ n\ n\ 2$ C_{2v}^{10}
	$35\ C\ m\ m\ 2$ C_{2v}^{11}	$36\ C\ m\ c\ 2_1$ C_{2v}^{12}	$37\ C\ c\ c\ 2$ C_{2v}^{13}	$38\ A\ m\ m\ 2$ C_{2v}^{14}	$39\ A\ b\ m\ 2$ C_{2v}^{15}
	$40\ A\ m\ a\ 2$ C_{2v}^{16}	$41\ A\ b\ a\ 2$ C_{2v}^{17}	$42\ F\ m\ m\ 2$ C_{2v}^{18}	$43\ F\ d\ d\ 2$ C_{2v}^{19}	$44\ I\ m\ m\ 2$ D_{2v}^{20}

(Continued)

Crystals and Crystal Structures.   Richard J. D. Tilley
© 2006 John Wiley & Sons, Ltd

(Continued)

Point group*	Space group**			
	45 $I\,b\,a\,2$ C_{2v}^{21}	46 $I\,m\,a\,2$ C_{2v}^{22}		
$mmm,$ D_{2h}	47 $P\,2/m\,2/m\,2/m$ $Pmmm,\ D_{2h}^1$	48 $P\,2/n\,2/n\,2/n$ $Pnnn,\ D_{2h}^2$	49 $P\,2/c\,2/c\,2/m$ $Pccm,\ D_{2h}^3$	50 $P\,2/b\,2/a\,2/n$ $Pban,\ D_{2h}^4$	51 $P\,2_1/m\,2/m\,2/a$ $Pmma,\ D_{2h}^5$
	52 $P\,2/n\,2_1/n\,2/a$ $Pnna,\ D_{2h}^6$	53 $P\,2/m\,2/n\,2_1/a$ $Pmna,\ D_{2h}^7$	54 $P\,2_1/c\,2/c\,2/a$ $Pcca,\ D_{2h}^8$	55 $P\,2_1/b\,2_1/a\,2/m$ $Pbam,\ D_{2h}^9$	56 $P\,2_1/c\,2_1/C\,2/n$ $Pccn,\ D_{2h}^{10}$
	57 $P\,2/b\,2_1/c\,2_1/m$ $Pbcm,\ D_{2h}^{11}$	58 $P\,2_1/n\,2_1/n\,2/m$ $Pnnm,\ D_{2h}^{12}$	59 $P\,2_1/m\,2_1/m\,2/n$ $Pmmn,\ D_{2h}^{13}$	60 $P\,2_1/b\,2/c\,2_1/n$ $Pbcn,\ D_{2h}^{14}$	61 $P\,2_1/b\,2_1/c\,2_1/a$ $Pbca,\ D_{2h}^{15}$
	62 $P\,2_1/n\,2_1/m\,2_1/a$ $Pnma,\ D_{2h}^{16}$	63 $C\,2/m\,2/c\,2_1/m$ $Cmcm,\ D_{2h}^{17}$	64 $C\,2/m\,2/c\,2_1/a$ $Cmca,\ D_{2h}^{18}$	65 $C\,2/m\,2/m\,2/m$ $Cmmm,\ D_{2h}^{19}$	66 $C\,2/c\,2/c\,2/m$ $Cccm,\ D_{2h}^{20}$
	67 $C\,2/m\,2/m\,2/a$ $Cmma,\ D_{2h}^{21}$	68 $C\,2/c\,2/c\,2/a$ $Ccca,\ D_{2h}^{22}$	69 $F\,2/m\,2/m\,2/m$ $Fmmm,\ D_{2h}^{23}$	70 $F\,2/d\,2/d\,2/d$ $Fddd,\ D_{2h}^{24}$	71 $I\,2/m\,2/m\,2/m$ $Immm,\ D_{2h}^{25}$
	72 $I\,2/b\,2/a\,2/m$ $Ibam,\ D_{2h}^{26}$	73 $I\,2_1/b\,2_1/c\,2_1/a$ $Ibca,\ D_{2h}^{27}$	74 $I\,2_1/m\,2_1/m\,2_1/a$ $Imma,\ D_{2h}^{28}$		

Tetragonal

Point group*	Space group**			
$4,\ C_4$	75 $P\,4$ C_4^1	76 $P\,4_1$ C_4^2	77 $P\,4_2$ C_4^3	78 $P\,4_3$ C_4^4	79 $I\,4$ C_4^5
	80 $I\,4_1$ C_4^6				
$\bar{4},\ S_4$	81 $P\,\bar{4}$ S_4^1	82 $I\,\bar{4}$ S_4^2			
$4/m,\ C_{4h}$	83 $P\,4/m$ C_{4h}^1	84 $P\,4_2/m$ C_{4h}^2	85 $P\,4/n$ C_{4h}^3	86 $P\,4_2/n$ C_{4h}^4	87 $I\,4/m$ C_{4h}^5
	88 $I\,4_1/a$ C_{4h}^6				
$422,\ D_4$	89 $P\,4\,2\,2$ D_4^1	90 $P\,4\,2_1\,2$ D_4^2	91 $P\,4_1\,2\,2$ D_4^3	92 $P\,4_1\,2_1\,2$ D_4^4	93 $P\,4_2\,2\,2$ D_4^5
	94 $P\,4_2\,2_1\,2$ D_4^6	95 $P\,4_3\,2\,2$ D_4^7	96 $P\,4_3\,2_1\,2$ D_4^8	97 $I\,4\,2\,2$ D_4^9	98 $I\,4_1\,2\,2$ D_4^{10}
$4mm,$ C_{4v}	99 $P\,4\,m\,m$ C_{4v}^1	100 $P\,4\,b\,m$ C_{4v}^2	101 $P\,4_2\,c\,m$ C_{4v}^3	102 $P\,4_2\,n\,m$ C_{4v}^4	103 $P\,4\,c\,c$ C_{4v}^5
	104 $P\,4\,c\,n$ C_{4v}^6	105 $P\,4_2\,m\,c$ C_{4v}^7	106 $P\,4_2\,b\,c$ C_{4v}^8	107 $I\,4\,m\,m$ C_{4v}^9	108 $I\,4\,c\,m$ C_{4v}^{10}
	109 $I\,4_1\,m\,d$ C_{4v}^{11}	110 $I\,4_1\,c\,d$ C_{4v}^{12}			
$\bar{4}2m,$ $\bar{4}m2,$	111 $P\,\bar{4}\,2\,m$ D_{2d}^1	112 $P\,\bar{4}\,2\,c$ D_{2d}^2	113 $P\,\bar{4}\,2_1\,m$ D_{2d}^3	114 $P\,\bar{4}\,2_1\,c$ D_{2d}^4	115 $P\,\bar{4}\,m\,2$ D_{2d}^5

(Continued)

(Continued)

Point group[*]	Space group[**]			
D_{2d}	116 $P\,\bar{4}\,c\,2$ D_{2d}^6	117 $P\,\bar{4}\,b\,2$ D_{2d}^7	118 $P\,\bar{4}\,n\,2$ D_{2d}^8	119 $I\,\bar{4}\,m\,2$ D_{2d}^9	120 $I\,\bar{4}\,c\,2$ D_{2d}^{10}
	121 $I\,\bar{4}\,2\,m$ D_{2d}^{11}	122 $I\,\bar{4}\,2\,d$ D_{2d}^{12}			
$4/mmm$, D_{4h}	123 $P\,4/m\,2/m\,2/m$ $P4/mmm$, D_{4h}^1	124 $P\,4/m\,2/c\,2/c$ $P4/mcc$, D_{4h}^2	125 $P\,4/n\,2/b\,2/m$ $P4/nbm$, D_{4h}^3	126 $P\,4/n\,2/n\,2/c$ $P4/nnc$, D_{4h}^4	127 $P\,4/m\,2_1/b\,2/m$ $P4/mbm$, D_{4h}^5
	128 $P\,4/m\,2_1/n\,2/c$ $P4/mnc$, D_{4h}^6	129 $P\,4/n\,2_1/m\,2/m$ $P4/nmm$, D_{4h}^7	130 $P\,4/n\,2_1/c\,2/c$ $P4/ncc$, D_{4h}^8	131 $P\,4_2/m\,2/m\,2/c$ $P4_2/mmc$, D_{4h}^9	132 $P\,4_2/m\,2/c\,2/m$ $P4_2/mcm$, D_{4h}^{10}
	133 $P\,4_2/n\,2/b\,2/c$ $P4_2/nbc$, D_{4h}^{11}	134 $P4_2/n\,2/n\,2/m$ $P4_2/nnm$, D_{4h}^{12}	135 $P\,4_2/m\,2_1/b\,2/c$ $P4_2/mbc$, D_{4h}^{13}	136 $P4_2/m\,2_1/n\,2/m$ $P4_2/mnm$, D_{4h}^{14}	137 $P4_2/n\,2_1/m2/c$ $P4_2/nmc$, D_{4h}^{15}
	138 $P\,4_2/n\,2_1/c\,2/m$ $P4_2/ncm$, D_{4h}^{16}	139 $I\,4/m\,2/m\,2/m$ $I4/mmm$, D_{4h}^{17}	140 $I\,4/m\,2/c\,2/m$ $I4/mcm$, D_{4h}^{18}	141 $I\,4_1/a\,2/m\,2/d$ $I4_1/amd$, D_{4h}^{19}	142 $I\,4_1/a\;\;2/c\;\;2/d$ $I4_1/acd$, D_{4h}^{20}

Trigonal

Point group	Space group			
3, C_3	143 $P\,3$ C_3^1	144 $P\,3_1$ C_3^2	145 $P\,3_2$ C_3^3	146 $R\,3$ C_3^4	
$\bar{3}$, C_{3i}	147 $P\,\bar{3}$ C_{3i}^1	148 $R\,\bar{3}$ C_{3i}^2			
3 2, 312, 321, D_3	149 $P\,3\,1\,2$ D_3^1	150 $P\,3\,2\,1$ D_3^2	151 $P\,3_1\,1\,2$ D_3^3	152 $P\,3_1\,2\,1$ D_3^4	153 $P\,3_2\,1\,2$ D_3^5
	154 $P\,3_2\,2\,1$ D_3^6	155 $R\,3\,2$ D_3^7			
$3m$, $3m1$, $31m$, C_{3v}	156 $P\,3\,m\,1$ C_{3v}^1	157 $P\,3\,1\,m$ C_{3v}^2	158 $P\,3\,c\,1$ C_{3v}^3	159 $P\,3\,1\,c$ C_{3v}^4	160 $R\,3\,m$ C_{3v}^5
	161 $R\,3\,c$ C_{3v}^6				
$\bar{3}m$, $\bar{3}1m$, $\bar{3}m1$, D_{3d}	162 $P\,\bar{3}\,1\,2/m$ $P\bar{3}1m$, D_{3d}^1	163 $P\,\bar{3}\,1\,2/c$ $P\bar{3}1c$, D_{3d}^2	164 $P\,\bar{3}\,2/m\,1$ $P\bar{3}m1$, D_{3d}^3	165 $P\,\bar{3}\,2/c\,1$ $P\bar{3}c1$, D_{3d}^4	166 $R\,\bar{3}\,2/m$ $R\bar{3}m$, D_{3d}^5
	167 $R\,\bar{3}\,2/c$ $R\bar{3}c$, D_{3d}^6				

Hexagonal

Point group	Space group			
6, C_6	168 $P\,6$ C_6^1	169 $P\,6_1$ C_6^2	170 $P\,6_5$ C_6^3	171 $P\,6_2$ C_6^4	172 $P\,6_4$ C_6^5
	173 $P\,6_3$ C_6^6				
$\bar{6}$, C_{3h}	174 $P\,\bar{6}$ C_{3h}^1				

(Continued)

(Continued)

Point group[*]	Space group[**]			
$6/m$, C_{6h}	$175\ P\ 6/m$ C_{6h}^1	$176\ P\ 6_3/m$ C_{6h}^2		
622, D_6	$177\ P\ 6\ 2\ 2$ D_6^1	$178\ P\ 6_1\ 2\ 2$ D_6^2	$179\ P\ 6_5\ 2\ 2$ D_6^3	$180\ P\ 6_2\ 2\ 2$ D_6^4	$181\ P\ 6_4\ 2\ 2$ D_6^5
	$182\ P\ 6_3\ 2\ 2$ D_6^6				
$6mm$, C_{6v}	$183\ P\ 6\ m\ m$ C_{6v}^1	$184\ P\ 6\ c\ c$ C_{6v}^2	$185\ P\ 6_3\ c\ m$ C_{6v}^3	$186\ P\ 6_3\ m\ c$ C_{6v}^4	
$\bar{6}m2$, $\bar{6}2m$, D_{3h}	$187\ P\ \bar{6}\ m\ 2$ D_{3h}^1	$188\ P\ \bar{6}\ c\ 2$ D_{3h}^2	$189\ P\ \bar{6}\ 2\ m$ D_{3h}^3	$190\ P\ \bar{6}\ 2\ c$ D_{3h}^4	
$6/mmm$, D_{6h}	$191\ P\ 6/m\ 2/m\ 2/m$ $P6/mmm$, D_{6h}^1	$192\ P\ 6/m\ 2/c\ 2/c$ $P6/mcc$, D_{6h}^2	$193\ P\ 6_3/m\ 2/c\ 2/m$ $P6_3/mcm$, D_{6h}^3	$194\ P\ 6_3/m\ 2/m\ 2/c$ $P6_3/mmc$, D_{6h}^4	

Cubic

Point group[*]	Space group[**]			
23, T	$195\ P\ 2\ 3$ T^1	$196\ F\ 2\ 3$ T^2	$197\ I\ 2\ 3$ T^3	$198\ P\ 2_1\ 3$ T^4	$199\ I\ 2_1\ 3$ T^5
$m\bar{3}$, T_h	$200\ P\ 2/m\ \bar{3}$ $Pm\bar{3}$, T_h^1	$201\ P\ 2/n\ \bar{3}$ $Pn\bar{3}$, T_h^2	$202\ F\ 2/m\ \bar{3}$ $Fm\bar{3}$, T_h^3	$203\ F\ 2/d\ \bar{3}$ $Fd\bar{3}$, T_h^4	$204\ I\ 2/m\ \bar{3}$ $Im\bar{3}$, T_h^5
	$205\ P\ 2_1/a\ \bar{3}$ $Pa\bar{3}$, T_h^6	$206\ I\ 2_1/a\ \bar{3}$ $Ia\bar{3}$, T_h^7			
432, O	$207\ P\ 4\ 3\ 2$ O^1	$208\ P\ 4_2\ 3\ 2$ O^2	$209\ F\ 4\ 3\ 2$ O^3	$210\ F\ 4_1\ 3\ 2$ O^4	$211\ I\ 4\ 3\ 2$ O^5
	$212\ P\ 4_3\ 3\ 2$ O^6	$213\ P\ 4_1\ 3\ 2$ O^7	$214\ I\ 4_1\ 3\ 2$ O^8		
$\bar{4}3m$, T_d	$215\ P\ \bar{4}\ 3\ m$ T_d^1	$216\ F\ \bar{4}\ 3\ m$ T_d^2	$217\ I\ \bar{4}\ 3\ m$ T_d^3	$218\ P\ \bar{4}\ 3\ n$ T_d^4	$219\ F\ \bar{4}\ 3\ c$ T_d^5
	$220\ I\ \bar{4}\ 3\ d$ T_d^6				
$m\bar{3}m$, O_h	$221\ P\ 4/m\ \bar{3}\ 2/m$ $Pm\bar{3}m$, O_h^1	$222\ P\ 4/n\ \bar{3}\ 2/n$ $Pn\bar{3}n$, O_h^2	$223\ P\ 4_2/m\ \bar{3}\ 2/n$ $Pm\bar{3}n$, O_h^3	$224\ P\ 4_2/n\ \bar{3}\ 2/m$ $Pn\bar{3}m$, O_h^4	$225\ F\ 4/m\ \bar{3}\ 2/m$ $Fm\bar{3}m$, O_h^5
	$226\ F\ 4/m\ \bar{3}\ 2/c$ $Fm\bar{3}c$, O_h^6	$227\ F\ 4_1/d\ \bar{3}\ 2/m$ $Fd\bar{3}m$, O_h^7	$228\ F\ 4_1/d\ \bar{3}\ 2/c$ $Fd\bar{3}c$, O_h^8	$229\ I\ 4/m\ \bar{3}\ 2/m$ $Im\bar{3}m$, O_h^9	$230\ I\ 4_1/a\ \bar{3}\ 2/d$ $Ia\bar{3}d$, O_h^{10}

[*]The first entry is the Hermann-Mauguin short symbol. The full symbol is given in Table 4.4 and in Appendix 3. The second entry is the Schoenflies symbol.

[**]The first entry is the space group number, the second is the full Hermann-Mauguin symbol, the third is the short Hermann-Mauguin symbol when it differs from the full symbol, or else the Schoenflies symbol, the forth entry, when it appears, is the Schoenflies symbol.

Appendix 5 Complex numbers

A complex number is a number written

$$z = a + ib$$

where i is defined as $i = \sqrt{-1}$, a is called the **real part** of the number and b the **imaginary part**. The **modulus** of a complex number $z = a + ib$ is written $|z|$ and is given by

$$|z| = \sqrt{a^2 + b^2}$$

An Argand diagram is a graphical method of representing a complex number, z, written as $a + ib$. The real part of the complex number, a, is plotted along the horizontal axis and the imaginary part, ib, along the vertical axis. The complex number z is then represented by a point whose Cartesian coordinates are (a, b), (Figure A5.1). (These diagrams are also referred to as *representations in the Gaussian plane* or *representations in the Argand plane*.)

The complex number

$$z = a + ib$$

can also be written in polar form, with **z** taken to represent a radial vector of length r, where

$$r = |z| = \sqrt{a^2 + b^2}$$

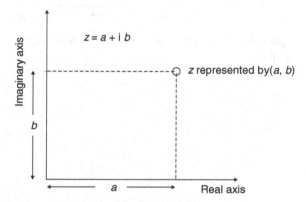

Figure A5.1 Representation of a complex number by an Argand diagram

r is a scalar, that is, a positive number with no directional properties, given by the modulus of vector **z**, written $|z|$, (Figure A5.2). The angle between the vector **z** and the horizontal axis, ϕ, measured in an anticlockwise direction from the axis, is called its **argument**. The real and imaginary parts of **z** are given by the projections of **z** on the horizontal and vertical axes, i.e.

$$\text{Real part} = a = r \cos \phi$$

$$\text{Imaginary part} = b = r \sin \phi$$

and $\mathbf{z} = r \cos \phi + i\, r \sin \phi = r(\cos \phi + i \sin \phi)$

Crystals and Crystal Structures.   Richard J. D. Tilley
© 2006 John Wiley & Sons, Ltd

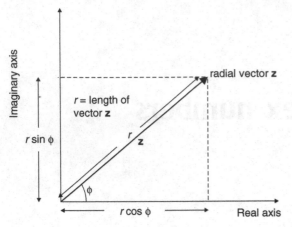

Figure A5.2   Representation of a vector as a complex number

The value of ϕ is given by

$$\phi = \arctan\left(r\sin\phi / r\cos\phi\right) = \arctan\left(b/a\right)$$

The **complex conjugate number** to z, written z^*, is obtained by replacing i with $-i$:

$$z = a + ib$$
$$z^* = a - ib$$

The product of a complex number with its complex conjugate is always a real number:

$$z\,z^* = (a + ib)(a - ib) = a^2 - i^2\,b^2 = a^2 + b^2$$

Appendix 6   Complex amplitudes

The addition of vectors that represent the scattering of a beam of radiation by several objects can be achieved algebraically by writing them in the form:

$$\mathbf{a} = a\,e^{i\phi} \text{ or } a\,e^{-i\phi}$$

where a is the scalar magnitude of the vector and ϕ is the phase. The vector $\mathbf{a}$ is called a **complex amplitude**. The value of i is taken as positive when the phase is ahead of a 'standard phase' and negative when the phase is behind the 'standard phase'. [The 'standard phase' in X-ray diffraction is set by the atom at the origin of the unit cell, (0, 0, 0)].

This terminology allows the scattering of X-rays by atoms in a unit cell to be added algebraically, by writing the scattering by an atom in a unit cell as a complex amplitude $\mathbf{f}$:

$$\mathbf{f} = f\,e^{i\phi}$$

where f is the scalar magnitude of the scattering factor and ϕ is the phase of the scattered wave. Using Euler's formula:

$$e^{i\phi} = \cos\phi + i\sin\phi$$

the scattering can be written

$$\mathbf{f} = f\{\cos\phi + i\sin\phi\}$$

This complex number can be represented in polar form as a vector $\mathbf{f}$ plotted on an Argand diagram.

$$\mathbf{f} = f\{\cos\phi + i\sin\phi\}$$

where f is the length of the scattering vector and ϕ is the argument (i.e. the phase angle) associated with $\mathbf{f}$, (Figure A6.1).

The advantage of using this representation is that algebraic addition is equivalent to vector addition. For example, suppose that it is necessary to add two vectors $\mathbf{f}_1$ with magnitude f_1 and phase ϕ_1 and $\mathbf{f}_2$ with magnitude f_2 and phase ϕ_2 to obtain to resultant vector $\mathbf{F}$, with magnitude F and phase θ, that is:

$$\mathbf{F} = F\,e^{i\theta} = f_1 e^{i\phi_1} + f_2 e^{i\phi_2}$$

Drawing these on a diagram in the Argand plane, (Figure A6.2) shows that the total displacement along the real axis, X, is:

$$X = x_1 + x_2 = f_1\cos\phi_1 + f_2\cos\phi_2$$

and along the imaginary axis, Y, is:

$$Y = y_1 + y_2 = f_1\sin\phi_1 + f_2\sin\phi_2$$
$$\mathbf{F} = (X + iY)$$

The length of $\mathbf{F}$ is then F:

$$F = \sqrt{X^2 + Y^2}$$

Crystals and Crystal Structures.   Richard J. D. Tilley
© 2006 John Wiley & Sons, Ltd

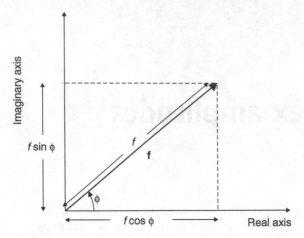

Figure A6.1 Representation of an atomic scattering factor as a complex amplitude vector

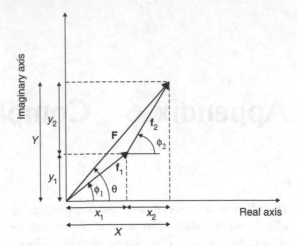

Figure A6.2 The addition of two complex amplitude vectors

and the phase angle is given by:

$$\tan \theta = Y/X$$

Clearly this can be repeated with any number of terms to obtain the algebraic sums:

$$X = x_1 + x_2 + x_3 \ldots$$
$$= f_1 \cos \phi_1 + f_2 \cos \phi_2 + f_3 \cos \phi_3 \ldots$$
$$= \sum_{n=1}^{N} x_n = \sum_{n=1}^{N} f_n \cos \phi_n$$
$$Y = y_1 + y_2 + y_3 \ldots$$
$$= f_1 \sin \phi_1 + f_2 \sin \phi_2 + f_3 \sin \phi_3 \ldots$$
$$= \sum_{n=1}^{N} y_n = \sum_{n=1}^{N} f_n \sin \phi_n$$

with $F = \sqrt{X^2 + Y^2}$

and

$$\tan \theta = Y/X$$

The intensity of a scattered beam, I, is given by multiplying $\mathbf{F}$ by its complex conjugate, $\mathbf{F}^*$ (see Appendix 5) to give a real number:

$$\mathbf{F} \times \mathbf{F}^* = (X + iY)(X - iY) = X^2 + Y^2 = F^2$$

or $\quad \mathbf{F} \times \mathbf{F}^* = F\,\mathrm{e}^{i\theta} \times F\,\mathrm{e}^{-i\theta} = F^2$

Answers to problems and exercises

Chapter 1

Quick Quiz

1c, 2b, 3b, 4a, 5b, 6c, 7b, 8c, 9c, 10a.

Calculations and Questions

1.1 Graphical, $a_H \approx 0.372$ nm; arithmetical, $a_H = 0.375$ nm, $c_H = 1.051$ nm.

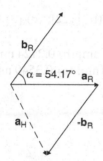

1.2

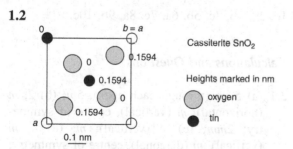

Cassiterite SnO$_2$

Heights marked in nm

Sn atoms at: 000; 0.2369 nm, 0.2369 nm, 0.1594 nm; O atoms at: 0.1421 nm; 0.1421 nm, 0; 0.3790 nm, 0.09476 nm, 0.1594 nm; 0.3317 nm, 0.3317 nm, 0; 0.09476 nm, 0.3790 nm, 0.1594 nm; Volume $= 0.0716$ nm^3.

1.3 Number of formula units, Z, $= 1$ SrTiO$_3$.

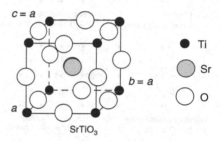

SrTiO$_3$

1.4 (a) 4270 kg m^{-3}; (b) 4886 kg m^{-3}.

1.5 0.7014 nm.

1.6 95.9 g mol^{-1}.

1.7 23.8 g mol^{-1}, (Mg).

1.8 2.

Chapter 2

Quick Quiz

1b, 2c, 3a, 4a, 5c, 6a, 7c, 8c, 9b, 10b.

Crystals and Crystal Structures. Richard J. D. Tilley
© 2006 John Wiley & Sons, Ltd

Calculations and Questions

2.1 (a) not a lattice; (b) rectangular; (c) hexagonal; (d) centred rectangular; (e) not a lattice; (f) oblique.

2.2

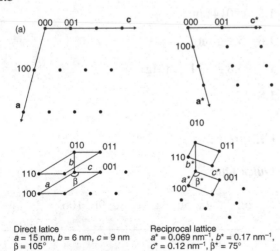

(a)

Direct lattice
$a = 8$ nm, $b = 12$ nm,
$\gamma = 110°$

Reciprocal lattice
$a^* = 0.125$ nm^{-1}, $b^* = 0.083$ nm^{-1}
$\gamma^* = 70°$

(b)

Direct lattice
$a = 10$ nm, $b = 14$ nm

Reciprocal lattice
$a^* = 0.10$ nm^{-1}, $b^* = 0.071$ nm^{-1}

(c)

Primitive direct lattice
$a = b = 17.2$ nm, $\gamma = 71°$

Reciprocal lattice
$a^* = b^* = 0.058$ nm^{-1}
$\gamma^* = 109$

2.3

(a)

Direct latice
$a = 15$ nm, $b = 6$ nm, $c = 9$ nm
$\beta = 105°$

Reciprocal lattice
$a^* = 0.069$ nm^{-1}, $b^* = 0.17$ nm^{-1},
$c^* = 0.12$ nm^{-1}, $\beta^* = 75°$

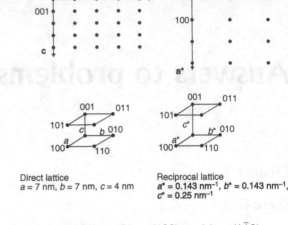

(b)

Direct lattice
$a = 7$ nm, $b = 7$ nm, $c = 4$ nm

Reciprocal lattice
$a^* = 0.143$ nm^{-1}, $b^* = 0.143$ nm^{-1},
$c^* = 0.25$ nm^{-1}

2.4 (a) (230); (b) (120); (c) $(1\bar{1}0)$ or $(\bar{1}10)$; (d) (310); (e) $(5\bar{3}0)$ or $(\bar{5}30)$.

2.5 (a) (110), $(11\bar{2}0)$; (b) $(1\bar{1}0)$, $(1\bar{1}00)$; (c) $(3\bar{2}0)$, $(3\bar{2}\bar{1}0)$; (d) (410), $(41\bar{5}0)$; (e) $(3\bar{1}0)$, $(3\bar{1}\bar{2}0)$

2.6 (a) [110]; (b) $[\bar{1}20]$; (c) $[\bar{3}\bar{1}0]$; (d) [010]; (e) $[\bar{1}\bar{3}0]$.

2.7 (a) [001]; (b) $[12\bar{2}]$; (c) $[01\bar{1}]$; (d) [101].

2.8 (a) 0.2107 nm; (b) 0.2124 nm; (c) 0.2812 nm; (e) 0.1359 nm; (e) 0.3549 nm.

Chapter 3

Quick Quiz

1c, 2b, 3b, 4c, 5b, 6a, 7c, 8a, 9b, 10c.

Calculations and Questions

3.1 (a) 5, *m* (through each apex); 5*m*; (b) 2, *m* (horizontal), *m* (vertical), centre of symmetry; 2*mm*; (c) *m* (vertical); *m*; (d) 4, *m* (vertical), *m* (diagonal), centre of symmetry; 4*mm*; (e) 2, *m* (horizontal), *m* (vertical), centre of symmetry; 2*mm*.

3.2 (a) 4-fold rotation axis at the centre of the pattern

(b) 2-fold rotation axis at the centre of the pattern plus a vertical mirror through the centre of the pattern

3.3

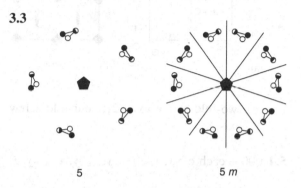

5   5 *m*

3.4 (a) *p2*; (b) *pg*; (c) *p2gg*.

3.5 (a) *p4*; (b) *p4mm*; (c) *p31m*.

3.6 (0.125, 0.475); (0.875, 0.525); (0.875, 0.475); (0.125, 0.525); (0.625, 0.975); (0.375, 0.025); (0.375, 0.975); (0.625, 0.025).

3.7 (0.210, 0.395); (0.790, 0.605); (0.605, 0.210); (0.395, 0.790); (0.290, 0.895); (0.710, 0.105); (0.895, 0.710); (0.105, 0.290).

Chapter 4

Quick Quiz

1b, 2a, 3c, 4b, 5c, 6b, 7b, 8b, 9a, 10c.

Calculations and Questions

4.1 (a) 2, 3, *m*, $\bar{4}$; $\bar{4}3m$; T_d

(b) 3, 2, *m* (vertical), *m* (horizontal); $\bar{6}m2$ or $\bar{6}2m$; D_{3h}

(c) 3, *m* (vertical); 3*m* (3*m*1 or 31*m*); C_{3v}

(d) 6, 2, $\bar{1}$, *m* (vertical), *m* (horizontal); 6/*mmm*; D_{6h}

(e) 4, $\bar{4}$, 3, $\bar{3}$, 2, $\bar{1}$, *m* (vertical), *m* (horizontal); $m\bar{3}m$; O_h

4.2 (a) 2/*m*, (b) *m*, (c) 2.

4.3 (a) *mmm*, (b) *mm2*, (c) 222.

4.4 (a) $\dfrac{2}{m}\dfrac{2}{m}\dfrac{2}{m}$, (b) $\dfrac{4}{m}\dfrac{2}{m}\dfrac{2}{m}$, (c) $\bar{3}\dfrac{2}{m}$, (d) $\dfrac{6}{m}\dfrac{2}{m}\dfrac{2}{m}$, (e) $\dfrac{4}{m}\bar{3}\dfrac{2}{m}$.

4.5

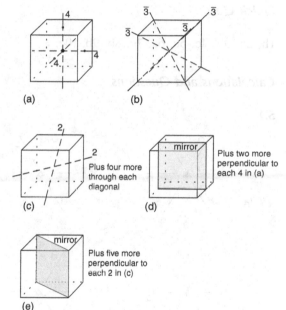

(a)   (b)

2₁

2

Plus four more through each diagonal

(c)

mirror

Plus two more perpendicular to each 4 in (a)

(d)

mirror

Plus five more perpendicular to each 2 in (c)

(e)

(a) Three 4-fold axes through the centre of each square face; four $\bar{3}$ axes along each body diagonal; six 2-fold axes along each diagonal; mirrors perpendicular to each 4 and 2 axis.

(b)
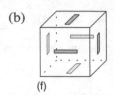

(f)

To go from $4/m\,\bar{3}\,2/m$ to $2/m\,\bar{3}$ it is necessary to remove *all* the tetrad (4) axes, (otherwise the first symbol would remain at 4 and the system would become tetragonal), and replace them with diad (2) axes.

4.6 (a) Cu, $m\bar{3}m$; (b) Fe, $m\bar{3}m$; (c) Mg, $6/mmm$

4.7 (a) cubic; (b) cubic; (c) hexagonal

Chapter 5

Quick Quiz

1b, 2a, 3b, 4c, 5c, 6a, 7c, 8b, 9b, 10c.

Calculations and Questions

5.1

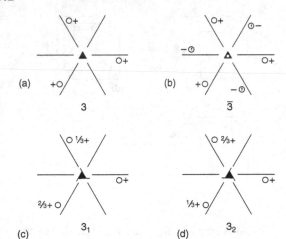

5.2 (a) $P\,1\,c\,1$; (b) $P\,2/m\,2/m\,2/m$; (c) $P\,4_2/n\,2/b\,2/c$; (d) $P\,6_3/m\,2/c\,2/m$; (e) $F\,4/m\,\bar{3}\,2/m$.

5.3 (a) tetrad axis; (b)

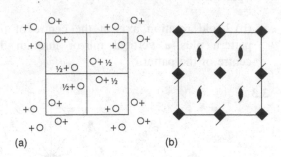

(c) two-fold screw axis, 2_1; four-fold screw axis, 4_2.

5.4 $000+$ each of: x, y, z; $\bar{x}, \bar{y}, z$; $\bar{y}, x, z$; $y, \bar{x}, z$;

$\frac{1}{2}\,\frac{1}{2}\,\frac{1}{2}+$ each of: x, y, z; $\bar{x}, \bar{y}, z$; $\bar{y}, x, z$; $y, \bar{x}, z$.

5.5 (a) tetrad axis parallel to the **c**-axis; mirrors with normals along [100], [010]; mirrors with normals along [110], [1$\bar{1}$0].

(b) diad axes, glide planes (see Figure 5.12).

(c) 2 Pu.

(d) Pu(1): 0, 0, 0.

  Pu(2): ½, ½, 0.4640.

(e) 4 S.

(f) S(1): 0, 0, 0.3670.

  S(2): ½, ½, 0.0970.

  S(3): ½, 0, 0.7320.

  S(4): 0, ½, 0.7320.

Chapter 6

Quick Quiz

1b, 2c, 3a, 4c, 5a, 6b, 7a, 8c, 9b, 10c.

Calculations and Questions

6.1 (111), 9.54°; (200), 11.04°; (220), 15.71°; (311), 18.51°; (222), 19.36°; (400), 34.68°.

6.2 (a) 0.2291 nm; (b) 10.2 nm.

6.3

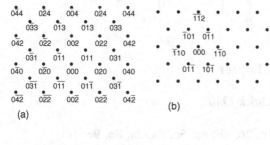

(a)

(b)

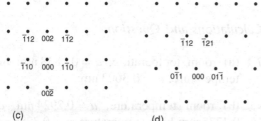

(c)

(d)

The value of λl will depend upon the scale of the figure. If the distance from 000 to 020 in (a) is, say, 20 mm, λl will be 3.30 mm nm.

6.4

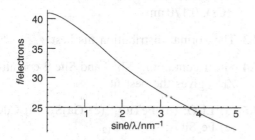

6.5

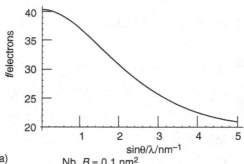

(a) Nb, $B = 0.1\ nm^2$

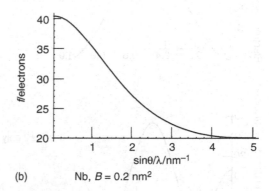

(b) Nb, $B = 0.2\ nm^2$

(a) 0.036 nm; (b) 0.050 nm.

6.6 $F_{110} = 37.46$, phase $= 0$;

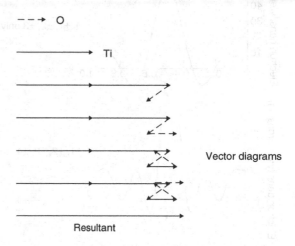

Vector diagrams

the vector diagram gives a value of 37.5 for F.

6.7 (a) 5138; (b) 124,318.

6.8

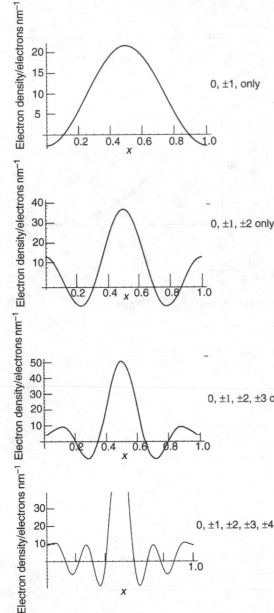

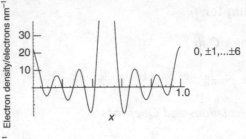

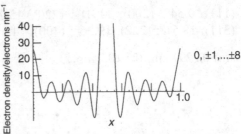

Chapter 7

Quick Quiz

1c, 2b, 3b, 4a, 5c, 6a, 7a, 8b, 9c, 10b.

Calculations and Questions

7.1 (a) room temperature, $a = 0.2838\,nm$; high temperature, $a = 0.3603\,nm$;

(b) room temperature, $a = 0.2924\,nm$, $c = 0.4775\,nm$; high temperature, $a = 0.3286\,nm$;

(c) room temperature, $a = 0.5583\,nm$; high temperature, $a = 0.4432\,nm$.

7.2 $r(Cl)$, $0.181\,nm$; $r(Li)$, $0.075\,nm$; $r(Na)$, $0.101\,nm$; $r(K)$, $0.134\,nm$; $r(Rb)$, $0.148\,nm$; $r(Cs)$, $0.170\,nm$.

7.3 The normal distribution fits best.

7.4 Site 1 containing Nb^{5+} and Site 2 containing Zr^{4+} gives the best fit.

7.5 (a) Li_2O; (b) $FeTiO_3$; (c) Ga_2S_3; (d) $CdCl_2$; (e) Fe_2SiO_4; (f) Cr_5O_{12}.

7.6

Metallic radii

Ionic radii

Covalent radii

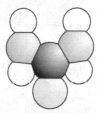

Van der Waals radii

Clearly ionic radii give the poorest description. Metallic radii are surprisingly similar to van der Waals radii, and so reasonable.

Chapter 8

Quick Quiz

1b, 2b, 3a, 4b, 5c, 6a, 7c, 8b, 9b, 10a.

Calculations and Questions

8.1 (a) f(mixed sites) = 22.72, F_{200} = 28.50; Intensity = 789.9; (b), F_{200} (pure TiO_2) = 13.80, Intensity (pure TiO_2) = 190.6; (c) F_{200} (pure SnO_2) = 61.5, Intensity (pure SnO_2) = 3778.6

8.2 Vegard's law is obeyed; (a) ≈1.2223 nm; (b) $x ≈ 0.27$, i.e. $(Y_{0.27}Tm_{0.73})_3Ga_5O_{12}$.

8.3 (a) vacancies, $La_{0.948}Mn_{0.948}O_3$, interstitials, $LaMnO_{3.165}$; (b) vacancies, 6594.2 kg m^{-3}, interstitials, 6956.9 kg m^{-3}.

8.4 (a)

15R, (32), (cchch)

(b)

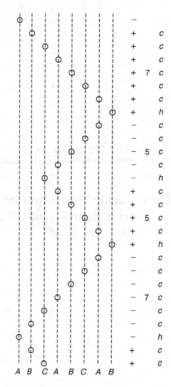

24H, (7557), $(hc_4)_2(hc_6)_2$

(c)

$-$		h
$+$		c
$+$	3	c
$+$		h
$-$		c
$-$	3	c
$-$		h
$+$		c
$+$	2	h
$-$		c
$-$	2	h
$+$		c
$+$	3	c
$+$		h
$-$		c
$-$	3	c
$-$		h
$+$		c
$+$	3	c
$+$		h
$-$		c
$-$	3	c
$-$		h
$+$		c
$+$	2	h
$-$		c
$-$	2	h

A B C A

$16H,(332233),(hcc)_4(hc)_2$

8.5 (a) The *CS* plane lies midway between (102) and (103), (see figure), (205), $W_nO_{3n-1.5}$; (b)

$A_{4.666}B_{3.666}O_{12}$, $n = 3.666$,   (3⅔);   (c). $Bi_2A_{1.5}B_{2.5}O_{10.5}$, $n = 2.5$, (2½).

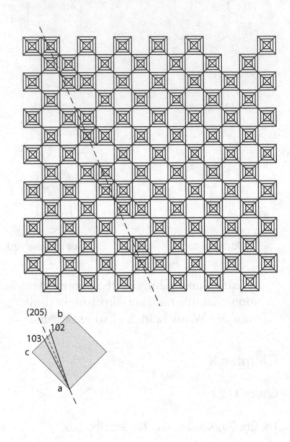

(205) b
102
103
c
a

8.6 (a) $Ba_{1.111}Fe_2S_4$, $c = 5.02$ nm   (average value); (b) $Ba_{14}(Fe_2S_4)_{13}$, $c = 7.14$ nm (average value);   (c) $Sr_8(TiS_3)_7$, $c = 2.092$ nm (average value).

Bibliography

The source for the structural information in this book is the EPSRC's Chemical Database Service at Daresbury. This can be accessed at: http://cds.dl.ac. uk/cds. See:

D. A. Fletcher, R. F. McMeeking and D. Parkin, J. Chem. Inf. Comput. Sci., **36**, 746–749 (1996).*

The Internet

For many people the internet will be the first port of call. Relatively few references to web sites have been included, as many tend to be ephemeral; usually produced in a wave of enthusiasm and then left to quietly moulder.

A problem with using the internet is the amount of information stored. For example, a search for 'quasicrystals' produced about a quarter of a million pages of reference, while 'protein crystallography' produced well over 2 million (mid-2005). Nevertheless, the internet is a valuable resource and will yield dividends when used thoughtfully. It is the probably the best starting point when the subject matter is new to the reader.

Crystal structures are best viewed as three-dimensional computer images that can be 'rotated' and viewed from any 'direction'. There are an increasing number of programs now available that can do this. In the first instance, search for '*crystallographic freeware*' or '*crystallographic shareware*' to obtain free downloads of suitable material. Searching on '*crystallographic software*' will also give information on commercially available software.

Crystal structures can be displayed, and downloadable programs for graphical presentation of crystal structures can be found at the EPSRC's Chemical Database Service at Daresbury. This can be accessed at: http://cds.dl.ac.uk/cds.

Protein structures are collated and can be viewed in various formats and rotated or transformed, via the Protein Data Bank: http://www.rscb.prg/pdb or other sites world wide, including http://pdb.ccdc.cam.ac.uk.

Downloadable software for a range of crystallographic applications, and much crystallographic information, is available at the CCP14 site: http://www.ccp14.ac.uk/

The values for Cromer-Mann coefficients used in the calculation of atomic scattering factors were taken from http://www-structure.llnl.gov. This site has excellent on-line tutorials that allow calculation of scattering factors, structure

---

*The format for all references to the scientific literature is: **Volume**, page numbers (year).

factors and diffraction intensities for one-dimensional atom arrays. Well worth several hours of study.

The International Union of Crystallography has information on all aspects of crystallography at: http://www.iucr.ac.uk/

This site contains a large number of 'Teaching Pamphlets' and other educational material.

Commercial software

The software used to produce many of the crystal structures displayed in this book was CaRIne v. 3.1. This is an easy to use program, most suited to inorganic crystal structures. A later version, CaRIne v 4.0, is under development.

A number of other software packages are available that are more suited to molecular modelling, all of which can be compared by searching for 'crystallographic software' using the internet.

General and introductory books:

Symmetry in nature is discussed lucidly by:

I. Stewart, *What Shape is a Snowflake?* Weidenfeld and Nicolson, London (2001).

For an introduction to organic chemistry, including descriptions of proteins, see:

J. McMurray, *Organic Chemistry*, 6th edition, International Student Editions, Thompson Brooks/Cole, (2004).

The point groups of molecules are introduced in:

D. F. Shriver, P. W. Atkins and C. H. Langford, *Inorganic Chemistry*, 2nd edition, Oxford University Press, Oxford, (1994).

P. W. Atkins, *Physical Chemistry*, Fifth edition, Oxford University Press, Oxford, (1994).

Atomic radii and size

Atomic radii are discussed by:

N. W. Alcock, *Structure and Bonding*, Ellis Horwood, Chichester, (1990). A large amount of information on atomic size and structure is to be found in many of the chapters in: M. O'Keeffe and A. Navrotsky (Eds.), *Structure and Bonding in Crystals*, Vol I and II, Academic Press, New York, (1981), especially:

M. O'Keefe and B. G. Hyde, Chapter 10, *The role of nonbonded forces in crystals*.

W. H. Baur, Chapter 15, *Interatomic distance predictions for computer simulation of crystal structures*.

R. D. Shannon, *Bond distances in sulfides and a preliminary table of sulfide crystal radii*.

Bond-valence

I. D. Brown, Chapter 14, *The bond-valence method: an empirical approach to chemical structure and bonding*, in: M. O'Keeffe and A. Navrotsky (Eds.), *Structure and Bonding in Crystals*, Vol I and II, Academic Press, New York, (1981).

I. D. Brown, *The Chemical Bond in Inorganic Chemistry*, International Union of Crystallography Monographs on Crystallography No. 12, Oxford University Press, Oxford, (2002).

N. Brese and M. O'Keeffe, *Bond-valence parameters for solids*. Acta Crystallogr., **B47**, 192–197 (1991).

P. Müller, S. Köpke and G. M. Sheldrick, *Is the bond-valence method able to identify metal atoms in protein structures?* Acta Crystallogr., **D59**, 32–37 (2003).

Crystallography, crystal chemistry and physics

The many aspects of crystallography and crystal structure description are covered, from a variety of standpoints, in the following books. Some are rather old but would be available in libraries, and all are well worth consulting.

W. L. Bragg, *The Development of X-ray Analysis*, Dover, New York, (1992).

C. Giacovazzo, H. L. Monaco, G. Artioli, D. Viterbo, G. Ferraris, G. Gilli, G. Zanotti and M. Catti, *Fundamentals of Crystallography*, 2nd edition, Oxford University Press, Oxford, (2002).

C. Hammond, *Introduction to Crystallography*, The Royal Microscopical Society Microscopy Handbooks 19, Oxford University Press, Oxford, (1990).

B. G. Hyde and S. Andersson, *Inorganic Crystal Structures*, Wiley-Interscience, New York, (1989).

W. B. Pearson, *The Crystal Chemistry and Physics of Metals and Alloys*, Wiley–Interscience, New York, (1972).

J-J. Rousseau, *Basic Crystallography*, Wiley, Chichester, (1998).

J. V. Smith, *Geometrical and Structural Crystallography*, Wiley, New York, (1982).

A. F. Wells, *Structural Inorganic Chemistry*, 5th edition, Oxford University Press, Oxford, (1984).

The determination of crystal structures is described by:

W. Clegg, *Crystal Structure Determination*, Oxford Chemistry Primers, No. 60, Oxford University Press, Oxford, (1998).

W. Massa (trans. R. O. Gould), *Crystal Structure Determination*, Springer, New York, (2001).

D. Viterbo, Chapter 6, *Solution and refinement of crystal structures*, in: C. Giacovazzo, H. L. Monaco, G. Artioli, D. Viterbo, G. Ferraris, G. Gilli, G. Zanotti and M. Catti, *Fundamentals of Crystallography*, 2nd edition, Oxford University Press, Oxford, (2002).

This chapter also contains descriptions of direct methods, including the 'Shake and Bake' technique.

Symmetry is covered at the most fundamental level in:

International Tables for Crystallography, Vol A, *Space-Group Symmetry*, Ed. T. Hahn, 2nd edition, International Union of Crystallography/D. Reidel, Dordrecht, (1987).

A good survey of Pauling's rules is in:

F. D. Bloss, *Crystallography and Crystal Chemistry*, Holt, Rinehart and Winston, New York, (1971).

Crystal structures in terms of anion centred polyhedra are discussed by:

M. O'Keeffe and B. G. Hyde, *An alternative approach to non-molecular crystal structures*, Structure and Bonding, **61**, 77–144 (1985).

The physical properties of crystals are covered by:

F. D. Bloss, *Crystallography and Crystal Chemistry*, Holt, Rinehart and Winston, New York, (1971).

M. Catti, Chapter 10, *Physical properties of crystals: phenomenology and modelling*, in: C. Giacovazzo, H. L. Monaco, G. Artioli, D. Viterbo, G. Ferraris, G. Gilli, G. Zanotti and M. Catti, *Fundamentals of Crystallography*, 2nd edition, Oxford University Press, Oxford, (2002).

Electron diffraction and microscopy

The most comprehensive overview of electron microscopy and diffraction, especially the geometry of electron diffraction patterns, is:

P. B. Hirsch, A. Howie, R. B. Nicholson, D. W. Pashley and M. J. Whelan, *Electron Microscopy of Thin Crystals*, Butterworths, London, (1965). Unfortunately this book is now hard to find.

D. B. Williams and C. B. Carter, *Transmission Electron Microscopy*, Kluwer Academic/Plenum, (1996).

Incommensurately modulated structures

C. Giacovazzo, Chapter 4, *Beyond ideal crystals*, in: C. Giacovazzo, H. L. Monaco, G. Artioli, D. Viterbo, G. Ferraris, G. Gilli, G. Zanotti and M. Catti, *Fundamentals of Crystallography*, 2nd edition, Oxford University Press, Oxford, (2002).

T. Janssen and A. Janner, *Incommensurabilty in crystals*, Adv. Physics, **36**, 519–624 (1987).

E. Makovicky and B. G. Hyde, *Incommensurate, two-layer structures with complex crystal chemistry: minerals and related synthetics*, Materials Science Forum, **100 & 101**, 1–100 (1992)

S. van Smaalen, *Incommenurate crystal structures*, Crystal. Reviews, **4**, 79–202 (1995).

G. A. Wiegers, *Misfit layer compounds: structures and physical properties*, Prog. Solid State Chemistry, **24**, 1–139 (1996).

R. L. Withers, S. Schmid and J. G. Thompson, *Compositionally and/or displacively flexible systems and their underlying crystal chemistry*, Prog. Solid State Chemistry, **26**, 1–96 (1998).

Fundamental information concerning incommensurate diffraction patterns is given in:

International Tables for Crystallography, Vol. B, *Reciprocal space*, 2nd edition, Ed. U. Schmueli, Kluwer, Dordrecht, (2001).

References to $Ba_xFe_2S_4$

I. E. Grey, *The structure of $Ba_5Fe_9S_{18}$*, Acta Crystallogr., **B31**, 45–48 (1975).

M. Onoda and K. Kato, *Refinement of structures of the composite crystals $Ba_xFe_2S_4$ (x = 10/9 and 9/8) in a four-dimensional formalism*, Acta Crystallogr., **B47**, 630–634 (1991).

and references therein.

References to Sr_xTiS_3

M. Saeki, M. Ohta, K. Kurashima, M. Onoda, *Composite crystal $Sr_{8/7}TiS_y$ with y = 2.84 – 2.97*, Materials Research Bulletin, **37**, 1519–1529 (2002).

O. Gourdon, V. Petricek and M. Evain, *A new structure type in the hexagonal perovskite family; structure determination of the modulated misfit compound $Sr_{9/8}TiS_3$*, Acta Crystallogr., **B56**, 409–418 (2000).

and references therein.

Modular structures

A. Baronet, *Polytypism and stacking disorder*, Chapter 7 in: Reviews in Mineralogy, **27**, Ed. P. R. Busek, Mineralogical Soc. America, (1992).

G. Ferraris, Chapter 7, *Mineral and inorganic crystals*, in: C. Giacovazzo, H. L. Monaco, G. Artioli, D. Viterbo, G. Ferraris, G. Gilli, G. Zanotti and M. Catti, *Fundamentals of Crystallography*, 2nd edition, Oxford University Press, Oxford, (2002).

G. Ferraris, E. Mackovicky and S. Merlino, *Crystallography of Modular Materials*, International Union of Crystallography Monographs on Crystallography No. 15, Oxford University Press, Oxford, (2004).

R. J. D. Tilley, *Principles and Applications of Chemical Defects*, Stanley Thornes, Cheltenham, (1998), especially Chapter 9.

D. R. Veblen, *Electron microscopy applied to nonstoichiometry, polysomatism and replacement reactions in minerals*, Chapter 6 in:

Reviews in Mineralogy, **27**, Ed. P. R. Busek, Mineralogical Soc. America, (1992).

D. R. Veblen, *Polysomatism and polysomatic series: a review and applications*, Am. Mineral., **76**, 801–826 (1991).

Nets

M. O'Keeffe and B. G. Hyde, *Plane nets in crystal chemistry*, Phil. Trans. Royal Soc. Lond., **295**, 553–623 (1980).

O. Delgardo-Friedrichs, M. D. Foster, M. O'Keeffe, D. M. Proserpio, M. M. J. Treacy and O. M. Yaghi, *What do we know about three-periodic nets?* J. Solid State Chemistry, **178**, 2533–2544 (2005).

O. Delgardo-Friedrichs and M. O'Keeffe, *Crystal nets as graphs: terminology and definitions*. J. Solid State Chemistry, **178**, 2480–2485 (2005).

The illustrations alone in these three papers are worth viewing.

See also the references in these papers, and many of the papers in: J. Solid State Chemistry, **152**, (2000).

Proteins

Protein chemistry and physics, including structural characterisation, are lucidly described in:

D. Whitford, *Proteins, Structure and Function*, Wiley, Chichester, (2005).

The early history of protein crystallography will be found at: http://www.umass.edu/microbio/rasmol/1st_xtls.htm

The solution of crystal structures, with much information on solving the phase problem for protein structures is given by:

D. Viterbo, Chapter 6, *Solution and refinement of crystal structures*, and G. Zanotti, Chapter 9, *Protein crystallography*, in: C. Giacovazzo, H. L. Monaco, G. Artioli, D. Viterbo, G. Ferraris, G. Gilli, G. Zanotti and M. Catti, *Fundamentals of Crystallography*, 2nd edition, Oxford University Press, Oxford, (2002).

The first reported protein structure was myoglobin:

J. C. Kendrew, G. Bodo, H. M. Dintzis, R. G. Parrish, H. Wyckoff and D. C. Phillips, *A three-dimensional model of the myoglobin molecule obtained by X-ray analysis*, Nature, **181**, 662–666 (1958).

Quasicrystals

The first report of a quasicrystal was:

D. Schechtman, I. Blech, D. Gratias and J. W. Cahn, *Metallic phase with long-range orientational order and no translational symmetry*, Phys. Rev. Letters, **53**, 1951–1953 (1984).

See also:

C. Giacovazzo, Chapter 4, *Beyond ideal crystals*, in: C. Giacovazzo, H. L. Monaco, G. Artioli, D. Viterbo, G. Ferraris, G. Gilli, G. Zanotti and M. Catti, *Fundamentals of Crystallography*, 2nd edition, Oxford University Press, Oxford, (2002).

D. R. Nelson, *Quasicrystals*, Scientific American, **255**, [2] p32, (1986).

P. W. Stephens and A. I. Goldman, *The structure of quasicrystals*, Scientific American, **264**, [4] p24, (1991).

A. Yamamoto, *Crystallography of quasiperiodic crystals*, Acta Crystallogr., **A52**, 509–560 (1996).

M. Senechal, *Quasicrystals and Geometry*, Cambridge U. P., Cambridge, (1995).

Tiling

Martin Gardiner, *Mathematical Games*, Scientific American, **233** [1] p112, 1975. [This was used to construct Figure 3.16.]

Martin Gardiner, *Mathematical Games*, Scientific American, **233** [2] p112, 1975.

Martin Gardner, *Mathematical Games*, Scientific American, **236** [1] p110, 1977. [This was used to construct Figures 3.17 and 3.18.]

C. Giacovazzo, Chapter 4, *Beyond ideal crystals*, in: C. Giacovazzo, H. L. Monaco, G. Artioli, D. Viterbo, G. Ferraris, G. Gilli, G. Zanotti and M. Catti, *Fundamentals of Crystallography*, 2nd edition, Oxford University Press, Oxford, (2002).

B. Grünbaum and G. C. Shephard, *Tilings and Patterns*, W. H. Freeman, New York, (1987).

Formula index

Page numbers in italic, e.g. *4*, refer to figures, and in bold, e.g. **2**, refer to entries in tables.

Crystals and Crystal Structures. Richard J. D. Tilley
© 2006 John Wiley & Sons, Ltd

Subject index

Crystals and Crystal Structures.   Richard J. D. Tilley
© 2006 John Wiley & Sons, Ltd